Protein Tailoring for Food and Medical Uses

Protein Tailoring for Food and Medical Uses

edited by

Robert E. Feeney
John R. Whitaker
Department of Food Science and Technology
University of California
Davis, California

MARCEL DEKKER, INC. New York and Basel

Library of Congress Cataloging-in-Publication Data

Protein tailoring for food and medical uses.

Based on an American Chemical Society symposium held Sept. 8-13, 1985 in Chicago.
Includes bibliographies and index.
1. Proteins--Derivatives--Congresses. 2. Proteins--Structure-activity relationships--Congresses. 3. Proteins--Industrial applications--Congresses. 4. Food-Protein content--Congresses. 5. Drugs--Design --Congresses. I. Feeney, Robert Earl. II. Whitaker, John R. III. American Chemical Society. [DNLM: 1. Dietary Proteins--Therapeutic use--congresses. 2. Food Handling--Congresses. QU 55 P96012 1985]
QP551.P6977326 1986 664 86-8867
ISBN 0-8247-7616-X

MARCEL DEKKER, INC.
270 Madison Avenue, New York, New York 10016

Current printing (last digit):
10 9 8 7 6 5 4 3 2 1

PRINTED IN THE UNITED STATES OF AMERICA

Preface

This American Chemical Society symposium is the third in a series held at five-year intervals. The first one was in Mexico City in 1975 at the First North American Congress of Chemistry and consisted of two half-day sessions covering only the modifications of food proteins [R. E. Feeney and J. R. Whitaker, Food Proteins: Improvement through Chemical and Enzymatic Modification, Adv. Chem. Series, Vol. 160, American Chemical Society, Washington, D.C. (1977)]. The second one was held five years ago at the Second North American Congress of Chemistry and consisted of three half-day sessions covering modification of proteins of interest in both foods and medicine (pharmacology) [R. E. Feeney and J. R. Whitaker, Modification of Proteins: Food, Nutritional, and Pharmacological Aspects, Adv. Chem. Series, Vol. 198, American Chemical Society, Washington, D.C. (1982)]. This current symposium further develops these subjects and attempts to show interrelationships between them. Its international flavor is evidenced by the home countries of the twenty-five authors of the fourteen articles: twelve authors from the United States, nine from Japan, two from Norway, and one each from Canada and Denmark.

Foods are eaten by people, and people take medicines and are treated with medical products. It is these interrelationships between modifications of proteins in foods and in medicine that this symposium delineates. Many of the techniques that are used in foods can also be used medically, but unfortunately many of the medical procedures are still unsuitable for foods. Part of this may be because the approval and use of medical products are usually under supervision of a physician, whereas no one knows exactly how, and to what extent, foods will be used once they are released. In addition, foods are used in much higher quantities, so chemical changes in them could have a more quantitative effect. Nevertheless, the general theme does exist.

Since the last symposium of 1980 two of the areas that have shown tremendous advances have been the basic area of mechanism-based inactivation of enzymes by the derivatives of substrates (sometimes called suicide reagents) and the logarithmically developing area of genetic engineering. These have been further aided by advances in the computerization of data and molecular imaging, and structural graphics of interacting systems. Along with this has been the more precisely engineered formation of protein conjugates with other substances and with other proteins. Again, most of these advances have been in the medical area rather than in the food area.

The food area, however, has not been neglected, as is evidenced by papers in this volume on the formation of peptide products with pharmacological action (opioids), of peptides with different flavor characteristics, of low phenylalanine products for dietetic purposes, and the covalent incorporation of essential amino acids to improve nutritional quality. But the bulk modification of proteins for human use is still far from adaptation. Continued studies in this area on the safety of these products should allow some of them to reach the market in the future. Much can be done to modify the proteins in foods to make them more acceptable, more useful, and more nutritious, certainly one of the main objectives of scientists trying to make better foods.

The rapid advancements in all of these areas have resulted in the coinage (and adoption) of new words and phrases that have different meanings to different individuals. In particular, the relations among "protein modification," "protein tailoring" (chosen by the editors as the title of this monograph), "protein engineering," and "recombinant proteins" may be confusing. "Protein modification" is a broad term that includes any physical or chemical change caused by treatment of the protein by chemical, enzymatic or physical means. The editors have chosen "protein tailoring" to imply a comparatively specific modification to be used for a specific purpose. "Protein engineering" is generally, but not always, used for those modifications caused by changes in the genetic code (usually by *in vitro* mutagenesis), while "recombinant proteins" applies specifically to proteins made by *in vitro* mutagenesis.

In the selection of the speakers for this symposium, it was necessary to choose from a large array of excellent individuals. This again is evidence of an increased interest and research in the field, and it is hoped that the selection of does cover the main areas where advances are being made. Certainly the omissions of many appropriate participants and some subjects were unavoidable.

The success of the symposium was greatly aided by financial support from the following: Abbott Laboratories, American Cyanamid Company, Anheuser-Busch Companies, Dart and Kraft Foundation, Frito-Lay, General Mills, Monsanto Company, Pfizer, Proctor & Gamble Company, Rohm and Haas Company, Stauffer Chemical Company, and Syntex USA.

Robert E. Feeney
John R. Whitaker

Contents

Contributors

JENS ADLER-NISSEN Enzymes Research and Development, Novo Industri A/S, Bagsvaerd, Denmark

SOICHI ARAI Department of Agricultural Chemistry, The University of Tokyo, Bunkyo-ku, Tokyo, Japan

WILLIAM F. BENISEK Department of Biological Chemistry, School of Medicine, University of California, Davis, California

HIDEO CHIBA Department of Food Science and Technology, Kyoto University, Sakyo-ku, Kyoto, Japan

ROBERT E. FEENEY Department of Food Science and Technology, University of California, Davis, California

MAUREEN HEARNE Cancer Research Institute, University of California, San Francisco, California

NORIKO HIRAO Department of Agricultural Chemistry, The University of Tokyo, Bunkyo-ku, Tokyo, Japan

RAFAEL JIMENEZ-FLORES Department of Food Science and Technology, University of California, Davis, California

YOUNG C. KANG Department of Food Science and Technology, University of California, Davis, California

HIROSHI MAEDA Department of Microbiology, Kumamoto University Medical School, Kumamoto, Japan

JON W. MARSH Laboratory of Molecular Biology, National Institute of Mental Health, Bethesda, Maryland

YASUHIRO MATSUMURA Department of Microbiology, Kumamoto University Medical School, Kumamoto, Japan

CLAUDE F. MEARES Department of Chemistry, University of California, Davis, California

DAVID M. NEVILLE, JR. Laboratory of Molecular Biology, National Institute of Mental Health, Bethesda, Maryland

TATSUYA ODA Department of Microbiology, Kumamoto University Medical School, Kumamoto, Japan

SJUR OLSNES Department of Biochemistry, Norsk Hydro's Institute for Cancer Research, Montebello, Oslo, Norway

MARK J. POZNANSKY Department of Physiology, University of Alberta, Edmonton, Alberta, Canada

THOMAS RICHARDSON Department of Food Science and Technology, University of California, Davis, California

KIRSTEN SANDVIG Department of Biochemistry, Norsk Hydro's Institute for Cancer Research, Montebello, Oslo, Norway

KAZUMI SASAMOTO Department of Microbiology, Kumamoto University Medical School, Kumamoto, Japan

RICHARD B. SILVERMAN Department of Chemistry and Department of Biochemistry, Molecular Biology, and Cell Biology, Northwestern University, Evanston, Illinois

MICHIKO WATANABE Department of Agricultural Chemistry, The University of Tokyo, Bunkyo-ku, Tokyo, Japan

RONALD WETZEL Biocatalysis Department, Genentech, Inc., South San Francisco, California

JOHN R. WHITAKER Department of Food Science and Technology, University of California, Davis, California

MASAAKI YOSHIKAWA Department of Food Science and Technology, Kyoto University, Sakyo-ku, Kyoto, Japan

Protein Tailoring for Food and Medical Uses

1

Tailoring Proteins for Food and Medical Uses: State of the Art and Interrelationships

Robert E. Feeney

Department of Food Science and Technology
University of California
Davis, California

I. INTRODUCTION

Today's highly sophisticated science of the modifications of proteins has ancient roots. Indeed, the tailoring of proteins for food and medicinal uses precedes the beginning of what we now call biochemistry. We see some early examples in foods that were prepared by mixing materials such as herbs, spices, oils, and fermentation products. Other examples include the ancient use of alkali for the processing of fish to produce the stabilized product called lutfisk in Scandinavian countries and for the preservation of corn meal by the Indians of the southern United States, Mexico, and South America. Various oxidizing agents such as bromates are used to produce a better performing flour for bread making, and enzymatic treatments are widely used in foods for beer making and for the preparation of certain nutritional products. Ancient medicine also has tales of the treatment of materials with acid and alkali. In the early part of this century, formaldehyde was developed for the formation of toxoids from bacterial toxins and is still in use today, albeit the method and the chemistry could be considered barbaric. Formaldehyde and glutaraldehyde are currently used today for the sterilization and cross-linking of connective tissue and heart valves for surgical prosthesis.

Chemical modification of proteins was pursued more precisely early in the twentieth century, mainly as an analytical procedure for side-chain amino acids. Somewhat later, starting in the late nineteen forties, methods were developed for specific inactivation of biologically active proteins and titration of their essential groups.

Enzymatic modifications have particularly developed in the past decade, as many more enzymes have become economically available. Various procedures have also been developed for stabilizing and immobilizing them. Both the food and medicinal industries employ these today.

Currently, methods that are used include not only the previously described chemical methods and enzymatic procedures but also what is popularly called genetic engineering. Some of today's chemical and enzymatic techniques for tailoring proteins are as follows:

Chemical

1. Derivatization (e.g., change physical, enzymatic, toxic properties)
2. Covalent attachment of proteins to other proteins and other biologically active substances
3. *In vitro* mutagenesis
4. Synthetic peptides and proteins

Enzymatic

1. Proteolysis
2. "Reverse proteolysis" (e.g., plastein)
3. Transamination
4. Removal of derivative (e.g., sialic acid)

Some of the active areas in foods include the use of these procedures to generate enzymes with tailored characteristics for special purposes in processing, while in medicinal chemistry proteins are targeted to specific sites in the body by connecting to drugs and other proteins such as toxins or hormones. Enzymes with newly designed characteristics are being made by site-directed mutagensis. However, an area for the future might well be chemical synthesis of proteins containing non-amino acid residues, thereby providing characteristics unattainable by site-directed mutagenesis.

As all these advances occur, the lines between the food and medicinal sciences blur. Studies on opioids from foods overlap with studies on brain peptides and peptide hormones. Studies on enzymes for food manufacture intermesh with studies on enzymes for medicinal uses. Such intermeshings occur not only at the basic level but also at the applied level.

II. ENZYMATIC AND CHEMICAL PROCEDURES

A. *ENZYMATIC REACTIONS*

Enzymatic reactions have been developed in the past decade and a half on at least two fronts: a) the development of specialized enzymes for particular purposes, either by the economical preparation of more specific enzymes for certain purposes or now by genetic engineering, and b) the use of enzymes under special circumstances to increase specificity or expected lifetimes. The latter has been achieved by the use of different physical conditions, such as the use of organic solvents at different temperatures. An example where chemical modifications also play a role is the use of immobilized enzymes, which has grown into a large industry.

Specialized enzymes have been used or developed for proteins, but most all of these are under research phases as of this time. However, under this general umbrella is the modification of enzymes, primarily the immobilized enzymes, to make them more suitable for a variety of purposes. Those related to nonprotein

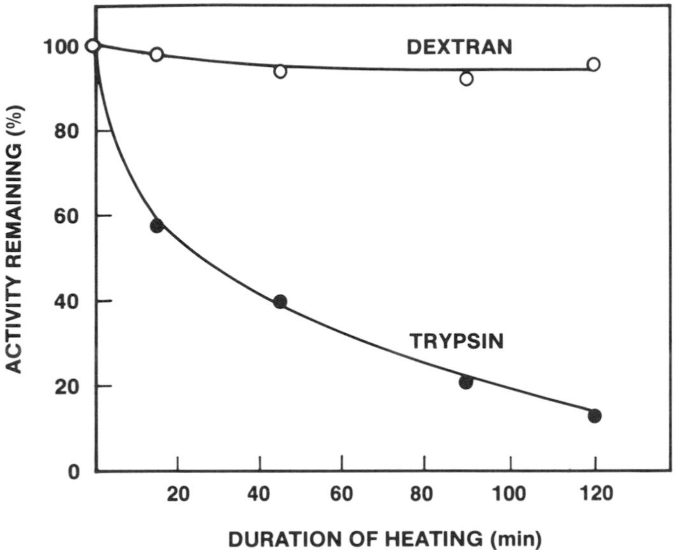

FIG. 1 *Loss of activity on heating trypsin (●) and its conjugate with dextran (○) at 37°C. Samples (125 μl) of native and modified enzyme (activity 2 U/ml measured towards p-toluene sulfonyl-L-arginine methyl ester) were heated for different lengths of time in 80 mM borate buffer pH 8.1, then the activity remaining determined. From Marshall and Rabinowitz (4).*

uses include ones for steroids or carbohydrates. The reader is referred to the numerous chapters in larger editions now available on immobilized enzymes (1,2) and the article by Benisek and Hearne in this volume (3). Other studies have concerned the preparation of enzymes with greater heat stability. These have been achieved by the attachment of such hydrophilic materials as dextrans (Fig. 1) (4). Now genetic engineering is also successfully used to make more heat stable enzymes (5).

Proteins have been tailored by using carboxypeptidases or aminopeptidases to change the carboxyl or amino terminals, and endopeptidases are used to make peptides for producing fragments of proteins that have properties different from the original native molecule. One example is the formation of opioids from casein as described by Chiba and Yoshikawa in this volume (6).

B. *GENERAL CHEMICAL PROCEDURES*

Many chemical procedures have been recently employed to modify active centers or to change the combining properties of enzymes. These are mainly applications to some of the fundamental enzyme chemistry studies, and a detailed coverage of them is not pertinent to this section. These have been extensively treated in several texts or monographs (7-13). Two examples of these types of studies from the author's laboratory were modifications of a relatively simple protein, the antifreeze glycoprotein (14,15) (Table 1), and much more complex proteins, the transferrins (19) (Table 2). With the glycoproteins, other than for scission of the polypeptide chain by proteolysis, the carbohydrate moiety was the susceptible portion. With the transferrins, two histidines, two tyrosines and one arginine per metal-binding site were shown to be essential for metal binding, while oxidation of over half of the methionines in ovotransferrin to the sulfoxides did not significantly impair the uptake of iron and synthesis of hemoglobin by chick reticulocytes (21).

There are also many other examples of modifications that alter, but do not destroy, activities. In the author's laboratory it was shown that the introduction of modifying groups that made the proteins more acid (lowered the isoelectric point) caused avian ovomucoids to react much more rapidly with proteolytic enzymes (22). Also, guanidination of lysine groups in lima bean trypsin inhibitor to produce homoarginine residues made it a considerably stronger inhibitor (23). Periodate-cyanoborohydride modification of the carbohydrate in the toxic plant protein ricin has been shown to decrease the toxicity *in vitro*, but to increase the toxicity *in vivo*, possibly by prolonging the lifetime of the modified protein in the body (24).

Of all the chemical reactions available for proteins, only a fraction (possibly less than 25%) of them are used for practical purposes. Some of the ones receiving current attention are listed in Table 3. These, in turn, can be further varied by changing the

TABLE 1 *Modifications*[1] *of a simple protein: antifreeze glycoprotein*[2]

Modification	Protein group modified	Effect on activity
1) N-Reductive methylation	*Amino*[3]	*Active*
2) N-Acetylation	*Amino*[3]	*Active*
3) Carbodiimide[3]	*Amino and carboxyl*[4]	*Active*
4) O-Acetylation	*Sugar hydroxyls*	*Inactive*
5) Periodate	*Sugar (scission)*	*Inactive*
6) Borate[5]	*Sugar hydroxyls*	*Inactive*
7) Enzyme: galactose oxidase	*Oxidation of C-6 hydroxyls to aldehyde*	*Active*
8) Bisulfite[5] *adduct of number 7 (oxidized enzyme)*	*Complex of aldehyde*	*Inactive*
9) I_2 *on number 7 (oxidized enzyme)*	*Oxidation of aldehyde to acid*	*Inactive*
10) Enzyme: subtilisin	*Peptide bonds*	*Inactive*

[1]Only modifications done in our laboratory (14,16) are described. For most other studies see reviews by A. L. DeVries (17,18).

[2]Repeating polymers of Ala-Ala-(β-galactosyl(1→3)-α-N-acetylgalactosamine)Thr, with some Pro following Thr in shorter components.

[3]Done on preparation with single terminal amino group blocked by N-dimethylation.

[4]Formation of a peptide bond between the amino terminal of one molecule with the carboxy terminal of a second (with blocked amino terminal) to form a dimer.

[5]These are reversible reactions.

TABLE 2 *Modifications[1] of a complex protein: transferrin[2]*

Modification	Primary group modified	Effect on Activity
1) *N-Reductive methylation*	*Lysine amino*	*None*
2) *N-Acetylation*	*Lysine amino*	*None*
3) *O-Acetylation[3]*	*Tyrosine*	*Inactive*
4) *Iodination* (I_3^-)[3]	*Tyrosine*	*Inactive*
5) *Periodate[3,4]*	*Tyrosine*	*Inactive*
6) *Diethyl pyrocarbonate[3]*	*Histidine*	*Inactive*
7) *Photooxidation*	*Histidine*	*Inactive*
8) *Phenylglyoxal[3]*	*Arginine*	*Inactive*

[1]Only modifications done in the author's laboratory are listed. A summary of modifications of other investigators as well as those of the author has been recently presented (19).

[2]The transferrins (serum, ovo-, and lacto-) are homologous proteins with a large single polypeptide chain (approx. 76 Kd) containing two iron-binding sites. The groups demonstrated to be involved in iron binding are the side-chain histidines and tyrosines and an externally provided bicarbonate (or carbonate) and possibly an arginine and a water molecule (19).

[3]These modifications do not inactivate the ternary iron-bicarbonate transferrin complex. Note, however, that photooxidation does inactivate the complex (20). Also, oxidation by relatively low concentrations of periodate acts like an affinity reagent, possibly binding where bicarbonate binds in the complex.

[4]Low concentrations of periodate on the iron-free protein also modifies approximately half of the methionines, which do not appear essential, because "all" of the methionine can be oxidized in a sample in 8 M urea, with nearly complete restoration of activity after renaturation.

TABLE 3 *Some commonly used reagents and reactions*

A. Alkylation

1) *Haloacetate*

$$\text{(P)}\text{-}NH_2 + ICH_2COO^- \xrightarrow{pH > 8.5} \text{(P)}\text{-}NHCH_2COO^- + I^- + H^+$$

2) *Reductive alkylation*

$$\text{(P)}\text{-}NH_2 + R\overset{O}{\overset{\|}{C}}R' \underset{+H_2O}{\overset{-H_2O}{\rightleftharpoons}} \text{(P)}\text{-}N{=}CRR' \xrightarrow[NaBH_4]{pH \sim 9} \text{(P)}\text{-}NH\text{-}CHRR'$$

3) *Michael addition*

$$\text{(P)}\text{-}SH + \text{N-ethylmaleimide } (NCH_2CH_3) \xrightarrow{pH\ 5} \text{(P)}\text{-}S\text{-succinimide } (NCH_2CH_3)$$

B. Amide Formation

1) *Acid anhydride*

$$\text{(P)}\text{-}NH_2 + O(\overset{O}{\overset{\|}{C}}\text{-}CH_3)_2 \xrightarrow{pH > 7} \text{(P)}\text{-}NH\overset{O}{\overset{\|}{C}}CH_3 + CH_3\overset{O}{\overset{\|}{C}}O^- + H^+$$

2) *N-hydroxysuccinimide esters*

$$\text{(P)}\text{-}NH_2 + R\text{-}\overset{O}{\overset{\|}{C}}\text{-}N\text{(succinimide)} \xrightarrow{pH\ 7\text{-}8} \text{(P)}\text{-}NH\text{-}\overset{O}{\overset{\|}{C}}\text{-}R + HO\text{-}N\text{(succinimide)}$$

TABLE 3 *Some commonly used reagents and reactions (continued)*

3) *Carbodiimide*

$$\text{(P)}-\overset{O}{\overset{\|}{C}}-O^- + R-N=C=N^+(H)-R' + H^+ \xrightarrow{pH \sim 5} \text{(P)}-\overset{O}{\overset{\|}{C}}-O-C(=\overset{+}{N}HR)(NHR')$$

$$\xrightarrow{HX} \text{(P)}-\overset{O}{\overset{\|}{C}}-X + O=C(NHR)(NHR') + H^+$$

C. O-Acylation

1) *N-acetylimidazole*

$$\text{(P)}-C_6H_4-O^- + H^+ + \text{N-acetylimidazole } (N-\overset{O}{\overset{\|}{C}}CH_3) \xrightarrow{pH\ 7.5} \text{(P)}-C_6H_4-O\overset{O}{\overset{\|}{C}}CH_3 + \text{imidazole (N-H)}$$

D. Oxidation

1) H_2O_2

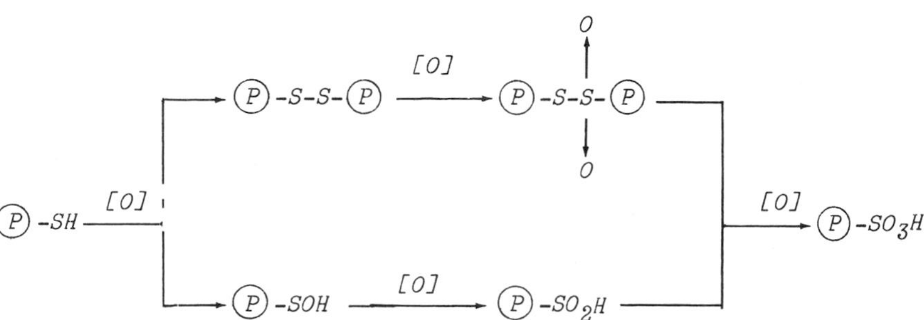

TABLE 3 *Reagents and reactions (continued)*

E. Stabilized Schiff Base Products

1) *Glutaraldehyde*

$$\text{Ⓟ}-NH_2 \quad \overset{H}{\underset{O}{\diagdown\!\!\!/\!/}}C-CH_2-CH_2-CH_2-C\overset{H}{\underset{O}{}} \longrightarrow \textit{Polymers with cyclic derivatives of glutaraldehyde}$$

F. Phosphorylation

$$\text{Ⓟ}-OH \quad + \quad POCl_3 \longrightarrow \textit{Many products on N- and HO- groups}$$

---

nonreactive part of the reagent, e.g., the length and hydrophobicity of the side chains.

Several more specialized adaptations for chemical modification have particular uses. One is a kinetic or statistical way for the determination of the numbers of a particular kind of side chain that can not be modified without loss of function of a protein. Examples of this were the finding that two histidines, two tyrosines and one arginine per metal-bindng site in transferrins appeared essential (19). Another covers a group of methods broadly classified as affinity labelling (25), a procedure whereby a particular reagent has a relatively high affinity for the active center of a biologically active protein. With the transferrins, periodate appeared to have this affinity (19). Affinity interactions are currently important for purification of modified proteins. Many variations of these exist. One of these is described in the article by Benisek and Hearne (3) who probe the active site of the enzyme, a steroid isomerase, with a solid phase photoaffinity reagent. The general area of mechanism-based enzyme inactivators (sometimes called suicide reagents) is described by Silverman (26). In this latter case, a principal practical interest is in the design of drugs.

TABLE 4 *Regeneration of 1-deoxyglycitolated bovine serum albumin with periodate*[1]

1-Deoxyglycitolated bovine serum albumin		Lysine recovered (%)[3,4] Periodate concentration (mM)	
Derivative	Glycitolation (%)[2]	10	20
Xylose	*86.1*	*65.3*	*86.8*
Arabinose	*81.0*	*63.7*	*86.8*
Ribose	*77.3*	*71.4*	*83.6*
Mannose	*70.8*	*60.0*	*84.6*
Galactose	*70.0*	*86.3*	*94.2*
Glucose	*39.0*	*88.3*	*93.6*
Lactose	*26.4*	*76.9*	*89.0*

[1]From Wong et al. (27).

[2]$\frac{\text{Modified lysine}}{\text{Total lysine}}$ X 100%.

[3]$\frac{\text{Unmodified lysine after } 10_4^-}{\text{Total lysine}}$ X 100%.

[4]Conditions for 10 mM periodate: 0.02 mM glycitolated bovine serum albumin, in 0.2 M sodium phsophate, pH 7.0, incubated for 15 min.; Conditions for 20 mM periodate: 0.02 mM glycitolated bovine serum albumin, in 0.2 M sodium borate, pH 8.6, incubated for 20 min.

A third type of modification is broadly classified as reversible. Some of these are truly reversible, while others regenerate the original protein, such as the periodate-mediated oxidative removal of sugars that had been attached to amino groups

TABLE 5 *Noncleavable homobifunctional reagents*[1]

Name	Formula	Max. link (Å)	Reaction specificity
Bisimidates	$CH_3\text{-}O\text{-}C(=NH_2^+Cl^-)\text{-}(CH_2)_n\text{-}C(=NH_2^+Cl^-)\text{-}O\text{-}CH_3$		*Amine*
Dimethyl suberimidate (DMS)	$n = 6$	11	
N-hydroxysuccinimide ester of suberic acid (NHS-SA)	$[\text{-}(CH_2)_3\text{-}C(=O)\text{-}O\text{-}N(\text{succinimide})]_2$	11	*Amine*
Glutaraldehyde	$[=C(CHO)\text{-}(CH_2)_2\text{-}CH=]_n$		*Nonspecific*
Formaldehyde	$H_2C=O$		*Nonspecific*

[1]From Ji (30).

as 1-deoxyglycitolyl groups by reductive alkylation (Table 4) (27). Another type is the covalent attachment of a protein to another substance, as used for immobilization (1,2,28) and cross-linking (29,30), an area that deserves special attention for its extensive uses in tailoring proteins.

C. *CROSS-LINKING REAGENTS*

Cross-linking reagents have many uses in the tailoring of proteins, and although most of these are either classical methods

TABLE 6 *Nonphotoactivable heterobifunctional reagents*[1]

Name	Formula	Specificity
N-Succinimidyl-3-(2-pyridyldithio)propionate	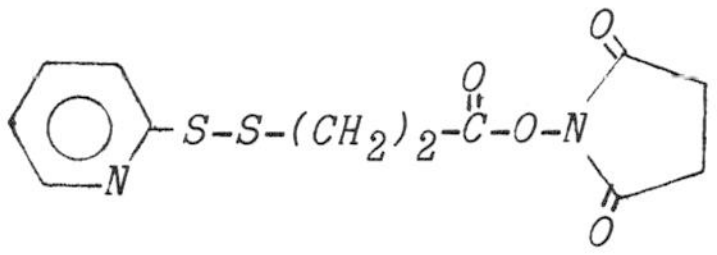 (pyridyl)-S-S-$(CH_2)_2$-C(=O)-O-N(succinimide)	*Amine, sulfhydryl*
Succinimidyl-4-(N-maleimidomethyl)-cyclohexane-1-carboxylate	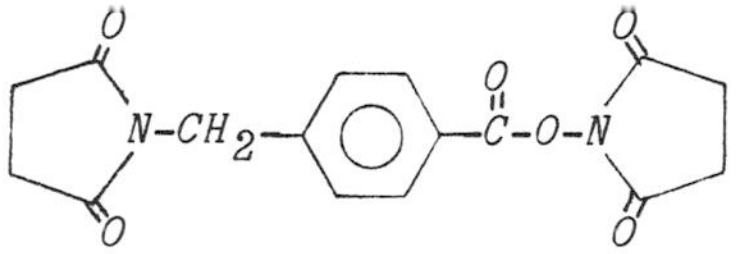 (succinimide)N-CH_2-(ring)-C(=O)-O-N(succinimide)	*Amine, sulfhydryl*
Succinimidyl-4-(p-maleimidophenyl)-butyrate	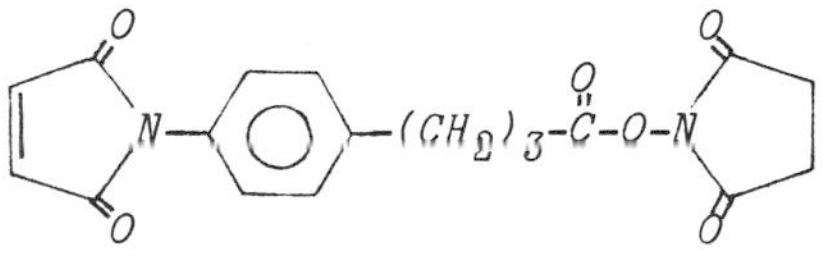 (maleimide)N-(phenyl)-$(CH_2)_3$-C(=O)-O-N(succinimide)	*Amine, sulfhydryl*
Carbodiimides	*R-N=C=N-R*	*Amine, carboxyl*

[1]Adapted from Ji (30).

already in use, or slight modifications thereof, a general description of the ones used should be considered. Some of the main cross-linking reactions that are used today are either homobifunctional (Table 5) or heterobifunctional (Table 6). These vary depending upon the properties of the protein that is being linked either with another protein or with another substance, such as a hormone, drug, or chelating agent. In all cases the importance of retaining the desired activities of both entities must be considered.

An example where a particular property of the protein can be used is seen in Fig. 2, where a sulfhydryl group is unessential for the properties concerned. Reagents can be attached to this

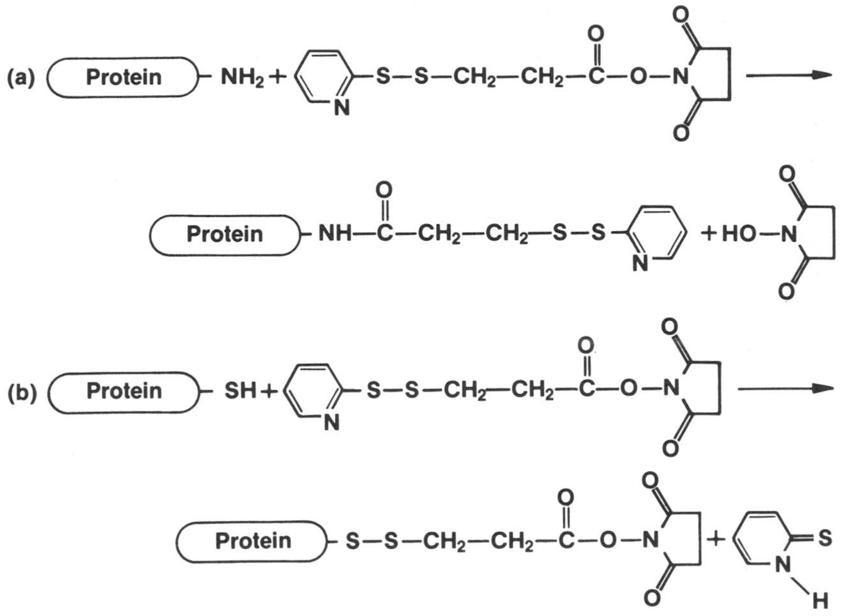

FIG. 2 *Modification of proteins by N-succinimidyl 3-(pyridyldithio)propionate. From Carlsson et al. (31).*

group. On the other hand, a sulfhydryl group may be introduced into a protein for the purpose of forming the cross-linkage, giving either a cleavable disulfide or a stable thioether. As in all chemical reactions involving proteins, a selection of the appropriate reagents is frequently an operational matter and must concern the conditions under which the reaction must be feasible.

As stated above, the bifunctional reagents for cross-linking can be either homobifunctional or heterobifunctional, depending upon the particular requirements of the conjugation to be done. Frequently heterobifunctional reagents can be used in a more discriminating manner because the reaction can be controlled better so as to avoid forming cross-linkages between the individual proteins; that is, intramolecular rather than intermolecular cross-linkages. An example of a heterobifunctional reagent first described by Carlsson et al. (31) is N-succinimidyl-

3-(2-pyridyldithio)propionate (SPDP). With this, 2-pyridyl disulfide groups are introduced into one of the proteins to be coupled, and on the other protein thio groups are introduced by pyridyldithiolation and subsequent reduction with dithiothreitol. On mixing, the two substituted proteins react to form a conjugate in which the linkage is a disulfide (Fig. 3).

An example in which a reactive group is introduced into a protein is one using an N-hydroxysuccinimidyl ester of iodoacetic acid. In this method the thio groups introduced into the protein are then reacted with iodoacetyl groups introduced into the other protein. This then gives the thioether linkage which is a stable linkage.

With hormone or drug conjugates (32) more strict attention must be paid to the diverse chemically reactive groups that are found in drugs and some hormones. For example, oxidation of some of the carbohydrate groups with periodate to the aldehydes and subsequent reduction of the Schiff base between these and amino groups can be used to form stable secondary amine products. Sometimes the particular properties of the interacting substances can also be used to obtain such reactions.

Many other types of cross-linking reagents have been developed but are not commonly used for tailoring proteins. These include ones with a cleavable group, an affinity group, and a photoaffinity group.

D. *GENETIC ENGINEERING TECHNIQUES*

Under this broad and general term can be classified all those procedures that somehow involve genetic material or the genetic process. Since these will be covered in detail in two other chapters (5,33), they will only be mentioned here briefly and simplistically.

One of the most well known and publicized procedures is not tailoring a protein into a new one, but rather making it available

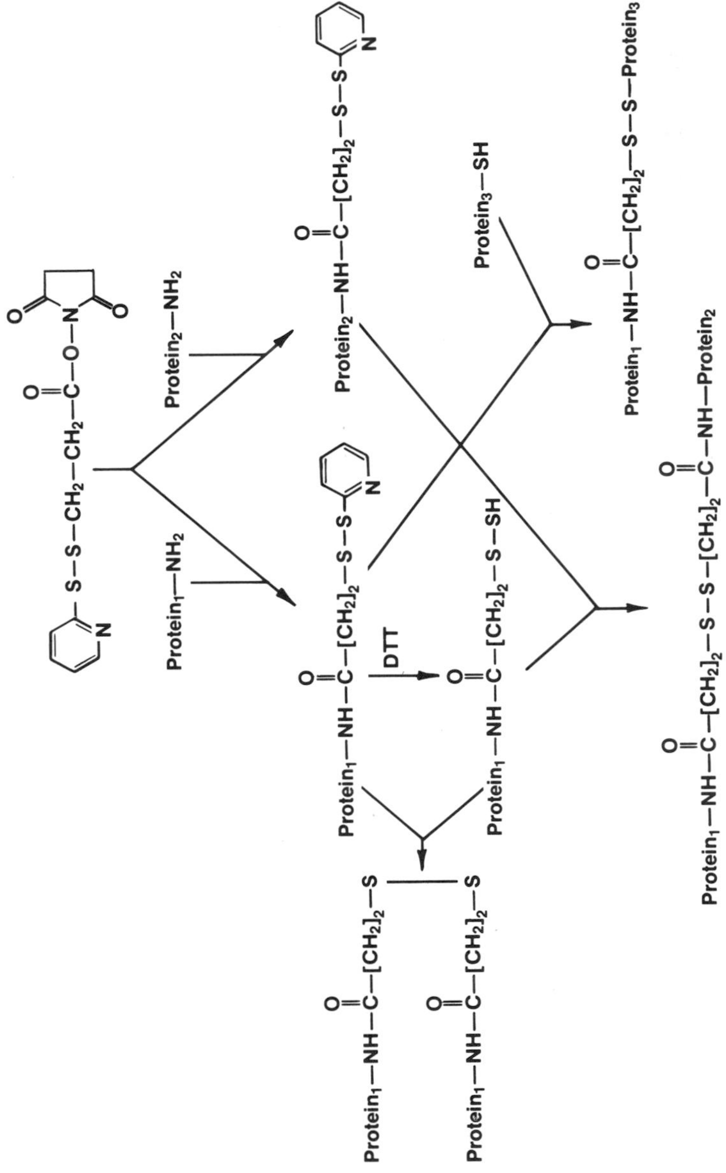

FIG.3 *Protein-Protein conjugation by N-succinimidyl 3-(2-pyridyldithio)propionate. From Carlsson et al (31).*

by inserting the correct gene into an organism that is capable of being cultivated in high yield to produce the desired protein.

However, site-directed mutagenesis is a procedure that does tailor proteins by making new ones. This exploding area is one for which there is great promise and high expectations. This new gene also must be inserted into an organism capable of producing it. Before the *in vitro* mutagenesis is done, however, extensive basic information must be compiled. Both the sequence and the general properties of the protein should be known. Then it is critical to have the cooperation of the molecular biologist, the computer graphologist for molecular modeling, and the x-ray structure physicist. Even then problems may exist such as pro- or preproproteins and the necessity for posttranslational modifications. Nevertheless, acquisition of knowledge of the relationship of structure to function is rapidly accelerating. For example, stabilization to environmental conditions has been achieved by several means (5). In subtilisin, the methionine 222 is susceptible to oxidation with concurrent inactivation of the enzyme; exchanging methionine 222 for a nonoxidizable residue greatly increased the enzyme's resistance to oxidation (34). Although there is presently insufficient information to predict conformation and function from just the sequence, this long-ago considered possibility (35) may not be too far in the future.

III. TAILORING OF FOOD PROTEINS

The modification of food proteins is a specialty area because of the large scope of the subject and the possibly very large quantities of material that need to be modified and that may be consumed by any one individual at a particular time. In addition, the modification of food proteins and the products that are produced from this are today readily seen by people and, as such, customs, fears and prejudices may be involved in their use. Nevertheless, so many modifications have been used that the public is generally accustomed to the general idea. However, the word they particularly fear is *chemical*.

There are many possible purposes for modifying food proteins. Some of the more common ones are listed below (36):

1. To retard deteriorative reactions (e.g., Maillard reaction)
2. To impart texture (e.g., texturized vegetable proteins)
3. To increase solubility (e.g., beverages)
4. To decrease solubility (e.g., cheeses)
5. To provide foaming and coagulation capacities (e.g., whipping agents, baking products)
6. To provide structural characteristics (e.g., texturized products)
7. To provide emulsifying capacities (e.g., mayonnaise)
8. To prevent interactions (e.g., encapsulation)
9. To remove off-flavors (e.g., as in soybean for tofu)
10. To remove toxic or inhibitory ingredients (e.g., beans and peas)
11. To attach chemicals covalently (e.g., colors or flavors)
12. To attach nutrients covalently (e.g., amino acids)

As can be seen, the reasons for modifying food proteins conveniently fall into three broad areas: 1) to form a more acceptable and utilizable product, 2) to form a product which is less susceptible to deterioration and 3) to form a product that is of nutritionally higher quality. All three of these categories must be considered and are usually interrelated in some way or another. There are two distinct divisions, and these are related to the quality of the food consumed and how certain foods are mixed with other foodstuffs. For example, a foodstuff like flour is consumed in large amounts, whereas another foodstuff may be used in relatively small amounts as a binding agent or as a decorative, a confectionary, or a "reward" type substance, such as a whipped topping to a dessert.

The second matter for consideration when modifying food proteins involves the concerns that must be examined. These could include such problems as: loss of nutritional value by formation of products whose peptide bonds are not split or whose hydrolyzed products may appear in the blood but not be utilized; formation of toxic or allergenic products; formation of undesirable organoleptic qualities (e.g., flavor, physical sensation); and interaction with other ingredients of the food product to cause many of the above mentioned effects. Here the possible difference between modified food proteins and modified proteins for medicinal purposes can be immediately seen. The food protein is used in relatively large amounts as compared to the medicinal protein, and it must necessarily be of a much lower cost. In addition, modification of the food proteins is done with relatively crude mixtures and purification is done, with only rare exceptions, because of the economics involved. In the case of medicinal products, the bad flavors can frequently be tolerated or can be circumvented by encapsulation of the material. With food proteins this is not possible. Since large quantities are consumed and they are always consumed by mouth, this also greatly increases the difficulty with the formation of unusual flavors. With food proteins the unusual is frequently just as much of a difficulty as the obnoxious because the unusual may be considered to be bad.

A. *ENZYMATIC MODIFICATION OF FOOD PROTEINS*

Enzymes have long been used for modifying food proteins and, as cited above, their use is many times more acceptable in the eye of the non-scientist than is the use of chemicals. Table 7 lists some classical uses of enzymes or their actions.

An enzymatically tailored food item for dietetic purposes is a low-phenylalanine product currently available in Japan. This was first ingeniously produced by partially hydrolyzing a soybean protein isolate (or a fish protein isolate), separating peptides, and resynthesizing by the plastein reaction products low in aromatic amino acids (38).

TABLE 7 *Proteolytic enzymes in food protein modification*[1]

Food	Purpose or action
Baked goods	*Softening action in doughs. Cut mixing time, increase extensibility of doughs. Improvement in texture, grain and loaf volume. Liberate β-amylase.*
Brewing	*Body, flavor, and nutrient development during fermentation. Aid in filtration and clarification. Chillproofing.*
Cereals	*Modify proteins to increase drying rate, improve product handling characteristics Production of miso and tufu.*
Cheese	*Casein coagulation. Characteristic flavor development during aging.*
Chocolate-cocoa	*Action on beans during fermentation.*
Egg, egg products	*Improve drying properties.*
Feeds	*Waste product conversion to feeds. Digestive aids, particularly for pigs.*
Fish	*Solubilization of fish protein concentrate. Recovery of oil and proteins from inedible parts.*
Legumes	*Hydrolyzed protein products. Removal of flavor. Plastein formation.*
Meats	*Tenderization. Recovery of protein from bones.*
Milk	*Coagulation in rennet puddings. Preparation of soybean milk.*
Protein hydrolysates	*Condiments such as soy sauce and tamar sauce. Bouillon. Dehydrated soups. Gravy powders. Processed meats. Special diets.*
Antinutrient factor removal	*Specific protein inhibitors of proteolytic enzymes and amylases. Phytate.*[2] *Gossypol.*[2] *Nucleic acid.*[2]
Wines	*Clarification.*[2]
In vivo processing[3]	*Conversion of zymogens to enzymes. Fibrinogen to fibrin. Collagen biosynthesis. Proinsulin to insulin. Macromolecular assembly.*

[1]From Whitaker (37).

[2]In large part caused by other than proteolytic enzymes.

[3]Representative examples given.

The use of modified enzymes for processing foods is developing rapidly. In each case the tailored proteins are used for processing purposes and not as part of the food, per se, such as immobilized enzymes.

B. *CHEMICAL MODIFICATION OF FOOD PROTEINS*

Chemical modification of food proteins has been done on a large scale, but most all of the possible applications are still at a laboratory research stage. Because of the particular uses of foods as described above, many of the chemical reactions that are used in fundamental research on proteins, as well as many of those that are used in the preparation of proteins for medicinal uses, cannot be used on food proteins at this time.

One of the principal areas studied in regard to food proteins has been the improvement of the physical (functional) properties. These have included methods to solubilize the protein, such as succinylation, or to increase the emulsifying properties by acylation with longer chain fatty acids. The latter has been done with milk proteins and other proteins (39); the laboratory of John E. Kinsella has been particularly active in the overall area (40,41).

Many of the modifications have been with plant proteins, because these are the ones that, when imbued with modified functional properties, can have a higher economic value and replace more expensive animal proteins. Figure 4 illustrates the shear modulus of protein gels formed by thermal denaturation as affected by the degree of acetylation of *Vicia faba* protein (42). Acylation with succinic anhydride or citraconic anhydride produces even more striking results insofar as solubility effects, because of the conversion of the amino group to a carboxyl group. In some of the plant proteins, increases in solubility such as obtained from succinylation have a particular value when attempting to make solutions of the normally insoluble proteins.

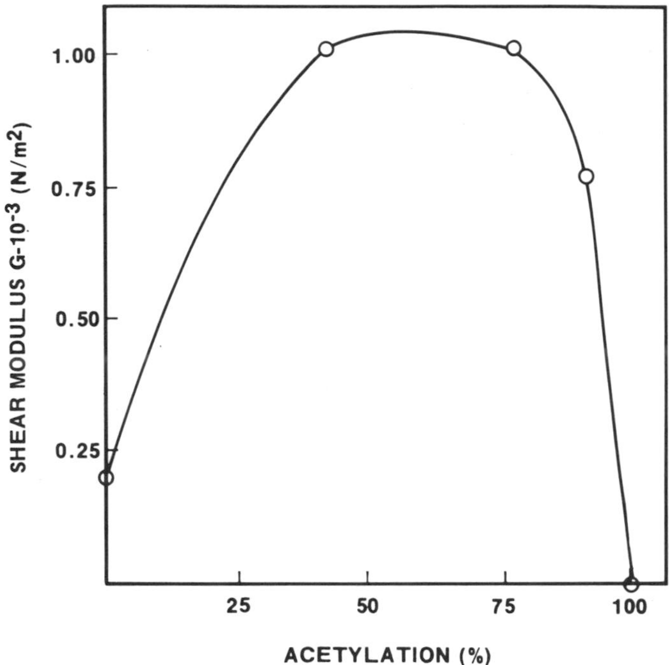

FIG. 4 *Shear modulus of protein gels (15%, pH 7) formed by thermal denaturation as dependent on the degree of acetylation of Vicia faba protein. From Schmandke et al. (42).*

In some cases attempts are made to introduce two new properties simultaneously, such as increased nutritional value and hydrophobicity. For the purpose of decreasing deteriorations, such as the Maillard reaction (reaction of sugar carbonyls with amino groups of lysines), attempts have been made to derivatize amino groups by acylation or by reductive alkylation (43). Because there was little information on the nutritional value of alkylated amino groups of lysine, it was necessary to study the effects of alkylation on the nutritional value in rats. It was found that a very high percentage of the lysine in casein substituted with dimethyl groups (50% substitution) would be nutritionally utilized, a condition which would make it possible to include rather large amounts of such modified proteins in the

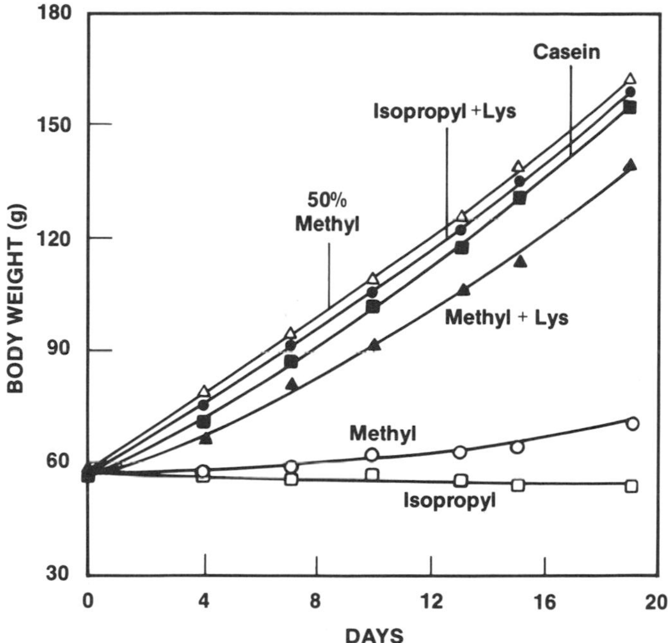

FIG. 5 *Growth of rats on N-methylated and isopropylated derivatives of casein. Diets and conditions are described in Lee et al. (43).*

diet. However, isopropylation did not support growth, although the rat was able to remove the isopropyl groups and excrete almost quantitatively the isopropyl lysine in the urine. In contrast, cyclopentylated casein was only partially digested down to the cyclopentyl lysine and partially excreted, and sugar derivatized casein was also unsatisfactorily utilized (44). Figure 5 shows the growth of rats on different substituted lysines of casein and the effects when 50% dimethylated casein was fed.

Two other studies have concerned the introduction of essential amino acids. The introduction of essential amino acids has been successfully achieved with tryptophan and methionine (45,46), and polymethionines have been put on (47). These have been described in more detail in this current volume (48).

Probably one of the most extensive uses of the modification procedure has been with alkali (49). The earlier studies with the Scandinavian lutfisk and the Indian corn meal for such dishes as tortillas have been more recently repeated with the use of alkali to form stabilized plant proteins primarily as meat substitutes or for the preparation of food bars (50). This process, however, requires careful quality control. There is a formation of cross-links from the production of lysinoalanine (49) from the β-elimination of cystine or glycosyl substituted serines or threonines and the subsequent condensation of these with the ε-amino group of lysine. Also formed is lanthionine, apparently from β-elimination of the cysteine. Such reactions not only produce undesirable compounds, particularly lysinoalanine which may accumulate in the kidney, but also cause losses of lysine. In addition, alkali causes racemization (49).

C. *GENETIC ENGINEERING IN FOOD PROTEINS*

At this time there appears to be little information available on genetically engineered food proteins with improved characteristics. There are, however, several examples where enzymes have been prepared that have superior characteristics for certain manufacturing processes. The most likely candidates at present would appear to be in the cheese industry, and these and other developments will be described in another article in this publication (33).

Laszlo P. Somogyi (51) has listed many new developments under way that are bringing new proteins to the food industry:

1. Recombinant DNA rennet
2. A new generation of destabilized microbial rennets with heat lability duplicating that of calf rennet
3. A bacterial β-amylase with improved heat stability for production of maltose syrup
4. Development of an enzyme system for accelerated ripening of cheese

5. The cloning of glucoamylase, a cellulase, and an oxidase
6. Cloning and expression via recombinant DNA in *Bacillus subtilis* of a neutral protease
7. Improved, less costly, and highly specific microbial lipases and esterases
8. An immobilized glucoamylase to convert dextrins in starch syrup to glucose
9. Genetically engineered enzymes for the hydrolysis of hemicellulose to pentose sugars followed by fermentation of pentoses to acetone and butanol (under development)

Some developmental products of specific organizations are: a neutral protease by Genex; a rennet by Gist-Brocades; a rennet by Dairyland Food Laboratories, Inc.; a "rennin" by Genencor, Inc.; and a glucoamylase, a pyranose-2-oxidase, and two cellulases (cellobiohydrolase and endogluconase) by Cetus Corporation.

IV. TAILORING OF MEDICINAL PROTEINS

Evidence for the tailoring of proteins for medicinal uses can be found scattered throughout historical writings. Today the tailoring of proteins is a burgeoning, rapidly expanding area. The following are examples of possible medicinal uses:

Encapsulation

Conjugates

1. Toxin: antibody
2. Enzyme: antibody
3. Chelating agent: antibody, other proteins
4. Anticancer proteins: synthetic polymer
5. Hormone: drug
6. Polylysine: drug
7. Dextran (and insulin): drug
8. Glycoprotein: drug

New enzyme

1. Enzyme: coenzyme conjugate
2. *In vitro* mutagenesis

Biologically active peptides

1. Synthesis
2. Enzyme hydrolysis

Peptides (for receptor blocking and immunogenicity)

Inactivated toxins (toxoids)

Some possible problems with tailored proteins for medicinal uses do exist, however. These include:

1. Heterogeneity of preparations
2. Lowered activity
3. Non-specificity
4. Cleared from system
5. Poor incorporation at target site
6. Accumulation in organs (e.g., kidney)
7. Immunological problems
8. Chemical or enzymatic breakdown

Nevertheless many of these difficulties are amenable to correction or avoidance by further research.

A. *ENZYMATIC MODIFICATIONS*

Enzymatic modifications of proteins for medicinal uses have been more or less restricted to nutritional preparations and for modification of nonproteins such as steroids. In some cases the use of enzymes for commercial processing of highly expensive items such as steroids has become an important phase of the pharmaceutical industry. The use of enzymes for such purposes, however, is not a part of the general topic of this review.

The use of enzymes for preparing peptides for research purposes has recently been employed for the formation of opioids from food proteins such as casein. This is covered extensively in the article by Chiba and Yoshikawa in this volume (6).

B. *CHEMICAL MODIFICATIONS*

Chemical modifications have long been used for modifying proteins in the pharmaceutical industry. The use of cross-linking agents such as formaldehyde for the conversion of toxins to toxoids is now over half a century old, and the more recent use of glutaraldehyde and formaldehyde for the fixation and preparation of animal heart valves has reached extensive industrial production (52). Perhaps up to 100,000 heart valves for human implantation have been made in one single year by these procedures.

A combined physical and chemical procedure for the conversion of proteins with a particular enzymatic activity to ones with a different enzymatic activity has been recently exploited by Keyes and Saraswathi (53). This is actually a procedure based on what was originally thought to be how antibodies were formed. In the older theories of antibody specificity it was believed that antibodies against certain substances were formed at the time of biosynthesis of the antibody by a template mechanism, i.e., the template influenced the folding and cross-linking by disulfides of the antibody. The recent work has altered protein structure by the effects of pH, temperature, solvent, or disulfide rearrangement and simultaneous equilibration with the template. These materials were then cross-linked with glutaraldehyde and purified. According to the authors, ribonuclease treated in the presence of indole-3-propionic acid produced two protein products with different esterase activity. Likewise, bovine serum albumin treated in the presence of indole also made an esterase, while α-amylase in the presence of cellobiose formed a β-amylase.

A different way of introducing a new enzymatic activity has been developed in the laboratory of E. T. Kaiser and has been recently reviewed (54). In essence, the method employs the introduction of a catalytic group at or on the periphery of the active site to give "a semi synthetic enzyme." In other words, the combining properties of the enzyme, the very important part of enzyme activity, are utilized, while the covalent addition of a prosthetic group can endow the protein with a catalytic activity that it did not have before.

In some instances a synthesis of part of a protein may have considerable activity in certain systems, particularly those which react with certain receptors on a cell or react with immunological systems. One such type is the synthesis of 20 amino acids which have a diptheria toxin-like function (55). In other instances a minor change in a small peptide can greatly aid in the function of an otherwise unusable compound for certain modes of entry to the body. A related example is an insertion of an unnatural amino acid (D-alanine) in enkaphalin pentapeptide which gave a stability against digestive enzymes for oral presentation.

An exotic way of delivering compounds to a particular site or organ is to tie them to a particular messenger addressed to that site or organ. Such messengers can be proteins or hormones. There are now many examples where materials have been attached to proteins or to peptides and have considerable target specificity (56). Table 8 shows the inhibition of RNA polymerase by, and the binding of the toxin amatoxin to, amatoxin specific antibodies by mixed anhydride coupling of the amatoxins to various proteins, polylysine or polyornithine. Figure 6 shows the inhibition of HeLa cell growth by 6-aminonicotinamide conjugated to polylysine. Other protein carriers are serum transferrin and glycoproteins of various types, particularly those having affinity for some lectins. Other chemical substances such as anti-cancer agents like neocarcinostatin (59) or chelating agents (60) have been attached to macromolecular carriers. Chelating agents can be used so effectively that they are currently under extensive clinical trials in this country and in Japan. The neocarcinostatin conjugate with different types of hydrophobic and hydrophilic materials can be directed at different organs of the body, while the chelating agents frequently attached to transferrins can bring particular radioactive elements to tumor-bearing organs either for imagery or for killing of the rapidly growing tissue.

Two or more proteins may be attached together or two proteins may be attached to other substances. When subunits of some

TABLE 8 *Carbodiimide coupling of amatoxins to various proteins*[1]

Compound[2]	Ratio	Carbodiimide used[3]	Increase in β-amanitin (βA) toxicity after coupling to proteins (free βA = 1)[4]
Bovine serum albumin- (βA)	*0.6*	*MCDI*	*6*
Bovine serum albumin-(βA)	*1.6*	*ECDI*	*50-100*
Calf asialofetuin-(βA)	*5.7*	*ECDI*	*9*
Calf asialofetuin-(αA)	*0.2*	*ECDI*	*70*

[1]From Faulstich and Fiume (57).
[2]βA = β-amanitin and αA = α-amanitin.
[3]MCDI = 1-cyclohexyl-3-[2-morpholinyl-(4)-ethyl]carbodiimide and ECDI = 1-ethyl-3-(dimethylaminopropyl)carbodiimide.
[4]The LD_{50} (ip) of free βA for mouse is 0.5 mg/kg body weight.

proteins are exchanged and covalently attached, they are popularly described as chimeric proteins. These will be covered in several other articles (in this publication) (29,61,62). Earlier studies on these were done in the exquisite researches of Dr. S. Olsnes and colleagues working with chimera of abrin and ricin. In these studies they were separating the two halves of the molecule connected by disulfide linkages; the *b* unit served as the recognition protein and the device for helping get the *a* unit into the cell, and the *a* unit bore the toxic enzyme to kill the cell. These studies have been extended to such important systems as

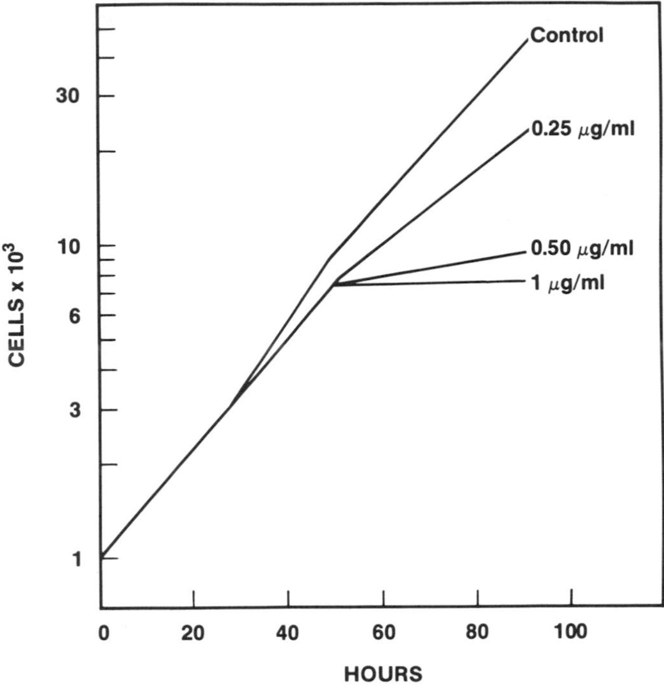

FIG. 6 *The inhibition of the HeLa cell growth by the methotrexate conjugate. The conjugate is 1:20 (w/w) methotrexate/poly(L-lysine) HBr and the poly(L-lysine) used is M_r 3000. Note the lack of inhibition until the second day. From Arnold (58).*

combining the diptheria toxin with the *b* units of ricin or abrin or with serum transferrin. This is an area currently receiving intense interest. A different conjugate of two proteins is formed by the attachment of a toxic protein to an antibody specific for certain sites (29). A still further elaboration of this is the covalent attachment of three proteins together, such as studied by Poznansky (61), in which the serum albumin is attached to a conjugate of an enzyme and a recognition substance such as an antibody or a hormone. In these instances the serum albumin apparently stabilizes the conjugate to destruction in the body.

C. *PRODUCTS OF GENETIC ENGINEERING*

One of the main thrusts today in the formation of proteins or peptides for medicinal use is their production in microorganisms by genetic transfer. Although this is not a modification or a tailoring of the protein in any sense, it represents a tailoring of the system and is a large and important program with items already appearing on the market.

The large revolution in this area is in the use of *in vitro* mutagenesis to form new proteins or polypeptides that will have novel activities in medicinal uses. This area, very valuable for both fundamental protein chemistry and for practical application, is accelerating at a logarithmic rate in many institutions and companies around the world. Since these also will be discussed in another article (5), they will only be mentioned here. The magnitude of the operations in the U.S.A. is seen in the extensive programs of some of the U.S. biotechnology companies (Table 9). On the international scale, the German Bayer Company has a large in-house program, cooperative programs with several other pharmaceutical companies, and agreements with several universities (64).

V. FURTHER POSSIBLE APPLICATIONS

Protein tailoring can have many further applications, depending upon economics as well as scientific advances. The general optimism of the future of biotechnology has been expressed in an editorial note by Daniel E. Koshland, Jr. (65), the editor of *Science*, and the articles published in that particular issue of the journal (66-72).

A. *FOOD*

Protein tailoring should have a rapidly expanding role in the general area of biotechnology in food production and processing (70). Certainly, an increase in the nutritional value of plant products always appears promising.

TABLE 9 *U.S. biotechnology companies*[1]

Company[2]	1983 Sales	1984 Sales	1990 Sales (projected)	Major products
Genentech	*42.4*	*69.8 (est)*	*633-950*	*HGH, IF, insulin, TNF, TPA*
Cetus	*18.5*	*35.9*	*313-470*	*Diagnostics, IF, IL-2, TNF*
Biogen	*18.4*	*31.4*	*313-470*	*IF, IL-2, TPA, TNF*
Hybritech	*16.0*	*30.3*	*173-260*	*Diagnostics, MAbs*
Genex	*11.1*	*34.8*	*214-300*	*IF, IL-2, L-Aspartate*
Collaborative research	*7.7*	*N/A*	*73-110*	*BCGF, IF, IL-1 IL-2, KPA*
Centocor	*7.4*	*12.8*	*113-170*	*Diagnostics, MAbs*
Molecular genetics	*6.9*	*N/A*	*73-110*	*Vaccines, diagnostics*

[1]From Dibner (63).

[2]The largest US companies with pharmaceutical emphasis based on 1983 sales from available data, in current $ millions. From *Chemical Week*, *Genetic Engineering News*, and *SCRIP*, 1983, 1984; N/A = data not available; est = estimate. Product abbreviations: BCGF, B Cell Growth Factor; HGH, Human Growth Hormone; IF, Interferon; IL, Interleukin; KPA, Kidney Plasminogen Activator; MAbs, Monoclonal antibodies; TNF, Tumor Necrosis Factor; TPA, Tissue Plasminogen Activator.

B. *AGRICULTURE*

Plant development, a cornerstone of agriculture, is an area under explosive expansion, and new horizons, such as finding the new plant vector (*Arabidopsis thaliana*) (66), are continually being cited. Disease resistant and chemical (e.g., herbicide) resistant strains are continually being sought.

C. *VETERINARY MEDICINE*

Although veterinary medicine has not been included in the present article, protein tailoring should have one of its most prominent roles in this general area. Extensive contributions should come from the development of superior animals by genetic engineering, as well as from the development and use of new veterinary medical products by genetic engineering, including the wider scope of biotechnology. Articles are continually appearing on companies involved in these subjects (73). Undoubtedly, the possibility of much more rapid clearances of applications for animal usage as compared to clearances for human usage will greatly speed up expansion in this area.

D. *MEDICINE*

Ways to prevent, diagnose and treat human illnesses will obviously continue to be a primary subject for research on protein tailoring. Genetic engineering to replace enzyme deficiencies or to form better enzymes are possibilities. Many other areas appear to be as yet unexplored. One such area could be the production of structured proteins designed for bioprosthesis.

E. *DEFENSE*

As with almost all areas of science, developments from research on the modifications of proteins could be used for chemical and biological warfare (CBW). Although there is no quotable evidence for any comprehensive programs on the use of protein modifications for CBW, it appears irresponsible to the author to omit at least a mention of the subject. The following are a few possible candidates for inclusion in such programs. Truly defensive ones could include: antidotes for CBW agents; inactivators for CBW agents (filters, decontaminants, using immobilized proteins); and detectors for CBW agents (enzymes or receptors capable of giving a signal). Offensive ones (whether used by simple release into an environment, such as an aerosol or in the water supply, or by more direct means) could include: targeted toxins; new toxins (chemical or enzymatic derivatives); new toxins (*in vitro* mutagenesis); new virulent pathogenic organisms or viruses; and anti-immune substances, including viruses.

At this time only defensive applications are internationally legal. However, one cannot ignore that some nations, or even relatively small groups, could manufacture proteins tailored to kill with comparatively little equipment and expense. But, hopefully, world-wide opinion, the difficulties in using offensive agents, and preventative measures against their use, would make them unsuitable.

F. *INDUSTRY*

Immobilized enzymes or recombinantly produced enzymes already have applications in both the food and pharmaceutical industries. Newly designed specificities and stabilities to different environments (e.g., temperature, solvents) should continue to appear. Perhaps recombinantly produced enzymes will be the critical component in what the popular press has called the "biochip" of the future.

VI. CONCLUSIONS

The modification of the structure of protein either by chemical reactions, enzymatic treatments, or *in vitro* mutagenesis, or a combination of these, is expanding exponentially. Although most all of these are still at the laboratory stage, several are now important products for food processing and medicine. Modification of food proteins for direct consumption, however, lags way behind. Only such established treatments as proteolysis, alkali or oxidizing agents are currently permitted. However, more modified food proteins should become acceptable as research results indicate their healthfulness and as the public develops acceptances for modified proteins for medical uses. Tailoring proteins for medical use is currently being studied in many laboratories, and many clinical tests are underway. New products for medical use are expected to appear each year from the many companies engaged in these areas.

ACKNOWLEDGMENTS

The author would like to thank Chris Howland for editorial assistance and Diana Melbourn for typing the manuscript.

REFERENCES

1. K. Mosbach (ed.), Immobilized Enzymes, *Methods in Enzymology*, Vol. 44, Academic Press, New York, p. 999 (1976).
2. W. H. Scouten (ed.), *Solid Phase Biochemistry. Analytical and Synthetic Aspects*, John Wiley & Sons, New York, p. 779 (1983).
3. W. F. Benisek and M. Hearne, Probing the active site of a steroid isomerase with a solid phase reagent, *Protein Tailoring and Reagents for Food and Medical Uses* (R. E. Feeney and J. R. Whitaker, eds.), Marcel Dekker, New York (1986).
4. J. J. Marshall and M. L. Rabinowitz, Enzyme stabilization by covalent attachment of carbohydrate, *Arch. Biochem. Biophys. 167*: 777 (1975).
5. R. Wetzel, Medical applications of protein engineering, *Protein Tailoring and Reagents for Food and Medical Uses* (R. E. Feeney and J. R. Whitaker, eds.), Marcel Dekker, New York (1986).
6. H. Chiba and M. Yoshikawa, Biologically functional peptides from food proteins, *Protein Tailoring and Reagents for Food and Medical Uses* (R. E. Feeney and J. R. Whitaker, eds.), Marcel Dekker, New York (1986).
7. C. H. W. Hirs (ed.), Enzyme Structure, *Methods in Enzymology*, Vol. 11, Academic Press, New York, p. 988 (1967).

8. G. Means and R. E. Feeney, *Chemical Modification of Proteins*, Holden-Day, San Francisco, p. 254 (1971).

9. A. N. Glazer, R. J. Delange and D. S. Sigman, *Chemical Modification of Proteins. Selected methods and analytical procedures*, Elsevier, Amsterdam p. 205 (1975).

10. C. H. W. Hirs and S. N. Timasheff (eds.), Enzyme Structure Part E, *Methods in Enzymology*, Vol. 47, Academic Press, New York, p. 668 (1977).

11. C. H. W. Hirs and S. N. Timasheff (eds.), Enzyme Structure Part I, *Methods in Enzymology*, Vol. 91, Academic Press, New York, p. 693 (1983).

12. R. L. Lundblad and C. M. Noyes, *Chemical Reagents for Protein Modification*, Vol. 1, CRC Press, Boca Raton, Florida, p. 180 (1984).

13. R. L. Lundblad and C. M. Noyes, *Chemical Reagents for Protein Modification*, Vol. 2, CRC Press, Boca Raton, Florida, p. 178 (1984).

14. R. E. Feeney and Y. Yeh, Antifreeze proteins from fish bloods, *Adv. Protein Chem. 32*: 191 (1978).

15. R. E. Feeney, T. S. Burcham and Y. Yeh, Antifreeze glycoproteins from polar fish blood, *Ann. Rev. Biophys. Biophys. Chem. 15*: in press (1986).

16. R. E. Feeney, Penguin egg-white and polar fish blood-serum proteins, *Int. J. Peptide Protein Res. 19*: 215 (1982).

17. A. L. DeVries, Antifreeze peptides and glycopeptides in cold-water fishes, *Ann. Rev. Physiol. 45*: 245 (1983).

18. A. L. DeVries, Role of glycopeptides and peptides in inhibition of crystallization of water in polar fishes, *Phil. Trans. R. Soc. Lond. B 304*: 575 (1984).

19. R. E. Feeney, D. T. Osuga, C. F. Meares, D. R. Babin and M. H. Penner, Studies on iron-binding sites of transferrin by chemical modification, *Structure and Function of Iron Storage and Transport Proteins* (I. Urushizaki, P. Aisen, I. Listowsky and J. R. Drysdale, eds.), Elsevier, Amsterdam, p. 231 (1983).

20. T. B. Rogers, R. A. Gold and R. E. Feeney, Ethoxyformylation and photooxidation of histidines in transferrins, *Biochemistry 16*: 2299 (1977).

21. M. H. Penner, D. T. Osuga, C. F. Meares and R. E. Feeney, Interaction of oxidized chicken ovotransferrin with chicken embryo red blood cells, *Biochim. Biophys. Acta 827*: 389 (1985).

22. M. M. Simlot and R. E. Feeney. Relative reactivities of chemically modified turkey ovomucoids, *Arch. Biochem. Biophys. 113*: 64 (1966).

23. R. Haynes and R. E. Feeney, Transformation of active-site lysine in naturally occurring trypsin inhibitors. A basis for a general mechanism for inhibition of proteolytic enzymes, *Biochemistry 7*: 2879 (1968).

24. P. E. Thorpe, S. I. Detre, B. M. J. Foxwell, A. N. F. Brown, D. N. Skilleter, G. Wilson, J. A. Forrester and F. Stirpe, Modification of the carbohydrate in ricin with metaperiodate-cyanoborohydride mixtures. Effects on toxicity and *in vivo* distribution, *Eur. J. Biochem. 147*: 197 (1985).

25. W. B. Jakoby and M. Wilchek (eds.), Affinity Labeling, *Methods in Enzymology*, Vol. 46, Academic Press, New York, p. 774 (1977).

26. R. B. Silverman, Mechanism-based enzyme inactivators for medical uses, *Protein Tailoring and Reagents for Food and Medical Uses* (R. E. Feeney and J. R. Whitaker, eds.), Marcel Dekker, New York (1986).

27. W. S. D. Wong, M. M. Kristjansson, D. T. Osuga and R. E. Feeney, 1-Deoxyglycitolation of protein amino groups and their regeneration by periodate oxidation, *Int. J. Peptide Protein Res. 26*: 55 (1985).

28. W. H. Scouten, *Affinity Chromatography. Bioselective adsorption on inert matrices*, John Wiley & Sons, New York, p. 348 (1981).

29. J. W. Marsh and D. M. Neville, Jr., Immunotoxins: chemical variables affecting cell killing efficiencies, *Protein Tailoring and Reagents for Food and Medical Uses* (R. E. Feeney and J. R. Whitaker, eds.), Marcel Dekker, New York (1986).

30. T. H. Ji, Bifunctional reagents, *Methods in Enzymology*, Vol. 91 (C. H. W. Hirs and S. N. Timasheff, eds.), Academic Press, New York, p. 580 (1983).

31. J. Carlsson, H. Drevin and R. Axén, Protein thiolation and reversible protein-protein conjugation. N-Succinimidyl 3-(2-pyridyldithiol)propionate, a new heterobifunctional reagent, *Biochem. J. 173*: 723 (1978).

32. J. M. Varga, Hormone-drug conjugates, *Methods in Enzymology*, Vol. 112 (K. J. Widder and R. Green, eds.), Academic Press, New York, p. 259 (1985).

33. R. Jimenez-Flores, Y. Kang and T. Richardson, Genetic engineering of enzymes and food proteins, *Protein Tailoring and Reagents for Food and Medical Uses* (R. E. Feeney and J. R. Whitaker, eds.), Marcel Dekker, New York (1986).

34. D. A. Estell, T. P. Graycar and J. A. Wells, Engineering an enzyme by site-directed mutagenesis to be resistant to chemical oxidation, *J. Biol. Chem. 260*: 6518 (1985).

35. R. E. Feeney and R. G. Allison, *Evolutionary Biochemistry of Proteins: Homologous and Analogous Proteins from Avian Egg Whites, Blood Sera, Milk, and Other Substances*, John Wiley & Sons, New York, p. 290 (1969).

36. R. E. Feeney and J. R. Whitaker, Chemical and enzymatic modification of plant proteins, *Seed Storage Proteins* (A. Altschul and H. L. Wilcke, eds.), *New Protein Foods*, Vol. 5, Academic Press, New York, p. 181 (1985).

37. J. R. Whitaker, Enzymatic modification of proteins applicable to foods, *Food Proteins: Improvement through Chemical and Enzymatic Modification* (R. E. Feeney and J. R. Whitaker, eds.), Adv. Chem. Series, Vol. 160, American Chemical Society, Washington, D.C., p. 95 (1977).

38. M. Yamashita, S. Arai and M. Fujimaki, A low-phenylalanine, high-tyrosine plastein as an acceptable dietetic food. Method of preparation by use of enzymatic protein hydrolysis and resynthesis, *J. Food Sci. 41*: 1029 (1976).

39. M. Watanabe and S. Arai, Proteinaceous surfactants prepared by covalent attachment of L-leucine n-alkyl esters to food proteins by modification with papain, *Modification of Proteins: Food, Nutritional, and Pharmacological Aspects* (R. E. Feeney and J. R. Whitaker, eds.), Adv. Chem. Series, Vol. 198, American Chemical Society, Washington, D.C., p. 199 (1982).

40. J. E. Kinsella, Texturized proteins: fabrication, flavoring, and nutrition, *CRC Crit. Rev. Food Sci. Nutr. 10*: 147 (1978).

41. J. K. Shetty and J. E. Kinsella, Reversible modification of lysine: separation of proteins and nucleic acids in yeast, *Modification of Proteins: Food, Nutritional, and Pharmacological Aspects* (R. E. Feeney and J. R. Whitaker, eds.), Adv. Chem. Series, Vol. 198, American Chemical Society, Washington, D.C., p. 169 (1982).

42. H. Schmandke, R. Maune, S. Schuhmann and M. Schultz, Contribution to the characterization of acetylated protein fractions of Vicia faba in view of their functional properties, *Nahrung 25*: 99 (1981).

43. H. S. Lee, L. C. Sen, A. J. Clifford, J. R. Whitaker and R. E. Feeney, Effect of reductive alkylation of the ε-amino group of lysyl residues of casein on its nutritive value in rats, *J. Nutr. 108*: 687 (1978).

44. H. S. Lee, L. C. Sen, A. J. Clifford, J. R. Whitaker and R. E. Feeney, Preparation and nutritional properties of caseins covalently modified with sugars. Reductive alkylation of lysines with glucose, fructose, or lactose, *J. Agric. Food Chem. 27*: 1094 (1979).

45. A. J. Puigserver, L. C. Sen, E. Gonzales-Flores, R. E. Feeney and J. R. Whitaker, Covalent attachment of amino acids to

casein. 1. Chemical modification and rates of in vitro enzymatic hydrolysis of derivatives, *J. Agric. Food Chem. 27*: 1098 (1979).

46. A. J. Puigserver, L. C. Sen, A. J. Clifford, R. E. Feeney and J. R. Whitaker, Covalent attachment of amino acids to casein. 2. Bioavailability of methionine and N-acetylmethionine covalently linked to casein. *J. Agric. Food Chem. 27*: 1286 (1979).

47. H. F. Gaertner and A. J. Puigserver, Covalent attachment of poly(L-methionine) to food proteins for nutritional and functional improvement, *J. Agric. Food Chem. 32*: 1371 (1984).

48. J. R. Whitaker, Covalent attachment of essential amino acids to proteins to improve their nutritional and functional properties, *Protein Tailoring and Reagents for Food and Medical Uses* (R. E. Feeney and J. R. Whitaker, eds.), Marcel Dekker, New York (1986).

49. J. R. Whitaker and R. E. Feeney, Chemical and physical modification of proteins by the hydroxide ion, *CRC Crit. Rev. Food Sci. Nutr. 19*: 173 (1983).

50. R. E. Feeney, R. B. Yamasaki and K. F. Geoghegan, Chemical modification of proteins: an overview, *Modification of Proteins: Food, Nutritional and Pharmacological Aspects* (R. E. Feeney and J. R. Whitaker, eds.), Adv. Chem. Series, Vol. 198, American Chemical Society, Washington, D.C., p. 3 (1982).

51. L. P. Somogyi, Trends, opportunities and strategies for entry, *Enzymes-Worldwide*, Dec. issue: 48 (1984).

52. A. Carpentier, A. Nashef, S. Carpentier, N. Goussef, J. Relland, R. J. Levy, M. C. Fishbein, B. El Asmar, M. Benomar, S. El Sayed and P. G. Donzeau-Gouge, Prevention of tissue valve calcification by chemical techniques, *Cardiac Bioprostheses. Proceedings of the Second International Symposium* (L. H. Cohn and V. Gallucci, eds.), Yorke Medical Books, New York, p. 320 (1982).

53. Semisynthetic enzymes formed from proteins, *Chem. Engr. News 62* (37): 33 (1984).

54. E. T. Kaiser, D. S. Lawrence and S. E. Rokita, The chemical modification of enzymatic specificity, *Ann. Rev. Biochem. 54*: 565 (1985).

55. J. Vane and P. Cuatrecasas, Genetic engineering and pharmaceuticals, *Nature 312*: 303 (1984).

56. K. J. Widder and R. Green (eds.), Drug and Enzyme Targeting Part A, *Methods in Enzymology*, Vol. 112, Academic Press, New York, p. 589 (1985).

57. H. Faulstich and L. Fiume, Protein conjugates of fungal toxins, *Methods in Enzymology*, Vol. 112 (K. J. Widder and R. Green, eds.), Academic Press, New York, p. 225 (1985).

58. L. J. Arnold, Jr., Polylysine-drug conjugates, *Methods in Enzymology*, Vol. 112 (K. J. Widder and R. Green, eds.), Academic Press, New York, p. 270 (1985).

59. H. Maeda, Y. Matsumura, T. Oda and K. Sasamoto, Cancer-selective macromolecular therapeusis: tailoring of an antitumor protein drug, *Protein Tailoring and Reagents for Food and Medical Uses* (R. E. Feeney and J. R. Whitaker, eds.), Marcel Dekker, New York (1986).

60. C. F. Meares, Attaching metal ions to antibodies, *Protein Tailoring and Reagents for Food and Medical Uses* (R. E. Feeney and J. R. Whitaker, eds.), Marcel Dekker, New York (1986).

61. M. J. Poznansky, Tailoring enzymes for more effective use as therapeutic agents, *Protein Tailoring and Reagents for Food and Medical Uses* (R. E. Feeney and J. R. Whitaker, eds.), Marcel Dekker, New York (1986).

62. K. Sandvig and S. Olsnes, Entry of protein toxins into cells, *Protein Tailoring and Reagents for Food and Medical Uses* (R. E. Feeney and J. R. Whitaker, eds.), Marcel Dekker, New York (1986).

63. M. D. Dibner, The pharmaceutical industry: impacts of biotechnology, *Trends Pharmacol. Sci. 6*: 343 (1985).

64. E. Truscheit, Gentechnologie: hoffnung im kampf gegen hunger und krankheiten, *Bayer-Berichte 51*: 2 (1984).

65. D. E. Koshland, Jr., Excursions in biotechnology, *Science 229*: 1191 (1985).

66. E. M. Meyerowitz and R. E. Pruitt, *Arabidopsis thaliana* and plant molecular genetics, *Science 229*: 1214 (1985).

67. S. L. Morrison, Transfectomas provide novel chimeric antibodies, *Science 229*: 1202 (1985).

68. B. Roizman and F. J. Jenkins, Genetic engineering of novel genomes of large DNA viruses, *Science 229*: 1208 (1985).

69. D. Botstein and D. Shortle, Strategies and applications of in vitro mutagensis, *Science 229*, 1193 (1985).

70. D. Knorr and A. J. Sinskey, Biotechnology in food production and processing, *Science 229*, 1224 (1985).

71. R. A. Smith, M. J. Duncan and D. T. Moir, Heterologous protein secretion from yeast, *Science 229*, 1219 (1985).

72. M. D. Dibner, Biotechnology in pharmaceuticals: the Japanese challenge, *Science 229*, 1230 (1985).

73. D. Webber, Agricultural biotech firm rides animal health products to market, *Chem. Engr. News 63* (43): 12 (1985).

2

Covalent Attachment of Essential Amino Acids to Proteins to Improve Their Nutritional and Functional Properties

John R. Whitaker

Department of Food Science and Technology
University of California
Davis, California

I. INTRODUCTION

Plant proteins, the major source of protein for most of the world's population, are usually deficient in one or more of the essential amino acids. For example, the limiting essential amino acid of legume seed protein is methionine while for cereal grains it is lysine. Tryptophan and threonine are often the second and/or third limiting essential amino acids. A deficiency in a limiting essential amino acid results in a lowered protein efficiency ratio (PER) value. For example, the PER value for most common beans is in the range of 0.9–1.0 while the addition of methionine to 3% (of dry bean weight) increased the value to about 2.6 (1). This is an important doubling of PER, thereby requiring half as much protein to meet the daily dietary requirement.

There are several potential and real solutions to improving the PER value of plant proteins. These include: (a) fortification of the protein source with the free limiting amino acid(s); (b) mixing proteins containing complementary amino acid compositions; (c) covalent attachment of essential amino acids to food proteins; (d) breeding of plants to improve the essential amino acid composition. This would include the use of recombinant DNA techniques; and (e) the fortification of proteins with polyamino acids.

The advantages of fortification of high protein foods with the free limiting amino acid(s) include the general ease of doing so (especially in flours and solutions) as well as economic considerations. However, there are a number of disadvantages. These include: (a) possible losses of the added amino acid during food preparation and processing; (b) side reactions of the amino acid occurring during processing (Maillard reaction, etc.); (c) undesirable effect on sensory properties of the food (such as methional formation as result of reduction of methionine); and (d) possible decreased biological utilization of free amino acids. (Free amino acids may arrive in the blood stream ahead of those released by proteolysis of proteins in the gut, thereby not being available simultaneously with the other essential amino acids for synthesis into required proteins and enzymes (2,3)).

Mixing of proteins with complementary amino acid compositions, if available, is an economical and appropriate way of meeting the daily nutritional needs of the essential amino acids. A mixture of rice and beans, for example, as used in some countries, has a high PER value. Unfortunately, the two protein sources are not always available simultaneously in developing countries.

Breeding, including recombinant DNA technology, has proved to be of only limited use so far. High lysine-containing corn is well known. Unfortunately, breeding to increase the lysine level has resulted in decreased yields and proteins with different functional properties. In common beans, very little work has been done in increasing the methionine content, in part because there is an incomplete understanding of the regulation of sulfur incorporation into methionine (and cystine and cysteine). As noted above, modification of amino acid composition (as a result of changing the percentage composition of one or more proteins) may result in changes in the functional properties of a high protein food as well as possible yield.

The remainder of this chapter will deal with the covalent attachment of essential amino acids to proteins, including some comments on the possible utility of polyamino acids fortification in improving the nutritional quality of proteins.

II. COVALENT ATTACHMENT OF ESSENTIAL AMINO ACIDS TO PROTEINS

Some of the purposes and advantages of covalent incorporation of limiting essential amino acids into high protein foods include: (a) improve nutritional quality; (b) prevent loss of added amino acid(s); (c) alleviate undesirable flavors of the free amino acid(s) (or products of the amino acid(s)); (d) possible changes in functional properties of the modified proteins; and (e) prevent deteriorative reactions (Maillard reaction, etc.) during processing and storage of the food. A fundamental advantage is to permit a thorough investigation of the *in vivo* digestibility of the isopeptide bond formed as result of covalent attachment of the amino acid(s) to proteins. Disadvantages, of course, are the additional costs of the process as well as in gaining approval of the product for food use.

Several laboratory-scale methods with potential for commercial scaleup have been developed by protein chemists for the covalent incorporation of amino acids into protein. These include: (a) water-soluble carbodiimide method; (b) activated ester method; (c) carboxyanhydride method; (d) $POCl_3$-activated incorporation; (e) fortification with polyamino acid polymers; (f) plastein reaction; (g) transglutaminase method; (h) sulfhydryl-disulfide interchange method; and (i) recombinant DNA method.

Table 1 *Types of bonds formed by covalent incorporation of amino acids into proteins*

Normal peptide bond

$$H_3\overset{\oplus}{N}\text{-}CHR_1\text{-}\boxed{\underset{\alpha\quad\alpha}{CO\text{-}NH}}\text{-}CHR_2\text{-}CO\text{-}(P)$$

$$(P)\text{-}NH\text{-}CHR_1\text{-}\boxed{\underset{\alpha\quad\alpha}{CO\text{-}NH}}\text{-}CHR_2\text{-}COO^{\ominus}$$

Isopeptide bonds

$$H_3\overset{\oplus}{N}\text{-}CHR_1\text{-}\boxed{\underset{\alpha\quad\epsilon}{CO\text{-}NH}}\text{-}(CH_2)_4\text{-}\underset{}{\overset{|}{\overset{NH}{}}}CH\text{-}CO\text{-}(P)$$

$$H_3\overset{\oplus}{N}\text{-}\underset{^{\ominus}OOC}{CH}\text{-}(CH_2)_x\text{-}\boxed{\underset{\beta,\gamma\quad\alpha}{CO\text{-}NH}}\text{-}CHR_1\text{-}CO\text{-}(P)$$

$$H_3\overset{\oplus}{N}\text{-}\underset{^{\ominus}OOC}{CH}\text{-}(CH_2)_x\text{-}\boxed{\underset{\beta,\gamma\quad\epsilon}{CO\text{-}NH}}\text{-}(CH_2)_4\text{-}\overset{NH}{CH}\text{-}CO\text{-}(P)$$

$$H_3\overset{\oplus}{N}\text{-}\underset{^{\ominus}OOC}{CH}\text{-}(CH_2)_4\text{-}\boxed{\underset{\epsilon\quad\beta,\gamma}{NH\text{-}CO}}\text{-}(CH_2)_x\text{-}\overset{HN}{CH}\text{-}CO\text{-}(P)$$

$$H_3\overset{\oplus}{N}\text{-}\underset{^{\ominus}OOC}{CH}\text{-}(CH_2)_4\text{-}\boxed{\underset{\epsilon\quad\alpha}{NH\text{-}CO}}\text{-}CHR_1\text{-}NH\text{-}(P)$$

$$^{\ominus}OOC\text{-}CHR_1\text{-}\boxed{\underset{\alpha\quad\beta,\gamma}{NH\text{-}CO}}\text{-}(CH_2)_x\text{-}\overset{HN}{CH}\text{-}CO\text{-}(P)$$

When amino acids are incorporated *in vitro* into proteins, isopeptide bonds as well as peptide bonds are usually formed. The nature of these bonds is shown in Table 1. Addition of an amino acid to either the α-amino group at the N-terminal end or the α-

carboxyl group at the C-terminal end of the protein chain results in a peptide bond. However, additions of the amino acid to the ϵ-amino group of a lysyl residue or to the β- or γ-carboxyl group of an aspartyl or glutamyl residue of a protein results in an isopeptide bond. The *in vitro* and *in vivo* digestibility of the isopeptide bond will be addressed below.

A. *CARBODIIMIDE METHOD*

The carbodiimide method, developed first by Khorana (4) and Sheehan and Hess (5), involves activation of the carboxyl group of a protein (or N-acyl amino acid) with a water soluble carbodiimide such as 1-ethyl-3-(3-dimethylaminopropyl)carbodiimide. The reaction is shown in Eq. (1).

$$\text{Ⓟ-COOH} + \text{RN=C=NR'} \xrightarrow{H^{\oplus}} \text{Ⓟ-}\overset{O}{\overset{\|}{C}}\text{O-C}\begin{matrix}\diagup NHR' \\ \diagdown\!\!\diagdown \overset{\oplus}{N}HR\end{matrix} \tag{1}$$

$$\downarrow H_2N\text{-}CHR_1\text{-}COOR_2$$

$$\text{Ⓟ-CO-NH-CHR}_1\text{-COOR}_2 + \text{RNH-CO-NHR'}$$

$$\downarrow OH^{\ominus}$$

$$\text{Ⓟ-CO-NH-CHR}_1\text{-COO}^{\ominus} + R_2OH$$

The carbodiimide-activated intermediate is quite reactive and undergoes two additional reactions: (a) hydrolysis to the original protein plus a substituted urea, R-NH-CO-NHR', or (b) O to N migration to give Ⓟ-CO-NR-CO-NHR'. These two side reactions must be minimized by control of the experimental conditions. The other major problem is the formation of the substituted urea, RNH-CO-NHR', in equivalent amount to peptide bond formation. This must be removed before use of the protein for food or feed.

Table 2 shows some data of Voutsinas and Nakai (6) for the covalent incorporation of methionine and tryptophan into soy

Table 2 *Covalent attachment of amino acids to proteins by the carbodiimide method*[a]

Protein	Methionine	Tryptophan
	(grams/16 grams nitrogen)	
Control soy protein	*0.94*	*0.95*
Soy protein + L-methionine	*5.92*	*0.90*
Soy protein hydrolyzate + L-methionine	*7.22*	*0.94*
Soy protein + L-tryptophan	*0.83*	*10.74*
Soy protein isolate + L-tryptophan	*0.94*	*17.05*

[a]*Ref. 6.*

protein by the water soluble carbodiimide method. The methionine and tryptophan enrichments of the intact protein were 6.3- and 11.3-fold, respectively. Increased enrichment was obtained by use of partially hydrolyzed soy protein, as a result of increase in carboxyl group availability. Either the carboxyl groups on the protein or the carboxyl group of the amino acid may be activated by the water soluble carbodiimide. If the latter reaction is used the amino group of the amino acid should be derivatized.

B. *ACTIVE ESTER METHOD*

The active ester method, first developed by Wieland et al. (7), Bodanszky (8) and Schwyzer et al. (9), involves the use of the N-hydroxysuccinimide derivative (10) of an amino acid in which the carboxyl group is activated. The amino group of the amino acid must be masked (derivatized) usually with a t-butoxy group (t-BOC; $(CH_3)_3$-CO-CO-) but see below.

The reaction used by Puigserver et al. (11) to modify casein with several amino acids is shown in Figure 1. The best conditions for efficient covalent attachment of amino acids to casein were a 5% casein solution in 15% dimethylformamide (to keep the active ester soluble) at pH 9.0 and 25°C to which the active-ester derivative of the amino acid was added (6:1 ratio of active ester:amino groups of the protein). After 4 hours, 90-95% of the α- and ϵ-amino groups of the protein were modified. The derivative of tyrosine, at the OH group (see Fig. 1), was removed by adjusting the solution to pH 9.0 and holding for several hours at 25°C. The t-butoxy group was removed by treatment of the modified protein with trifluoroacetic acid. A number of modified caseins were prepared by this method and their chemical, physical, nutritional and functional properties determined (11-14). The method has also been used to covalently incorporate methionine into whole beans (15).

There are several problems with the method as described above. First, the N,N'-dimethylformamide is toxic to rats (15,16) and must be removed by exhaustive dialysis. This was easy to accomplish for the modified caseins but proved to be more difficult in modified whole beans (15). Second, the removal of the t-butoxy group with trifluoroacetic acid, while a facile and good reaction for peptide synthesis, results in a gummy protein during treatment. We found that N-acetyl amino derivatives can be used more effectively, in that the acetyl group can be removed *in vivo*, at least by rats (13).

C. *CARBOXYANHYDRIDE METHOD*

The carboxyanhydride method of covalently attaching amino acids to each other or to proteins was first proposed by Katchalski and Sela (17). In our opinion, it is superior to the carbodiimide and active-ester methods for attaching free amino acids to proteins in that the chemistry is easier and the reactions can be done at 4°C.

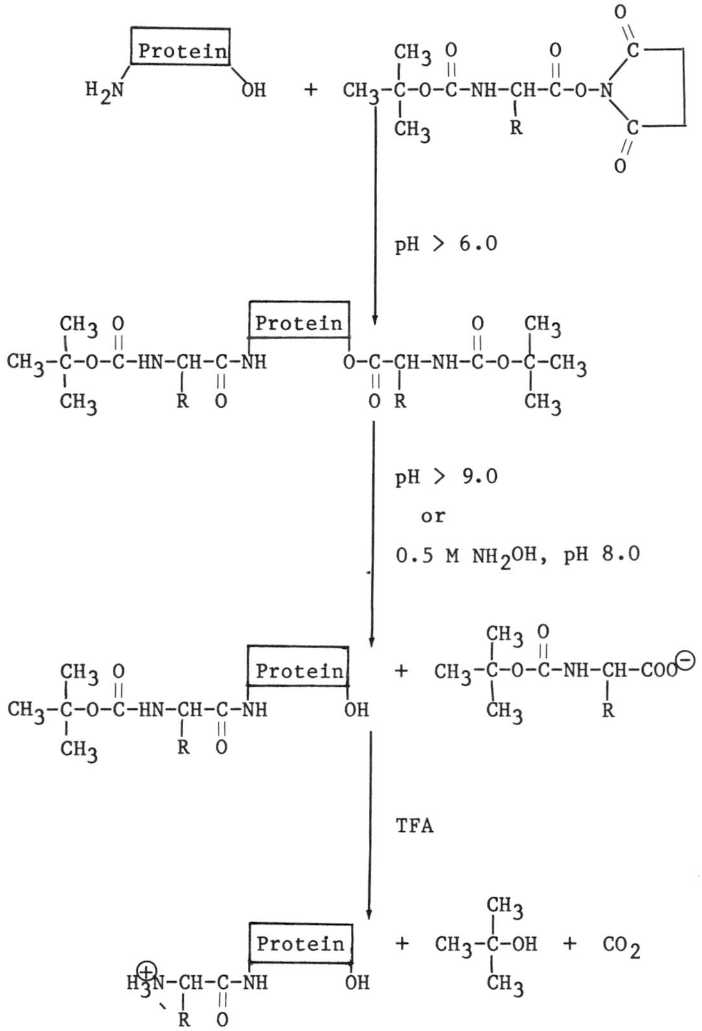

H_2N–[Protein]–OH + $(CH_3)_3C-O-CO-NH-CH(R)-CO-O-N$(succinimide)

pH > 6.0

$(CH_3)_3C-O-CO-HN-CH(R)-CO-NH$–[Protein]–$O-CO-CH(R)-NH-CO-O-C(CH_3)_3$

pH > 9.0

or

0.5 M NH_2OH, pH 8.0

$(CH_3)_3C-O-CO-HN-CH(R)-CO-NH$–[Protein]–OH + $(CH_3)_3C-O-CO-NH-CH(R)-COO^{\ominus}$

TFA

$H_3\overset{\oplus}{N}-CH(R)-CO-NH$–[Protein]–OH + $(CH_3)_3C-OH$ + CO_2

FIG. 1 *General scheme for covalent attachment of an amino acid to a protein by the active-ester method. TFA = trifluoroacetic acid.*

The chemistry of the carboxyanhydride method is shown in Eq. (2) and (3). First, one makes the N-carboxy-α-amino acid anhydride by reaction of the amino acid with phosgene ($COCl_2$; caution: very toxic, use highly efficient hood and other precautions) in tetrahydrofuran (THF) and dioxane as solvents (Eq. 2).

$$\text{R-}\underset{\text{NH}_2}{\text{CH}}\text{-COOH} + \text{COCl}_2 \xrightarrow[\text{Dioxane}]{\text{THF}} \text{R-}\underset{\text{NHCOCl}}{\text{CH}}\text{-COOH} + \text{HCl} \longrightarrow \left[\text{R-}\underset{\text{N=C=O}}{\text{CH}}\text{-COOH}\right] + \text{HCl} \dashrightarrow \begin{array}{c}\text{R-CH—CO}\\ \text{HN} \quad\quad \text{O}\\ \text{CO}\end{array} + \text{HCl} \quad (2)$$

After preparation, the N-carboxy-α-amino acid anhydride is added to another amino acid or to a protein and allowed to react as shown in Eq. (3) in order to covalently link the amino acid via a peptide or isopeptide bond.

$$\begin{array}{c}\text{R}_1\text{-CH—CO}\\ \text{HN} \quad\quad \text{O}\\ \text{CO}\end{array} + \text{H}_2\text{N-CHR}_2\text{-CO-}\text{Ⓟ} \xrightarrow[4^\circ\text{C}]{\text{pH 10.2}} \underset{\text{COO}^{\ominus}}{\text{HN}}\text{-CHR}_1\text{-CO-NH-CHR}_2\text{-CO-}\text{Ⓟ} \quad (3)$$

The completed reaction is then adjusted to pH 3-4 and warmed to 20°C to eliminate (as CO_2) the $-COO^{\ominus}$ group attached to -NH. The pH can then be readjusted to 10.2, the solution cooled to 4°C and another aliquot of the same or a different N-carboxy-α-amino acid anhydride added to permit coupling of another amino acid residue. This step-wise process can be repeated to build a homopolymer or heteropolymer chain of amino acids or amino acids coupled to the amino groups of a protein.

Alternatively, and preferably when a protein is to be enriched with only one limiting essential amino acid, the coupling can be performed in a single step when the reaction is performed at pH 6-7 and 4°C (see Eq. (4); Ref. 18).

$$
\begin{array}{l}
R_1\text{-CH—CO} \\
\quad \text{HN} \quad \text{O} \quad \text{(ring: HN—CO—O)} + H_2N\text{-CHR}_2\text{-CO-}Ⓟ \xrightarrow[4^\circ C]{\text{pH } 6\text{-}7} \underset{\text{COOH}}{\text{HN}}\text{-CHR}_1\text{-CO-NH-CHR}_2\text{-CO-}Ⓟ \\
\qquad\qquad\qquad\qquad\qquad\qquad\qquad\qquad \downarrow -CO_2 \qquad\qquad (4) \\
H\text{-(NH-CHR}_1\text{-CO)}_n\text{-NH-CHR}_2\text{-CO-}Ⓟ \longleftarrow \longleftarrow H_2N\text{-CHR}_1\text{-CO-NH-CHR}_2\text{-CO-}Ⓟ
\end{array}
$$

A number of experimental parameters affected the rate and extent of the single step covalent attachment of methionine to proteins (18). The highest number of methionine residues (54 moles/mol casein) was covalently attached to casein at pH 6.5 with the modification of 7 of the 12 amino groups of casein. By comparison, at pH 9.0, 20 moles of methionine per mol of casein were bound with modification of 4 of the 12 amino groups. The maximum efficiency of incorporation of methionine into casein (~75%) occurred at 6-8% casein concentration (pH 7.0, 4°C, 3:1 ratio of N-carboxy-α-amino anhydride to amino groups of casein). The extent of incorporation of methionine into casein increased linearly and proportionally with increase in the ratio of N-carboxy-α-amino acid anhydride to amino groups of casein between 0 and 8. At a ratio of 8, almost 80 residues of methionine were covalently bound to casein with modification of ~9 amino groups. L-Methionyl-casein was quite soluble up to incorporation of ~25 moles of methionine per mole of casein. Higher levels of modification resulted in marked decreases in solubility, as a result of attachment of this hydrophobic amino acid.

The carboxyanhydride method (either stepwise or single step method) can be used effectively to modify food proteins other than casein (Table 3). The efficiency of grafting was 68-80% of the

Table 3 *Modification of several food proteins by N-carboxy-L-methionine anhydride using the single step method*[a]

Protein	Methionine content (g/100 g protein)		Modified lysyl residues	Efficiency of grafting reaction[b]
	before	after	No. (% of total)	(%)
Casein	*3.0*	*22.0*	*3.9 (43)*	*70*
β-Lacto-globulin	*3.2*	*32.0*	*5.0 (42)*	*80*
Pea protein	*0.4*	*18.0*	*2.4 (40)*	*73*
Bean protein	*0.4*	*19.0*	*2.5 (41)*	*75*
Soy protein	*1.3*	*21.0*	*2.5 (40)*	*68*

[a]*Ref. 18.*

[b]*Percentage of added N-carboxy-L-methionine anhydride covalently attached to protein.*

added reagent covalently incorporated into the proteins, with a modification of about 40-43% of the amino groups. The methionine content was increased from 7-fold for casein to as much as 45- and 48-fold for pea protein and bean protein, respectively.

The question might be raised as to why modify the proteins so extensively when increasing the methionine much less is all that is required to increase the PER to the maximum value possible by adding methionine (for example). In the case of casein (Table 3) the methionine content was increased 7-fold (3.0 to 22.0 g/100 g protein) while an increase of 0.33-fold (3.0 to 4.0 g/100 g protein) would provide maximum improvement in PER. The justification is that it is far more economical to extensively modify a protein and then use it to fortify unmodified protein than to modify all the protein by the required amount. For example, 18% of the extensively modified L-methionyl-casein described above added to 82% of unmodified casein would result in maximum improvement in PER.

D. *POLYAMINO ACID POLYMERS*

The fortification of foods with polyamino acid polymers has not been described but appears to be worthy of research. Homo- or heteropolymers of amino acids can be made by the carboxyanhydride method described above and are available from suppliers of biochemical and organic compounds for research. The present price of the highly purified polymers is very high at the moment. Poly-L-lysine bromide in several molecular size ranges from 1000-4000 to 150,000-300,000 is $150 per gram from Sigma (February 1985). Poly-L-methionine ranging in molecular weight from 30,000-50,000 to 100,000-200,000 is $107 per gram. Poly-L-tryptophan ranging from 1000-5000 to 15,000-30,000 is $106 per gram. By comparison, pure L-lysine is $0.05/g, L-methionine is $0.60/g (DL-methionine is $0.03/g) and L-tryptophan is $3.65/g. Less rigorously controlled size and purity preparations could be made much more cheaply and may be suitable for fortification of high protein plant foods.

To the author's knowledge no one has seriously investigated the possibility of improving the PER of high protein plant foods by the incorporation of polyamino acid polymers. Newman et al. (19) have studied the nutritional quality of poly-L-lysine for weanling rats, comparing it to the nutritional quality of free lysine and proteins containing known quantities of lysine (Table 4). The weanling rats were able to use the lysine from poly-L-lysine at least as well, if not better, than free L-lysine.

For reasons stated at the beginning of this chapter there could be advantages of incorporating poly-L-amino acid polymers into foods for fortification rather than the free amino acids. Use of poly-L-amino acid polymers could be advantageous over covalent attachment of limiting essential amino acids to proteins in that it should be easier to control the process and purity of the product, perhaps resulting in easier approval for use in foods.

Table 4 *Nutritional quality of poly-L-lysine for weanling rats*[a]

Lysine source	Lysine content of diet (%)	PER
No lysine	*0.00*	*-2.05*
L-Lysine	*1.29*	*2.71*
Poly-L-lysine	*0.65*	*2.67*
Casein	*1.33*	*2.39*
Soybean meal	*0.94*	*2.18*

[a]*Ref. 19. Complete diet used in all cases except for lysine.*

E. *PHOSPHORUS OXYCHLORIDE ($POCl_3$) METHOD*

Proteins can be phosphorylated by a number of chemical reagents (20). These reagents include the general phosphorylating compounds phosphorus oxychloride ($POCl_3$), phosphoric acid in the presence of trichloroacetonitrile (21), phosphorus pentoxide dissolved in phosphoric acid (22) and sodium trimetaphosphate (23). More specific phosphorylating reagents, with decreased extent of modification, include monophenyl phosphodichloride (24), phosphoramidate (25,26), and diphosphoimidazole (27), and acetylphosphate, carbamoyl phosphate and 1,3-biphosphoglycerate (28).

Most phosphorylation reactions are performed in mixtures of organic and aqueous solvents because of rapid hydrolysis of the reagent by H_2O. For example, $POCl_3$ is often dissolved in an organic solvent (usually carbon tetrachloride) and added in small portions to aqueous protein solutions. The pH is controlled by adding sodium hydroxide and the reaction is performed in an ice bath.

Phosphorus oxychloride has been used more extensively than any of the other reagents to phosphorylate several proteins (20). Depending on the protein and the specific reaction

Table 5 *Simultaneous covalent attachment of lysine, tryptophan and phosphate to zein, catalyzed by* $POCl_3$[a]

Phosphate (moles/mol zein)		Tryptophan (moles/mol zein)		Lysine (moles/mol zein)	
In reaction	bound	In reaction	bound	In reaction	bound
200	0.74	5	1.6	10	5.9
200	0.67	5	1.2	20	2.0
200	0.67	10	0.75	40	3.6

[a]*Ref. 30, unpublished data. Performed in ice bath in 60% acetone solution.*

conditions, the phosphorus can be attached to hydroxyl oxygen (of serine and threonine, but rarely tyrosine), amino nitrogen and imidazole nitrogen. The phosphorylated nitrogen derivatives are quite acid labile while the phosphorylated oxygen derivatives are quite stable in acid but undergo β-elimination at alkaline pHs (29).

Recently, Sitohy et al. (30) have phosphorylated zein in 60% acetone to improve its water solubility. When free amino acids are also added to the acetone-water solution of protein prior to addition of small aliquots of $POCl_3$ dissolved in acetone, some of the amino acid(s) are covalently incorporated into the zein. Table 5 shows some of the results of incorporating lysine and tryptophan together into zein.

Preliminary experiments using *Tetrahymena thermophili* have shown that the PER value of zein is increased from ~0 to ~1.2 by the modification. A complete characterization of the modified zein, including detailed rat-feeding experiments, is presently underway.

F. *ENZYME-CATALYZED COVALENT INCORPORATION OF AMINO ACIDS*

Several enzyme-catalyzed reactions have been used to covalently incorporate amino acids into proteins. Because of the high specificity of enzyme-catalyzed reactions, the products are more readily predictable and fewer in number. For this reason, covalent incorporation of amino acids into proteins by enzymatic catalysis should present fewer complications in gaining approval by regulatory agencies. However, there are disadvantages. Presently, the economics of enzyme-catalyzed incorporation of amino acids into proteins is not favorable, the required enzymes may not be available in sufficient quantities and as a result of narrow specificity, the extent of modification of the protein is usually less (1-6 moles/mol protein) than by nonenzyme-catalyzed methods.

1. Plastein reaction As has been shown by Arai and coworkers (31-33), proteins such as soy protein can be partially hydrolyzed by papain or chymotrypsin, the hydrolyzate concentrated to about 35% solids and further incubated, with the result that the enzyme

Table 6 *Incorporation of methionine into proteins by "novel one step process" using papain*[a]

Protein	Methionine content (g/100 g protein)	
	original	modified
Soy protein isolate	*1.25*	*8.80 (8.85)*[b]
Soy flour	*1.30*	*8.21*

[a]*Ref. 33.*

[b]*Duplicate experiment.*

will resynthesize some of peptide bonds (plastein reaction). The net overall synthesis appears to be due to a combination of factors, the high substrate concentration of peptides, the driving force of insolubility of peptides containing an increased content of hydrophobic amino acids and the nucleophilic attack of the amino groups of the peptides on the acyl enzyme intermediate formed. This last reaction is shown in Eq. (5), using papain as the enzyme.

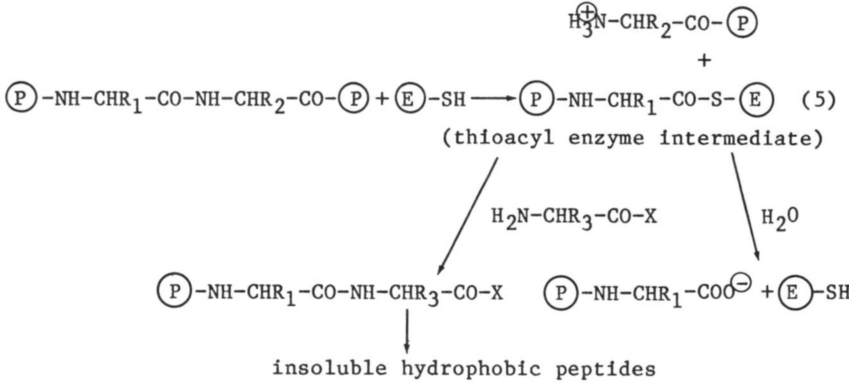

When the reaction is performed in the presence of an amino acid ester, such as L-methionine ethyl ester, some of the amino acid ester is covalently incorporated into the protein as shown by Eq. (5) where $H_2N\text{-}CHR_3\text{-}CO\text{-}X$ could be the added amino acid ester.

The results of experiments showing the incorporation of methionine into soy protein isolate and soy flour are shown in Table 6. There is a 6- to 7-fold increase in the amount of methionine in the final product compared to the starting material. The methionine content is 3-4 times that required to achieve maximum PER value of the protein.

2. *Transglutaminase reaction* Transglutaminase is present in blood and some other organs of animals. The role of transglutaminase in blood is to covalently cross-link the fibrin

molecules (formed via a complex cascade of proteolytic enzyme reactions terminating in the proteolysis of fibrinogen to fibrin) so that the clot is stable, thus preventing continued bleeding of a wound.

Transglutaminase has specificity for the glutamine side chain(s) of proteins. The second substrate of the reaction, a nucleophile, can be a large number of compounds including free amino acids, the lysyl side chains of proteins, etc. The reaction, shown in Eq. (6), uses L-methionine ethyl ester as the nucleophile.

$$\underset{\substack{| \\ CO \\ | \\ (P)}}{\overset{\substack{(P) \\ | \\ NH \\ |}}{HC}}-CH_2-CH_2-\overset{\substack{O \\ \|}}{C}-NH_2 + H_2N-\underset{\substack{| \\ (CH_2)_2 \\ | \\ S \\ | \\ CH_3}}{\overset{\substack{O \\ \| \\ COCH_2CH_3 \\ |}}{C}}-H \xrightarrow[\text{glutaminase}]{\text{trans-}} \underset{\substack{| \\ CO \\ | \\ (P)}}{\overset{\substack{(P) \\ | \\ NH \\ |}}{HC}}-CH_2-CH_2-CO-NH-\underset{\substack{| \\ (CH_2)_2 \\ | \\ S \\ | \\ CH_3}}{\overset{\substack{O \\ \| \\ C-OCH_2-CH_3 \\ |}}{CH}} + NH_3 \qquad (6)$$

The ethyl alcohol group can be removed from the product readily by holding at pH 8-9 for a few hours.

The data of Table 7 demonstrates the incorporation of lysine and methionine into proteins by the transglutaminase-catalyzed reaction. The methionine content of the four proteins used was increased about two-fold, while the lysine content of wheat gluten was increased five-fold.

Ikura et al. (35,36) have used the transglutaminase-catalyzed reaction to cross-link proteins for texturization and changes in functional properties. For example, 7S and 11S proteins from soy protein can be cross-linked by this reaction to produce chimeric proteins (See Chapter 1). In the covalent incorporation of

Table 7 *Methionine and lysine covalent incorporation into proteins by the transglutaminase-catalyzed reaction*[a]

Protein	Amino acid content (g/100 g protein)	
	Control	Enzyme-treated
Methionine incorporation		
α_{s1}-*Casein*	2.7	5.4
β-*Casein*	2.9	4.4
Soybean 7S proteins	1.1	2.6
Soybean 11S proteins	1.0	3.5
Lysine incorporation		
Wheat gluten	1.5	7.6

[a]*Ref. 34.*

[b]*L-Methionine ethyl ester or L-lysine was 20-68 mM in reaction containing 23 μg/ml transglutaminase, 1.5-10 mg/ml protein,* $CaCl_2$ *and dithiothreitol at pH 7.5 and 37°C.*

essential amino acids into proteins as described above, some interchain cross-linking of the protein also occurs.

The main source of transglutaminase for the above experiments was guinea pig liver. Obviously, if this reaction is to have any commercial utility, another source of transglutaminase, probably produced by recombinant DNA technology, must be found.

3. *Sulfhydryl-disulfide interchange reaction* The sulfhydryl-disulfide interchange reaction, shown in Eq. (7), is readily performed both nonenzymatically and enzymatically.

$$\text{Ⓟ-S-S-Ⓟ} + \text{RSH} \rightleftharpoons \text{Ⓟ-S-S-R} + \text{ⓅSH} \qquad (7)$$

The direction of the nonenzyme-catalyzed reaction is dependent on the relative concentrations of the reactions and products and their redox potentials. Two types of enzymes act upon disulfide bonds. A thiol-disulfide interchange enzyme (protein disulfide-isomerase; EC 5.3.4.1; other name, S-S- rearrangase) was first described in 1963 (37) and was purified from beef liver (38,39). More recently, the enzyme has been purified from *Candida claussenii* (40). This enzyme catalyzes the rearrangement of random "incorrect" pairs of half-cysteine residues to the native disulfide bonds in several proteins. Traces of a thiol compound are required for the reaction.

The second type of enzyme reduces disulfide bonds in proteins. Protein-disulfide reductase [NAD(P)H:protein-disulfide oxidoreductase; EC 1.6.4.4] purified from pea seeds (41) catalyzes the reaction shown in Eq. (8).

$$2\ \text{NAD(P)H} + \text{Ⓟ-disulfide} \rightleftharpoons 2\ \text{NAD(P)} + \text{Ⓟ-dithiol} \quad (8)$$

Another protein-disulfide reductase [glutathione:protein-disulfide oxidoreductase; EC 1.8.4.2] has been reported in hepatic tissue (42). It catalyzes the reaction shown in Eq. (9).

$$\underset{\text{(reduced)}}{2\ \text{Glutathione}} + \text{Ⓟ-disulfide} \rightleftharpoons \underset{\text{(oxidized)}}{\text{Glutathione}} + \text{Ⓟ-dithiol} \quad (9)$$

It is relatively easy to produce mixed disulfides by the reaction shown in Eq. (7). Therefore, the covalent incorporation of cysteine into proteins containing disulfide bonds is quite easy. Since cysteine is not an essential amino acid, although it has a sparing effect on methionine, no one appears to have examined this method for improving the cysteine/cystine content of proteins.

Use of the sulfhydryl-disulfide interchange method for covalent incorporation of essential amino acids into proteins is less straightforward. One way would be to use dipeptides containing the essential amino acid attached to L-cysteine. The dipeptide could be readily incorporated into the protein. It may be useful to test the feasibility of this method. Another potential method would be the attachment of -SH groups to the protein via pyridyldithiolation (43; See Chapter 1), followed by binding of the desired amino acid.

III. HYDROLYSIS OF PRODUCTS FORMED BY COVALENT ATTACHMENT OF AMINO ACIDS

As described at the beginning of this chapter, the covalent attachment of amino acids to proteins by both enzymatic and nonenzymatic methods is expected to give substantial amounts of isopeptide bonds. The results show that the products of these reactions can be utilized by proteolytic enzymes. The evidence for this is several fold. (a) The covalently attached amino acid is found free at higher levels initially in the blood when the modified protein is fed; (b) the PER value of the modified protein

Table 8 *Plasma concentration of some free amino acids in rats fed 10% protein diets*[a]

Proteins	Lys	Thr	Ser	Gly	Met
	(μmoles/100 ml plasma)				
Control casein	*101*	*19*	*34*	*32*	*5*
L-Methionyl-casein[b]	*96*	*17*	*33*	*27*	*39*

[a]*Ref. 12.*

[b]*>90% of the 12 lysyl residues of casein were modified. The methionine content of L-methionyl-casein was 16 moles methionine per mol casein compared to 5 moles methionine per mol casein for the control.*

is increased in proportion to the amount of limiting essential amino acid bound; and (c) *in vitro* studies show that certain enzyme systems can hydrolyze peptides and proteins containing some isopeptide bonds.

The effect of feeding L-methionyl-casein to rats on the initial free methionine level in the blood plasma is shown in Table 8. On feeding L-methionyl-casein the lysine content of the plasma was about the same as the control, while the methionine content was some 8 times higher. Both results indicate that the

Table 9 *Nutritional values of L-methionyl- and N-acetyl-L-methionyl-caseins for rats*[a]

Diet[b]		Weight gain (g)	PER
Control:	*10% commercial casein*	*72.6*	*2.46*
	10% commercial casein + 0.2% free L-methionine	*106.8*	*3.15*
Control:	*5% commercial casein + 5% control casein*[c] *+ 0.2% free L-methionine*	*66.7*	*2.97*
	5% commercial casein + 5% L-methionyl-casein	*60.5*	*2.92*
Control:	*3% commercial casein + 7% control casein + 0.26% N-acetyl-L-methionine*	*107.0*	*3.18*
	3% commercial casein + 7% N-acetyl-L-methionyl-casein	*89.8*	*2.95*

[a]*Ref. 12.*

[b]*Complete diet except for protein to which the indicated additions were made. The 10% commercial casein contained 0.26% methionine. The remainder of the diets contained 0.46% methionine.*

[c]*Control casein is commercial casein subjected to all the steps and conditions used in preparing the L-methionyl and N-acetyl-L-methionyl-caseins except no active-ester derivative of methionine was added.*

lysine and methionine are released from the protein by proteases as free amino acids.

Improvement of the PER of casein by covalent attachment of methionine to casein is shown in Table 9. Rat-feeding experiments showed that both the covalently-bound methionine (and N-acetyl methionine) and the lysyl residues to which it was primarily attached by isopeptide bonds were available to the rat (12,13). Addition of 0.2% free L-methionine to the 10% commercial casein-containing diet (increase in methionine from 0.26 to 0.46% of diet), resulted in an increase in PER from 2.46 to 3.15, showing that methionine is a limiting essential amino acid in casein at the 10% protein level.

A diet containing 5% commercial casein + 5% L-methionyl-casein (0.46% total methionine) was about as effective in improving the PER as the control diet containing 5% commercial casein + 5% control casein + 0.2% free L-methionine (0.46% total methionine; PER of 2.92 vs 2.97). A diet containing 3% commercial casein + 7% N-acetyl-L-methionyl-casein (0.46% total methionine) was as effective, based on PER, as the diet containing 5% commercial casein + 5% L-methionyl-casein (0.46% total methionine; PER of 2.95 vs 2.92) but somewhat less effective than the control diet for this experiment of 3% commercial casein + 7% control casein + 0.26% N-acetyl-L-methionine (0.46% total methionine; PER of 2.95 vs 3.18).

Therefore, as concluded above, the covalently bound methionine, whether the amino group is modified by an acetyl group or not, and the lysyl residues of casein to which it is covalently attached by isopeptide bonds is nutritionally available to rats.

The rate of *in vitro* proteolysis of modified caseins by chymotrypsin, rat bile pancreatin and bovine pancreatin was slower than unmodified casein. The extent of digestion was also lower, especially for acetyl-L-methionyl-casein and N-t-BOC-L-methionyl casein (see Table 10).

Table 10 *Relative extent of proteolysis of casein containing amino acids covalently attached by the active-ester method*[a]

Substrate	Degree of modification (%)	Rat bile pancreatin[b]	Bovine pancreatin[c]
		(relative extent of proteolysis)	
Control casein	*0*	*100*	*100*
Acetyl-L-methionyl-casein	*42*	*71*	*84*
BOC[d] *L-methionyl-casein*	*59*	*54*	*67*
L-Methionyl-casein	*59*	*95*	*90*

[a]*Ref. 13. 0.1% substrate, pH 7.0, 38°C, 48 h.*
[b]*1:100 (v/v).*
[c]*1:2 (v/v).*
[d]*BOC = t-butoxy group.*

In vitro studies have shown that the newly formed peptide and isopeptide bonds can be hydrolyzed by proteases present in the digestive system. Table 11 shows the hydrolysis of the isopeptide bond by tissue homogenates. While the substrates can be hydrolyzed by proteases present in all three tissue homogenates, the intestine homogenate has higher activity than the kidney and liver homogenates. The majority of the enzyme(s) is membrane bound since it is present primarily in the pellet rather than the supernatant.

Pepsin and the proteases in pancreatic juice (trypsin, chymotrypsin, elastase, carboxypeptidases) had very little activity on the isopeptide [^{14}C-methyl]-ϵ-N-L-methionyl-L-lysine (Table 12). However, the aminopeptidase of the intestinal membrane of hogs and rats rapidly hydrolyzed the isopeptide (44).

Gaertner and Puigserver (18) synthesized several isopeptides

Table 11 *Effect of tissue homogenates on hydrolysis of the isopeptide bond of two substrates*[a]

Tissue	Substrate	
	L-Methionyl-casein	ε-N-[^{14}C]-L-methionyl lysine
	(μmoles of methionine released)[b]	
Kidney homogenate	*1560*	*1340*
Liver homogenate	*2820*	*1920*
Intestinal homogenate	*5290*	*5980*
supernatant	*1240*	*560*
pellet	*3640*	*5160*

[a]*Ref. 13.*

[b]*Extent of hydrolysis (μmoles of free methionine released) was determined after incubation of substrate with tissue homogenates containing about 20 aminopeptidase units per assay for 15 h at 37°C.*

Table 12 *Hydrolysis of [^{14}C-methyl]ε-N-L-methionyl L-lysine by the proteolytic enzymes of the digestive tract*[a]

Enzymes	E/S (w/w)	Extent of hydrolysis[b] (%)
Hog		
Pepsin	*1:1*	*2*
Pancreatic juice	*1:6*	*3*
Solubilized aminopeptidase	*1:500*	*95*
Membrane-bound aminopeptidase	*1:500*	*91*
Rat		
Membrane-bound aminopeptidase	*1:500*	*93*

[a]*Ref. 44.*

[b]*Determined after a 3 h incubation period at 37°C. The concentration of isopeptide was 5 mM in 0.05 M phosphate buffer, pH 7.0.*

containing oligo(methionine) chains varying in length (2-5 residues) and covalently linked primarily to the ϵ-amino group of L-lysine. The bonds between methionyl-methionine units were normal peptide bonds, while the methionyl-lysine bond was an isopeptide bond. As determined by HPLC, pepsin cleaved the methionine oligomers to tripeptides (45). Chymotrypsin preferentially hydrolyzed the methionyl-methionine bond preceding the isopeptide bond. Cathepsin C released dimethionyl units from the oligomers. Intestinal aminopeptidase effectively hydrolyzed both the peptide and isopeptide bonds of the oligomers.

In vitro studies have shown that proteins modified by covalent attachment of amino acids can be hydrolyzed by proteases, although at a somewhat slower rate and to a lower extent than the control proteins. As shown by the data of Table 13, the initial rates of proteolysis and the extent of proteolysis of L-alanyl-casein and L-tryptophyl-casein were less than those for the control. The t-BOC group, attached to the amino group of the modifying amino acid, decreased both the initial rates and extents of proteolysis. There is some small effect of extent of amino group modification on the extent of proteolysis of the two proteins (without a t-BOC group). Even though the initial rate of proteolysis of L-tryptophyl-casein was essentially zero, the extent of proteolysis after 48 hours was 67-80% that of the control casein.

The data of Table 14 indicate that several extensively modified caseins are hydrolyzed by α-chymotrypsin, and the enzymes of rat bile pancreatic juice and bovine pancreatin. The relative extent of proteolysis ranged from 60% of control casein value for L-aspartyl-casein to 84% for L-alanyl-casein. It should be noted that a value of 100% (control casein) for relative extent of hydrolysis does not mean that all the peptide bonds of the protein are hydrolyzed, resulting in amino acids as products. Rather, it compares the number of susceptible bonds, depending on specificity of the enzyme(s), to the actual extent of hydrolysis.

Table 13 *Effect of extent of modification and derivatized amino group on in vitro hydrolysis of modified caseins*[a]

Protein	Amino group modification[b] (%)	Relative initial rate[c] (%)		Relative extent of hydrolysis[d] (%)	
		t-BOC[e]	NH_2 unmodified[f]	t-BOC	NH_2 unmodified[f]
Casein (control)	*0*	--	*100*	--	*100*
L-Alanyl-casein	*44*	*49*	*43*	*64*	*92*
	79	*18*	*43*	*44*	*85*
	88	*10 (34)*[g]	*43 (68)*[g]	*55*	*84*
L-Tryptophyl-casein	*54*	*0*	*0*	*57*	*80*
	87	*0*	*0*	*36*	*68*
	95	*0*	*0*	*35*	*67*
	97	*0*	*0*	*32*	*77*

[a]*Ref. 12. Digestion was performed with α-chymotrypsin except in cases noted as ( )*[g]*.*

[b]*Percent of amino groups of casein modified as estimated with trinitrobenzene sulfonic acid method (46).*

[c]*Reactions contained 0.1% protein in 0.1 M phosphate buffer, pH 7.0, for bovine pancreatin (1:60, enzyme to protein (w/w)) at 38°C or 0.1% protein in 0.02 M borate buffer, pH 8.2 for α-chymotrypsin (1:3000, enzyme to protein (w/w)) at 38°C.*

[d]*Reaction conditions as in c for chymotrypsin except used 1:300 enzyme to protein (w/w) and incubation time of 48 hours.*

[e]*Modified protein in which the new covalently attached amino acid has NH_2 group blocked with t-butoxy group. Rate and extent relative to unmodified casein.*

[f]*Modified protein in which the new covalently attached amino acid has the t-BOC group removed by treatment with trifluoroacetic acid.*

[g]*Digestion with bovine pancreatin.*

Table 14 *Extent of hydrolysis of modified caseins*[a]

Proteins	Amino group modification[c] (%)	α-chymotrypsin[d]	rat bile pancreatic juice[e]	bovine pancreatin[e]
		(relative extent of hydrolysis) (%)		
Casein (control)	*0*	*100*	*100*	*100*
Glycyl-casein[b]	*90*		*80*	*81*
L-Alanyl-casein[a]	*88*	*84*		
L-Aspartyl-casein[a]	*83*	*60*		
L-Methionyl-casein[b]	*90*		*71*	*78*
L-Tryptophyl-casein[a]	*97*	*77*		

[a]*Ref. 12.*

[b]*Ref. 13.*

[c]*Percent of amino groups of casein modified as estimated with trinitrobenzene sulfonic acid method (46).*

[d]*See footnotes c and d of Table 13.*

[e]*Reaction mixtures containing 0.1% protein in 0.1 M phosphate buffer, pH 7.0, were incubated at 38°C with 1:2 (w/w, enzyme to protein) bovine pancreatin and 1:100 (v/v) rat bile pancreatic juice. See Ref. 13 for further details.*

Several interesting conclusions can be drawn from the data of Table 15. First, α-chymotrypsin and trypsin hydrolyze the peptide bonds of the modified caseins at about the same relative initial rates, perhaps an unexpected result. The ε-amino groups of lysyl residues of the casein are modified by the covalent attachment of methionyl residues, thereby presumably preventing hydrolysis of the peptide bonds involving these modified lysyl residues. There is no modification of the tyrosyl, phenylalanyl or tryptophyl residues of casein, the specificity sites for chymotrypsin.

Table 15 *Effect of extent of modification on rate of hydrolysis of casein modified with L-methionine*[a]

Protein	Met[b]	Amino group modification (%)	Average methionyl chain length	Ratio isopeptide/peptide bonds added[c]	Initial rates, %[d]	
					α-chymotrypsin	trypsin
Casein (control)	*0*	*0*	--	--	*100*	*100*
L-Methionyl-casein	*12*	*95*	*1.1*	*19*	*52*	*53*
Poly-L-methionyl-casein-1	*11*	*35*	*2.6*	*0.62*	*32*	*34*
Poly-L-methionyl-casein-2	*37*	*60*	*5.1*	*0.24*	*18*	*14*
Poly-L-methionyl-casein-3	*25*	*95*	*2.2*	*0.84*	*8*	*11*

[a]*Ref. 18.*

[b]*In addition to the 5 methionine residues already present.*

[c]*The attachment of first methionyl residue to ε-amino group of a lysyl residue gives an isopeptide bond. Attachment of additional methionyl residues to the first methionyl residue is via normal peptide bonds.*

[d]*Reactions containing 0.1% protein in 0.02 M borate buffer, pH 8.2, were incubated at 38°C with 1:2000 (w/w, enzyme to protein) α-chymotrypsin or trypsin.*

Second, there is no correlation between the average methionyl chain length or the ratio of isopeptide bonds/peptide bonds added to the casein and the initial rates of hydrolysis. For example, with methionyl-casein, the average methionyl chain length is 1.1 and the ratio of isopeptide bonds/peptide bonds is 19. However, the relative initial rates of hydrolysis by α-chymotrypsin and trypsin (52 and 53%, respectively) are higher than for polymethionyl-casein-1 where the average methionyl chain length is 2.6 and the ratio of isopeptide bonds/peptide bonds is 0.62. This most likely indicates the hydrolysis of the isopeptide bond is not rate-limiting.

In summary, from the data presented one can conclude that the newly added amino acid residues to proteins, whether attached by isopeptide or peptide bonds, can be hydrolyzed by proteases *in vitro* or *in vivo*. There appears to be some decrease in the relative initial rates of hydrolysis *in vitro* and there is a small decrease in the extent of hydrolysis. Aminopeptidase(s) of the small intestinal mucosa appear to be largely responsible for hydrolysis of the isopeptide bonds, although the major pancreatic endoproteases, trypsin and chymotrypsin, can still hydrolyze these derivatives at least to 60-97% of that of control casein.

Rats can obtain full nutritional benefit of covalently attached methionine and the lysine to which it is covalently attached, by isopeptide bonds, as shown by feeding experiments.

IV. FUNCTIONAL PROPERTIES OF MODIFIED PROTEINS

Relatively little data are available on changes in the functional properties of casein (and other proteins) to which amino acids are added by covalent attachment. Some data on the physical properties of modified caseins are shown in Table 16. Adding 10-11 residues of amino acids to the ϵ-amino group of lysyl residues of casein had little effect on the relative viscosity, relative solubility and relative fluorescence (of tryptophan residues) of L-alanyl-casein, L-methionyl-casein, L-threonyl-casein and L-asparaginyl-casein. On the other hand, adding 10-11 residues of aspartyl or tryptophyl groups to casein had a very large effect on the relative solubility and relative fluorescence of L-aspartyl-casein and L-tryptophyl-casein. The results for L-tryptophyl-casein are expected, those for L-aspartyl-casein are not.

Table 16 *Physical properties of some modified caseins*[a]

Protein	Amino group modification (%)	Relative viscosity	Relative solubility	Relative fluorescence
Casein (control)	*0*	*1.00*	*1.00*	*1.00*
L-Alanyl-casein	*88*	*1.01*	*1.04*	*1.02*
L-Methionyl-casein	*89*	--	--	*1.02*
L-Threonyl-casein	*~80*	--	*1.02*	--
L-Asparaginyl-casein	*~80*	--	*1.00*	--
L-Aspartyl-casein	*83*	*1.06*	*0.85*	*0.68*
L-Tryptophyl-casein	*92*	*0.99*	*0.47*	*0.58*

[a]*Ref. 14. The original reference should be consulted for details of the experiments.*

V. CONCLUSIONS

Amino acids can be attached readily to food proteins by some modification of methods developed by protein chemists for peptide synthesis. As a result of using masking groups to block reaction of certain groups, the protein chemist has restricted reaction to α-carboxyl or α-amino groups during synthesis. Treatment of food proteins with activated derivatives of amino acids leads to substantial modification at the ϵ-amino groups of lysyl residues and β,γ-carboxyl groups of aspartyl and glutamyl residues, giving isopeptide bonds. Fortunately, the proteases of animals, primarily aminopeptidase(s), can hydrolyze these isopeptides. This results in release of the newly added amino acid and its availability, along with the amino acid to which it was attached (lysine for example).

Our work, and that of others show that limiting essential amino acids can be attached efficiently to food proteins in the laboratory, either enzymatically or nonenzymatically, with improvement in the PER of the protein, proportional to the amount of essential amino acid incorporated. Still to be determined is whether these methods can be scaled up to commercial size production, whether the economics will be favorable and whether the modified proteins will be safe for human and animal consumption. These present opportunities to the industrial and governmental scientists as well as the academicians.

ACKNOWLEDGMENT

The author thanks Virginia DuBowy for assistance in reference verification and in typing the manuscript.

REFERENCES

1. P. L. Antunes, V. C. Sgarbieri, and R. S. Garruti, Nutrification of dry bean (*Phaseolus vulgaris* L.) by methionine infusion, *J. Food Sci. 44:* 1302 (1979).
2. D. M. Matthews, Rates of peptide uptake by small intestine, In "Peptide Transport in Bacteria and Mammalian Gut", Assoc. Sci. Publ., Amsterdam, p. 71 (1972).
3. A. M. Ugolev, N. M. Timofeeva, L. F. Smirnova, P. De Laey, A. A. Gruzdkov, N. N. Iezuitova, N. M. Mityuskova, G. M. Roshchina, E. G. Gurman, V. M. Gusev, V. A. Tsvetkova, and G. G. Shcherbakov, Membrane and intracellular hydrolysis of peptides: differentiation, role and interrelations with transport, In "Peptide Transport and Hydrolysis", Ciba Found. Symp. 50: 221 (1977).
4. For a review see H. G. Khorana, The chemistry of carbodiimides, *Chem. Revs. 53:* 145 (1953).
5. J. C. Sheehan and G. P. Hess, A new method of forming peptide bonds, *J. Am. Chem. Soc. 77:* 1067 (1955).
6. L. P. Voutsinas and S. Nakai, Covalent binding of methionine and tryptophan to soy protein, *J. Food Sci. 44:* 1205 (1979).
7. I. Wieland, W. Schafer, and E. Bokelmann, Peptide syntheses. V. A convenient method for the preparation of acylthiophenols and their application in the synthesis of amides and peptides, *Ann. 573:* 99 (1951).

8. M. Bodanszky, Synthesis of peptides by aminolysis of nitrophenyl esters, *Nature 175:* 685 (1955).

9. R. Schwyzer, B. Iselin, and M. Feurer, Uber aktivierte Ester. 1. Aktivierte Ester der Hippursäure und ihre Umsetzungen mit Benzylamin, *Helv. Chim. Acta 38:* 69 (1955).

10. G. W. Anderson, J. E. Zimmerman, and F. M. Callahan, The use of esters of N-hydroxysuccinimide in peptide synthesis, *J. Am. Chem. Soc. 86:* 1839 (1964).

11. A. J. Puigserver, L. C. Sen, E. Gonzales-Flores, R. E. Feeney, and J. R. Whitaker, Covalent attachment of amino acids to casein. 1. Chemical modification and rates of in vitro enzymatic hydrolysis of derivatives, *J. Agric. Food Chem. 27:* 1098 (1979).

12. A. J. Puigserver, L. C. Sen, A. J. Clifford, R. E. Feeney, and J. R. Whitaker, A method for improving the nutritional value of food proteins: covalent attachment of amino acids, In "Nutritional Improvement of Food and Feed Proteins", M. Friedman, ed., Plenum Publ. Corp., p. 587 (1978).

13. A. J. Puigserver, L. C. Sen, A. J. Clifford, R. E. Feeney, and J. R. Whitaker, Covalent attachment of amino acids to casein. 2. Bioavailability of methionine and N-acetylmethionine covalently linked to casein, *J. Agric. Food Chem. 27:* 1286 (1979).

14. A. J. Puigserver, H. F. Gaertner, L. C. Sen, R. E. Feeney, and J. R. Whitaker, Covalent attachment of essential amino acids to proteins by chemical methods: nutritional and functional significance, In "Modification of Proteins: Food, Nutritional, and Pharmacological Aspects", R. E. Feeney and J. R. Whitaker, eds., *Adv. Chem. Ser. 198:* 149 (1982).

15. G. Matheis, L. C. Sen, A. J. Clifford, and J. R. Whitaker, Attachment of N-acetyl-L-methionine into whole soybeans and the nutritional consequences for rats, *J. Agric. Food Chem. 33:* 39 (1985).

16. R. E. Gosselin, H. C. Hodge, R. P. Smith, and M. N. Gleason, "Clinical Toxicity of Commercial Products. Acute Poisoning", 4th ed., Williams and Wilkins, Baltimore, MD, p. 135 (1976).

17. E. Katchalski and M. Sela, Synthesis and chemical properties of poly-α-amino acids, *Adv. Protein Chem. 13:* 243 (1958).

18. H. F. Gaertner and A. J. Puigserver, Covalent attachment of poly(L-methionine) to food proteins for nutritional and functional improvement, *J. Agric. Food Chem. 32:* 1371 (1984).

19. C. W. Newman, J. M. Maynes, and D. C. Sands, Poly-l-lysine, a nutritional source of lysine, *Nutr. Rep. Int. 22:* 707 (1980).

20. G. Matheis and J. R. Whitaker, Chemical phosphorylation of food proteins: an overview and a prospectus, *J. Agric. Food Chem. 32:* 699 (1984).

21. F. Cramer and G. Weimann, Trichloracetonitril, ein Reagenz zur selektiven Veresterung von Phosphorsäuren, *Chem. Ber. 94:* 996 (1961).

22. R. E. Ferrel, H. S. Olcott, and H. Fraenkel-Conrat, Phosphorylation of proteins with phosphoric acid containing excess phosphorus pentoxide, *J. Am. Chem. Soc. 70:* 2101 (1948).

23. H. -Y. Sung, H. -J. Chen, T. -Y. Liu, and J. -C. Su, Improvement of the functionalities of soy protein isolate through chemical phosphorylation, *J. Food Sci. 48:* 716 (1983).

24. E. Bourland, M. Pacht, and P. Grabar, Phosphorylation of amino acids and of gelatin by means of monophenylphosphorus dichloride, *Proc. Int. Congr. Biochem., 1st:* 115 (1949).

25. T. Müller, T. Rathlev, and Th. Rosenberg, Special cases of nonenzymatic transphosphorylation, *Biochim. Biophys. Acta 19:* 563 (1956).

26. T. Rathlev and Th. Rosenberg, Non-enzymic formation and rupture of phosphorus to nitrogen linkages in phosphoramido derivatives, *Arch. Biochem. Biophys. 65:* 319 (1956).

27. G. C. Taborsky, Phosphorylated ribonuclease: a study on the structural basis of enzymatic activity, *R. Trav. Lab. Carlsberg, Ser. Chim. 30:* 309 (1958).

28. F. Dallocchio, M. Matteuzzi, and T. Bellini, Nonenzymic protein phosphorylation. Phosphorylation of 6-phosphogluconate dehydrogenase by acyl phosphates, *Biochem. J. 203:* 402 (1982).

29. G. Matheis, M. H. Penner, R. E. Feeney, and J. R. Whitaker, Phosphorylation of casein and lysozyme by phosphorus oxychloride, *J. Agric. Food Chem. 31:* 379 (1983).

30. M. Sitohy, J. -M. Chobert, and J. R. Whitaker, Unpublished data, University of California, Davis (1985).

31. M. Fujimaki, S. Arai, and M. Yamashita, Enzymatic protein degradation and resynthesis for protein improvement. In "Food Proteins. Improvement Through Chemical and Enzymatic Modification", R. E. Feeney and J. R. Whitaker, eds., *Adv. Chem. Ser. 160:* 156 (1977).

32. M. Watanabe and S. Arai, Proteinaceous surfactants prepared by covalent attachment of L-leucine n-alkyl esters to food proteins by modification with papain. In "Modification of Proteins: Food, Nutritional and Pharmacological Aspects", R. E. Feeney and J. R. Whitaker, eds., *Adv. Chem. Ser. 198:* 199 (1982).

33. M. Yamashita, S. Arai, Y. Amano, and M. Fujimaki, A novel one-step process for enzymatic incorporation of amino acids

into proteins: application to soy protein and flour for enhancing their methionine levels, *Agric. Biol. Chem. 43:* 1065 (1979).

34. K. Ikura, M. Yoshikawa, R. Sasaki, and H. Chiba, Incorporation of amino acids into food proteins by transglutaminase, *Agric. Biol. Chem. 45:* 2587 (1981).

35. K. Ikura, T. Kometani, M. Yoshikawa, R. Sasaki, and H. Chiba, Crosslinking of casein components by transglutaminase, *Agric. Biol. Chem. 44:* 1567 (1980).

36. K. Ikura, T. Kometani, R. Sasaki, and H. Chiba, Crosslinking of soybean 7S and 11S proteins by transglutaminase, *Agric. Biol. Chem. 44:* 2979 (1980).

37. C. J. Epstein, R. F. Goldberger, and C. B. Anfinsen, The genetic control of tertiary protein structure: studies with model systems, *Cold Spring Harbor Symp. Quant. Biol. 28:* 439 (1963).

38. F. De Lorenzo, R. F. Goldberger, E. Steers, D. Givol, and C. B. Anfinsen, Purification and properties of an enzyme from beef liver which catalyzes sulfhydryl-disulfide interchange in proteins, *J. Biol. Chem. 241:* 1562 (1966).

39. S. Fuchs, F. De Lorenzo, and C. B. Anfinsen, Studies on the mechanism of the enzymic catalysis of disulfide interchange in proteins, *J. Biol. Chem. 242:* 398 (1967).

40. R. Kurane and Y. Minoda, Disulfide reduction and sulfhydryl oxidation by microbial enzyme. III. Purification of thiol-disulfide interchange enzyme from *Candida claussenii*, *Agric. Biol. Chem. 39:* 1417 (1975).

41. M. D. Hatch and J. F. Turner, A protein disulphide reductase from pea seeds, *Biochem. J. 76:* 556 (1960).

42. H. M. Katzen, F. Tietze, and D. Stetten, Further studies on the properties of hepatic glutathione-insulin transhydrogenase, *J. Biol. Chem. 238:* 1006 (1963).

43. J. Carlsson, H. Drevin, and R. Axén, Protein thiolation and reversible protein-protein conjugation. N-succinimidyl 3-(2-pyridyldithiol)propionate, a new heterobifunctional reagent, *Biochem. J. 173:* 723 (1978).

44. A. J. Puigserver, L. C. Sen, R. E. Feeney, and J. R. Whitaker, Hydrolyse enzymatique et stabilite en milieu acide des liaisons isopeptidiques de la L-methionyl-caseine, *Ann. Biol. Anim. Biochim. Biophys. 19:* 749 (1979).

45. H. Gaertner and A. J. Puigserver, Enzymatic hydrolysis of the model isopeptides N^{ϵ}-oligo(L-methionyl)-L-lysine, *Eur. J. Biochem. 145:* 257 (1984).

46. R. Fields, The rapid determination of amino groups, *Methods Enzymol. 25:* 464 (1972).

3

Modification to Change Physical and Functional Properties of Food Proteins

Soichi Arai, Michiko Watanabe, and Noriko Hirao

Department of Agricultural Chemistry
The University of Tokyo
Bunkyo-ku, Tokyo, Japan

I. INTRODUCTION

It is well known that among a variety of edible proteins available on earth, only a limited number are utilized, the rest still being disregarded. In most cases, the unutilized proteins are characterized by lack or insufficiency of the important physical and functional properties that determine many processing and organoleptic parameters of foods. To improve the utilization of these proteins the use of sophisticated technologies of modifying their chemical structures is necessary. Even high-quality proteins of nutritional importance may have scope for modification into products of increased added value which could be used as ingredients or additives for making foods with more acceptable characteristics.

In recent years, a special attention has been paid to the surface behaviors of food proteins, since the surface activity of a protein is a most important parameter related to its functional properties [1]. Generally, highly surface-active proteins have unique hydrophilic-hydrophobic structures of several types. These comprise the categories illustrated in Fig. 1.

FIG. 1 *Schematic diagrams depicting structures of more surface-active proteins (A - E) and those of less surface-active proteins (F and G). The diagrams represent hydrophilic regions with filled columns and hydrophobic regions with open columns.*

Casein components are good examples of such proteins. α_{s1}-Casein by itself possesses a reasonable degree of surface activity to emulsify oil finely. However, when this protein is modified with a protease to cut off its C-terminal structure, a more surface-active fragment with a typical amphiphilic structure (Fig. 1-A) is formed [2]. β-Casein per se has a clear amphiphilic structure (Fig. 1-B) contributing to its emulsifying activity [3]. This structure of β-casein is reflected in its function to form a casein micelle structure. On the other hand, κ-casein is characterized by its original hydrophilic-hydrophobic structure (Fig. 1-C) [4].

Chemical modification techniques are available which enhance hydrophobicity of a hydrophilic protein by attachment of, e.g., a long-chain alkyl group and hydrophilicity of a hydrophobic protein by attachment of, e.g., a sugar and the like. These modifications would possibly lead to formation of the structures shown in Fig. 1-A and Fig. 1-D and the structures shown in Fig. 1-B and Fig. 1-C, respectively.

Biological systems also provide interesting examples. Most biological cells contain preproproteins or preproteins that function at their inner surfaces. Chemically, these proteins bear hydrophobic signal peptide sequences constituting their N-terminal structures (Fig. 1-A). Looking for another example of a typical hydrophilic-hydrophobic structure, we find statherin, a small protein occurring in human saliva which is supersaturated with respect to calcium phosphates [5]. This protein stabilizes the supersaturation by adsorbing onto the surface of calcium phosphate crystals as they form. It is speculated that statherin has the structure as shown in Fig. 1-C and functions with its hydrophilic region adsorbed on the solid phase and its hydrophobic region exposed to the liquid phase. A more interesting example may be a series of antifreeze glycoproteins (AFGP) existing in the blood of winter polar fish [6]. These proteins, though greatly different from each other with respect to molecular weight, have a similar structure characterized by a repeated Ala-Ala-Thr sequence. Higher-molecular-weight species of these proteins, especially AFGP-4 with a molecular weight of ca. 17,000 daltons, are able to adsorb at the liquid water-ice crystal interface to inhibit the ice crystal growth [7]. Thus, the proteins protect polar fish from freezing to death at a subzero temperature around -2°C. It has been demonstrated that AFGP-4 has a unique hydrophilic-hydrophobic conformation (Fig. 1-E) which is responsible for such an interfacial function [8].

Most proteins have plural numbers of hydrophilic and hydrophobic regions in their molecules (Fig. 1-F). In general, these proteins, even though unfolded, are less surface-active than the proteins with the clear amphiphilic structures (Fig. 1-A through Fig. 1-E). Also, many glubular proteins constituted with their hydrophilic regions exposed and hydrophobic regions buried (Fig.1-G) are less surface-active, unless denatured to expose parts of the hydrophobic regions. Sometimes, enzymatic modification techniques are applied to these proteins for their partial hydrolysis to induce a similar conformational change. However, it is beyond the scope of this article to discuss the techniques used for such a particular purpose.

Our group carried out a study to produce a surface-active protein of the type shown in Fig. 1-D. For this purpose we have developed an enzymatic process that simulates the plastein reaction [9].

In the present paper we describe this enzymatic process and its application to succinylated α_{s1}-casein to understand the mechanism involved and to gelatin to produce a highly surface-active protein for food and industrial uses. The paper also deals with evaluation of this enzymatically modified gelatin in terms of its chemical, physical and functional properties in general and its freezing-retardation activity in particular.

II. ENZYMATIC PROCESS

A. *GENERAL ASPECT*

The plastein reaction, a unique protease-catalyzed process leading to formation of a plastein product from a protein hydrolysate, has been studied extensively for maximizing nutritive values of food proteins [10]. A great advantage of this reaction may be its ability to covalently incorporate amino acids (ester form) which have been added to the reaction system [10].

Recently, we succeeded in modifying the classical plastein reaction into a new process which would permit the use of a protein itself, instead of a protein hydrolysate, as a material to be improved [9]. Since this process is designed with the particular aim to incorporate an amino acid ester with greater efficiency, it is necessary to make the reaction drive in an alkaline system whereby an added amino acid ester is activated to a more effective nucleophile. The enzyme, papain (EC 3·4·22·2), although its activity for peptide bond hydrolysis is maximized in a weakly acidic environment, has the potential to catalyze this type of incorporation reaction most effectively at a pH value of 8 - 10 [11]. Another requirement of the process is that the substrate concentration in the medium be set at the highest possible level in order to minimize the water concentration. The replacement of a significant part of the water by an organic solvent such as acetone could fulfil the purpose [12]. A major reason for the existence of this requirement can be explained by the competitive hydrolysis-aminolysis kinetics:

$$E + S \rightleftarrows ES \rightarrow ES' + P_1 \quad \begin{cases} \xrightarrow{+H_2O} E + P_2 \\ \underset{}{\overset{+N}{\rightleftarrows}} ES'N \rightarrow E + P_3 \end{cases} \tag{1}$$

where the symbols stand for the following terms: E, enzyme (e.g., papain); substrate (protein); ES, Michaelis complex; ES', acyl- (or peptidyl-) enzyme; N, nucleophile (amino acid ester); P_1 and P_2, products formed from the substrate by simple hydrolysis; P_3 (S'N), product formed from the peptidyl-enzyme by aminolysis with the nucleophile. The process is understood to be primarily a competitive reaction involving the hydrolysis of the peptidyl-enzyme and its aminolysis leading to formation of the third product (P_3) which has covalently attached the amino acid ester (N) at the C-terminal. Thus, the lower the water concentration in the reaction system, the more efficient the formation of the product.

Using this process we undertook an experiment, with the expectation that the covalent attachment of a highly hydrophobic amino acid ester to a sufficiently hydrophilic protein would afford a product having the amphiphilic structure shown in Fig. 1-D.

The following section describes this topic in some detail.

B. APPLICATION

Bovine α_{s1}-casein (variant B) was selected as a well-defined substrate. It was prepared from fresh milk [13] and immediately succilylated [14]; the succinylation effected the enhancement of the hydrophilicity of this protein. As a hydrophobe to be attached, L-leucine *n*-dodecyl ester was used, α-methylene of which had been labelled with ^{13}C. It was confirmed that the ester gave a distinct CMR signal at a distance of 65.2 ppm from the signal of tetramethylsilane [15].

The enzymatic reaction of Eq. (1) was carried out under the following conditions: medium, 20 % acetone in 1 M carbonate (pH 9) containing 10 mM L-cysteine; concentration of succinylated α_{s1}-casein in the medium, 20 %; concentration of leucine 1-^{13}C-dodecyl ester in the medium, 0.25 M; concentration of papain (recrystal-

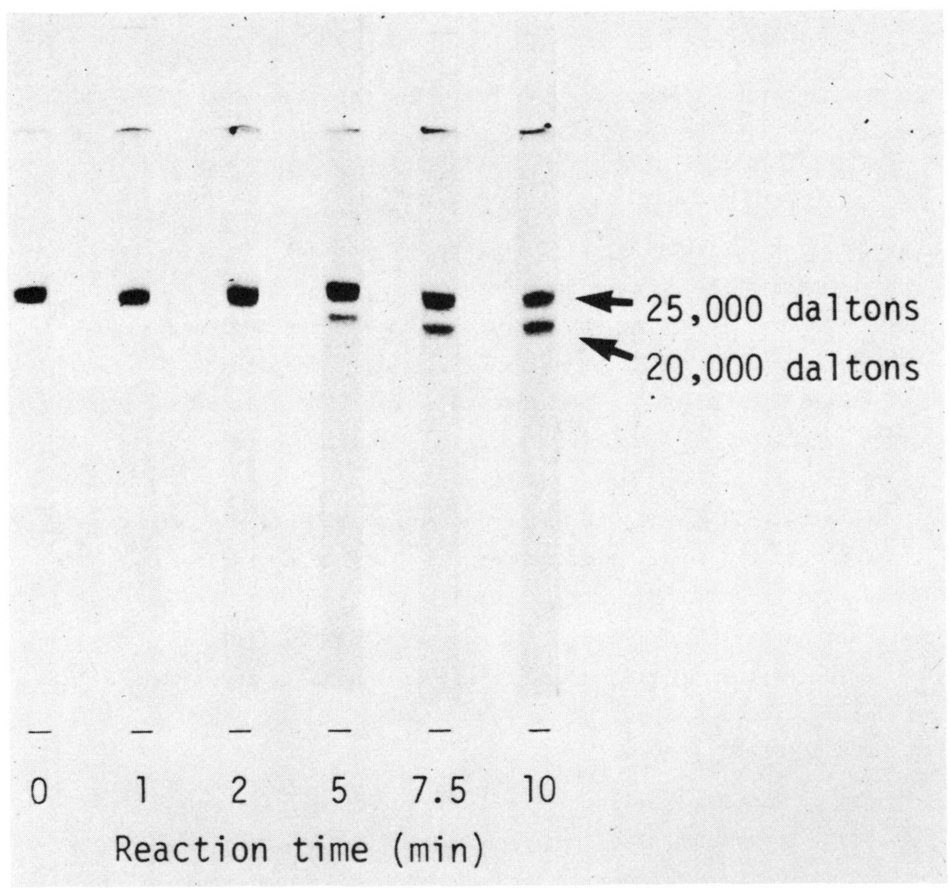

FIG. 2 *Polyacrylamide gel electrophoresis showing time-course degradation and formation of the substrate (ca. 25,000 daltons) and the product (ca. 20,000 daltons), respectively.*

lized preparation) in the medium, 0.02 %; incubation temperature, 37°C; and incubation time up to 60 min [16].

The reaction was followed by polyacrylamide gel electrophoresis (PAGE) in the presence of sodium dodecylsulfate (SDS). The SDS-PAGE demonstrated that during the reaction the substrate, 25,000 daltons, was degraded into a macropeptide having an approximate molecular weight of 20,000 daltons (Fig. 2). We then treated the 5 min incubation product with peptroleum ether/ 1 M carbonate (pH 9) to obtain the 20,000-dalton product by purification at the interface. The achievement of this surface purification indicated that the product had aquired a potent surface-active character [16]. CMR measurement demonstrated that the purified product gave a clear 65-ppm signal assignable to 1-^{13}C-methylene.

We conducted various analyses to find the chemical structure of the 20,000-dalton macropeptide and reached the conclusion that the substrate, succinylated α_{s1} casein, underwent the first degra-

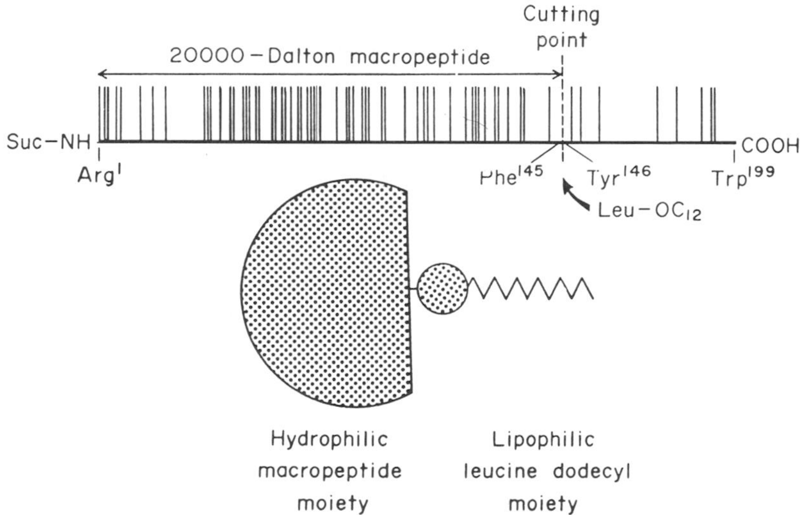

FIG. 3 *Covalent incorporation of L-leucine n-dodecyl ester into succinylated* α_{s1}*-casein by modification with papain and the resulting formation of a 20,000-dalton macropeptide with an amphiphilic structure. The position of hydrophilic amino acid residues on the protein molecule are marked with vertical bars.*

dation at the peptide bond involving Phe^{145} and Tyr^{146}, with the subsequent incorporation of the nucleophile, L-leucine *n*-dodecyl ester, into the same position to form a new C-terminal (Fig. 3). It follows that an amphiphilic structure is formed as illustrated in Fig. 3 [16]. Probably, the structure is responsible for the aquired surfactancy of the product.

III. ENZYMATICALLY MODIFIED GELATIN

A. *PRODUCTION*

Gelatin is a hydrophilic protein derived from collagen by heating under an acidic or alkaline condition. Collagen is characterized by its repeated X-Y-Gly sequences and therefore has no particular hydrophobic regions in its molecule. This character is considered to be retained in gelatin.

We used a commercial preparation of "alkali gelatin" and modified it with papain in the presence of L-leucine *n*-dodecyl ester as a hydrophobe to be attached. The conditions used in this case was similar to those described in the previous section (II-*B*). The reaction was stopped by acidifying to pH 1 with 1 N HCl and the acidified mixture dialyzed in running water. Lyophilization of the non-diffusible fraction gave a proteinaceous product in powder form which was purified further by treatment with a sufficient amount of hot acetone. HPLC analysis showed that these purification procedures were effective in removing low-molecular-weight species including unreacted leucine dodecyl ester (Fig. 4). The product resulting from the purification was named EMG-12.

B. *GENERAL PROPERTIES*

We conducted various experiments with EMG-12 to evaluate its chemical, physical and functional properties. TABLE 1 collates the data obtained from these experiments.

EMG-12, being a mixture of polypeptides, has a molecular weight distribution in the range of 2,000 - 40,000 daltons, with an average ca. 7,500 daltons [17]. The amount of the leucine dodecyl ester

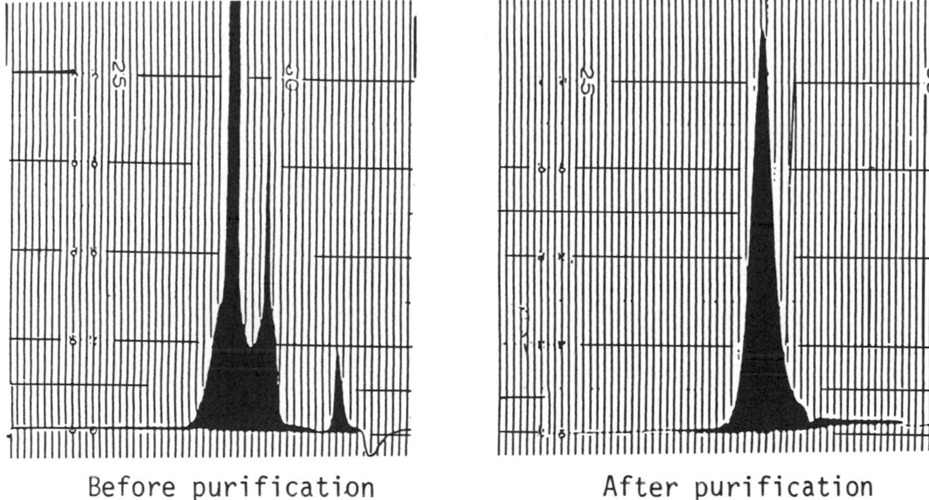

FIG. 4 *High performance liquid chromatography on Shodex OH PAK-B 800P (0.1 M phosphate, pH 8.0) of the enzymatically modified gelatin before and after purification.*

attached approximated 1 mol/7,500 g [17]. Probably as a result of its attachment, a great increase in hydrophobicity occurred with this protein [18].

Both electroconductivity and surface tension measurements with aqueous dispersions of EMG-12 estimated its critical micelle concentration at 0.02 - 0.04 % [19]. When EMG-12 was added to pure water at 20°C, its surface tension began to decrease rapidly from 72.75 dyn/cm to approximately 35 dyn/cm. In this respect, EMG-12 was well comparable with chemically synthesized surfactants for industrial uses. It was estimated that EMG-12, when dispersed in water at a concentration of 30 - 60 %, could form a thermotropic liquid crystal structure. This estimation was based on observing a birefringent phenomenon occurring when such a high concentration of an EMG-12 dispersion in water was exposed to polarized light [20].

With respect to emulsion properties in terms of emulsifying activity, emulsion stability against coalescence, and emulsion stability against creaming, EMG-12 was comparable with or superior to chemically synthesized surfactants, for example, sorbitan mono-

TABLE 1 *Chemical, physical, and functional properties of EMG-12 compared to gelatin and its hydrolysate (EMG-0)*

Item	Gelatin	EMG-0	EMG-12
General properties			
Molecular weight	*Higher than 30,000*	*av. 7,500*	*av. 7,500*
Dodecyl moiety content	—	—	*ca. 1 mol/7,500 g*
Effective hydrophobicity	*33*	*11*	*240*
Dispersion properties			
Critical micelle concentration	*No micelle formed*	*No micelle formed*	*0.02 - 0.04 %*
Decrease in surface tension	*Down to 60 dyn/cm*	*Down to 60 dyn/cm*	*Down to 35 dyn/cm*
Phase characteristic	*No phase transition observed*	*No phase transition observed*	*Partial formation of liquid crystals*
Emulsion properties			
Emulsifying activity	*Inferior to Tween-80*	*Inferior to Tween-80*	*Superior to Tween-80*
Emulsion stability	*Inferior to Tween-80*	*Inferior to Tween-80*	*Superior to Tween-80*
Interfacial molecular area	—	—	*48* $Å^2$
Cryophysical properties			
Supercooling of an o/w emulsion	—	—	*Stable at -10°C*
Supercooling of an aqueous dispersion	*Not stable*	*Not stable*	*Stable at -7°C*
Interaction with silver iodide crystals	*Not observed*	*Not observed*	*Adsorption for anti-nucleation*

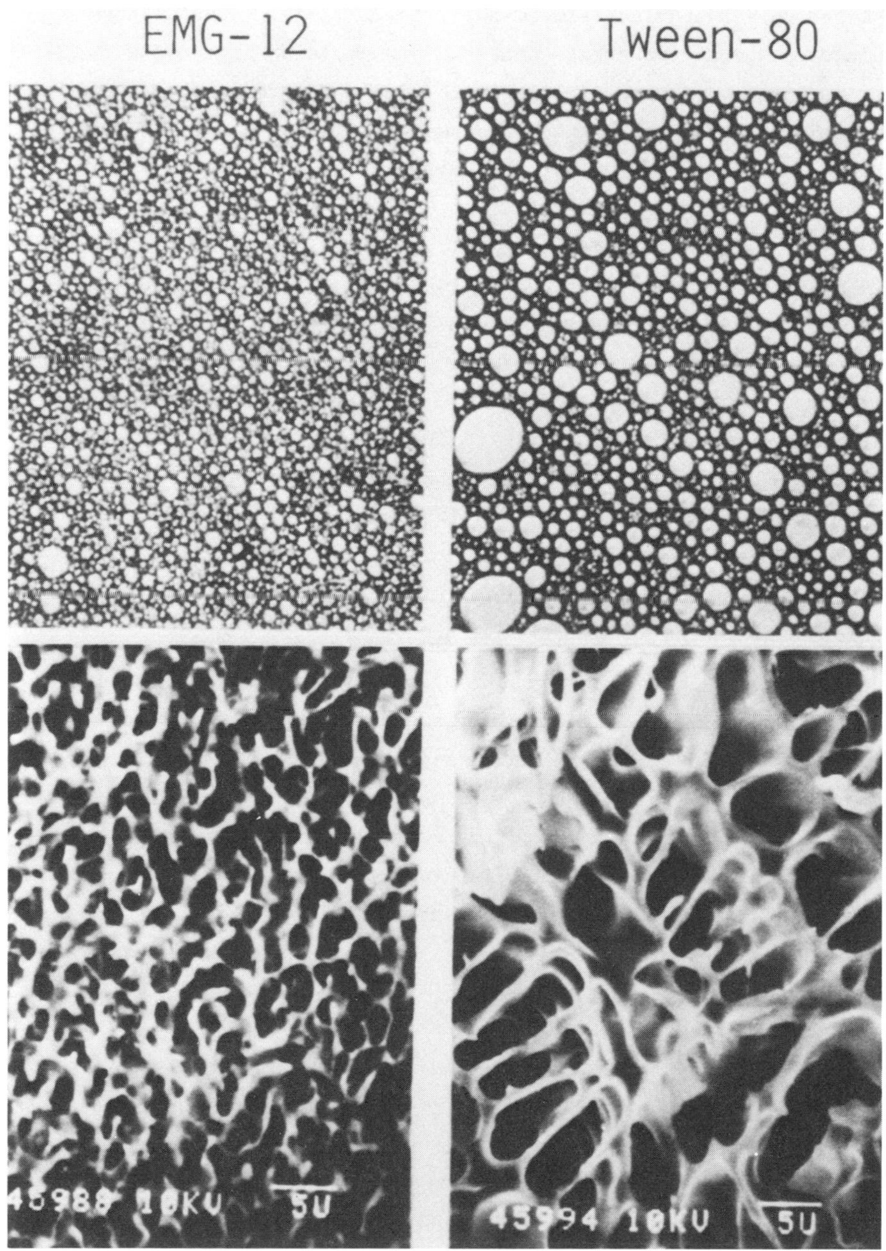

FIG. 5 *Observation by ordinary microscopy (upper) and scanning electron microscopy (lower) of an o/w emulsion produced with EMG-12 and that produced with Tween-80.*

oleate polyoxyethylene (Tween-80) [21][22][23]. FIGURE 5 shows pictures which demonstrate that the o/w emulsion produced with EMG-12 is finer in average oil particle size and more homogeneous in overall structure than that produced with Tween-80. Other properties of surface-chemical interest have also been evaluated as tabulated.

The reader can find how EMG-12 differs from gelatin or its hydrolysate with respect to almost all the parameters investigated. It is clear that the enzymatic modification caused such differences.

Recently, we found that EMG-12 added to water was able to retard freezing by supercooling. A similar phenomenon was observed for an o/w emulsion produced with EMG-12. It was thus expected that this proteinaceous surfactant could function as an antifreeze agent as well as an emulsifier. The next section describes this particular function of EMG-12 in detail.

C. *PARTICULAR FUNCTION*

For the crystallization of substances in general, the existence of nuclei is indispensable. The same holds true in the particular case of freezing water. Without any nucleus, water remains liquid even at subzero temperature, keeping a state of supercooling. The freezing of water starts with ice crystal formation which is initiated by ice nucleation. This is classified into the two categories: homogeneous nucleation in which water molecules themselves act as nuclei and heterogeneous nucleation which takes place with the aid of exogenous substances in terms of "motes". It is well known that silver iodide crystals can be heterogeneous nuclei with a potent nucleating activity. In the presence of even small pieces of this compound, it is generally difficult for water to supercool. In such a case, therefore, water freezes at the temperature corresponding to the melting point of ice.

We have often used crystallized silver iodide in order to evaluate how EMG-12 functions as an antifreeze agent despite the presence of this nucleus. Also, in every case we carried out experiments on a bulk water scale so that the so-called microcompartization effect leading to an extraordinary decrease in freezing temperature is excluded.

1. Antifreeze emulsion Since most o/w emulsions often lose their stability through the formation of ice crystals from the bulk water during freezer storage, it would be of benefit to develop an agent effective in inhibiting ice nucleation. It is preferable in this case to develop a high-molecular-weight agent which would act in a non-colligative manner, because any colligative decrease in freezing temperature is exclusively accompanied by an osmotic problem.

We produced o/w emulsions with EMG-12 and with the three control agents, gelatin hydrolysate (EMG-0) with an average molecular weight of ca. 7,500 daltons, polyglycerol stearate (PGS) and Tween-80. Each emulsion was placed in a sample tube fittable for pulsed

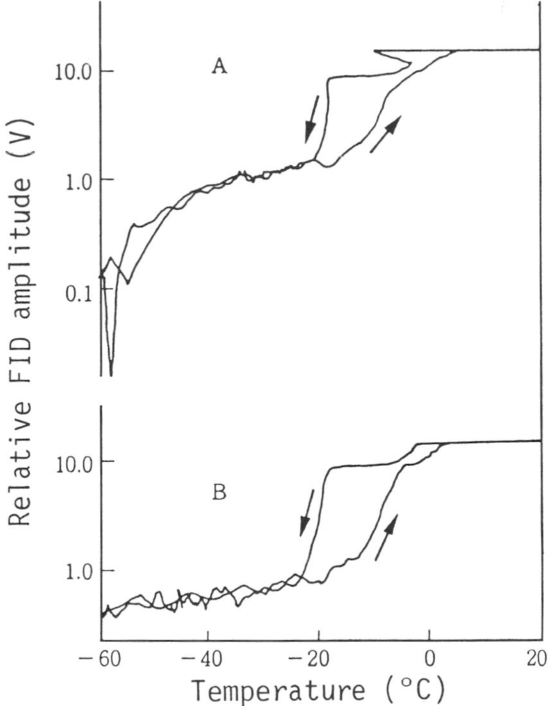

FIG. 6 *Freezing-thawing curves observed for an o/w emulsion produced with EMG-12 (A) and for that produced with Tween-80 (B).*

NMR measurement and cooled gradually down to -10°C while the FID amplitude was recorded automatically. For the emulsions produced with EMG-0 and Tween-80, the recorder indicated that the bulk water began to freeze when its temperature reached -4°C (Fig. 6). However, both of the emulsions produced with EMG-12 and PGS resisted freezing even at -10°C (Fig. 6), maintaining a state of supercooling for 20 hr or longer [24][25].

Another experiment was conducted to investigate what happened when ice nucleation was induced by adding silver iodide crystals to the emulsions. For this experiment, an EMG-12 dispersion (3 ml) in water was mixed with linoleic acid (3 ml) and a small amount (3 mg) of the crystals was added. Subsequently, the mixture was emulsified by ultrasonication at room temperature and the resulting emulsion allowed to stand at -10°C. The result showed that the emulsion produced with EMG-12 resisted freezing until its temperature reached -8.6 ± 0.7°C [26].

In order to find why supercooling in the emulsion produced with EMG-12 was stable even in the presence of added silver iodide crystals, we observed its structure by scanning electron microscopy. The observation indicated that no crystal pieces existed in the bulk water phase. By magnifying a picture of an oil particle mass that had constituted the emulsion, it was found that the crystal pieces were fixed to the oil particle mass at the surface (Fig. 7). It is therefore speculated that, as a result of such fixation of silver iodide, its ice-nucleating activity was lost or attenuated to a great extent.

2. *Antifreeze dispersion* EMG-12 was found effective in retarding the freezing of pure water as well. In this case, the effect was dependent critically on the amount of EMG-12 used. At a concentration of 0.01 % or lower, the water never supercooled. For stable supercooling it was necessary to use EMG-12 at 0.03 % or higher. Apparently, a critical zone existed in-between (Fig. 8) and, interestingly, it accorded with the critical micelle concentration of EMG-12 (Table 1)[24].

The degree of supercooling observed for an EMG-12 dispersion

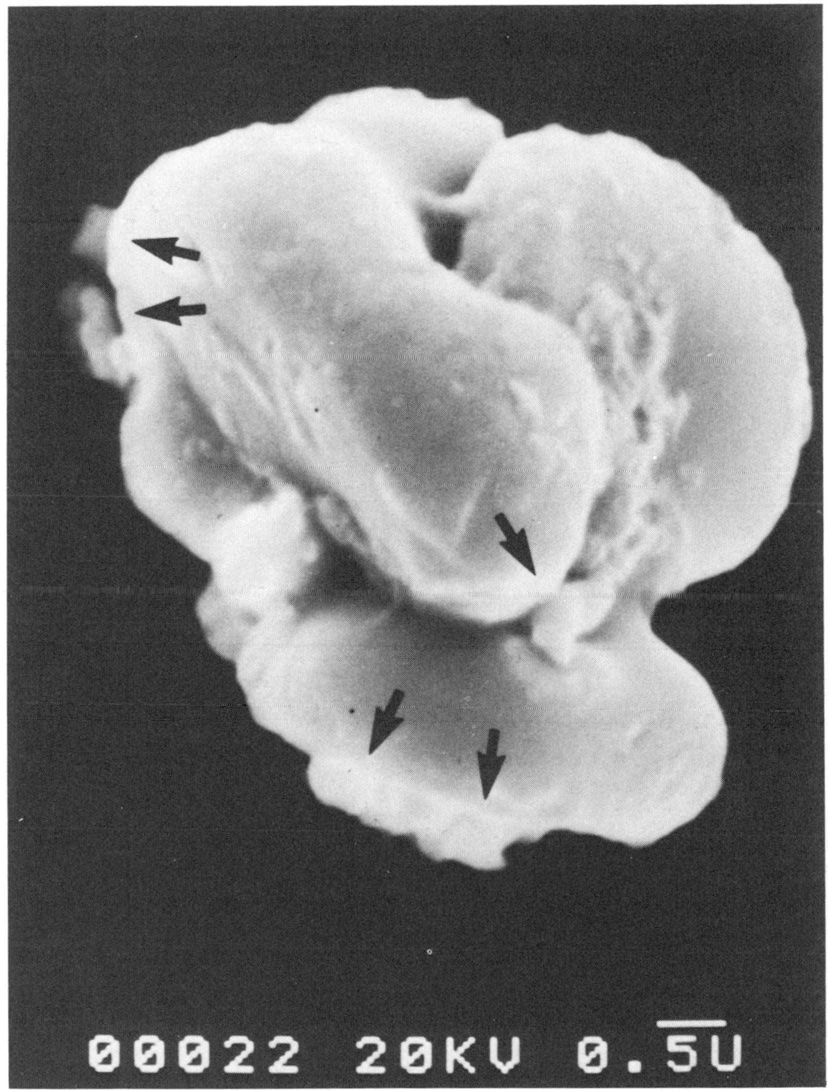

FIG. 7 *Scanning electron microgram showing silver iodide crystal pieces (arrows) adsorbed onto the surface of an oil particle mass.*

was around 7°C, significantly greater than that observed for an EMG-0 dispersion, although polyvinylpyrrolidone (PVP) used as another control resembled EMG-12 in respect of a supercooling

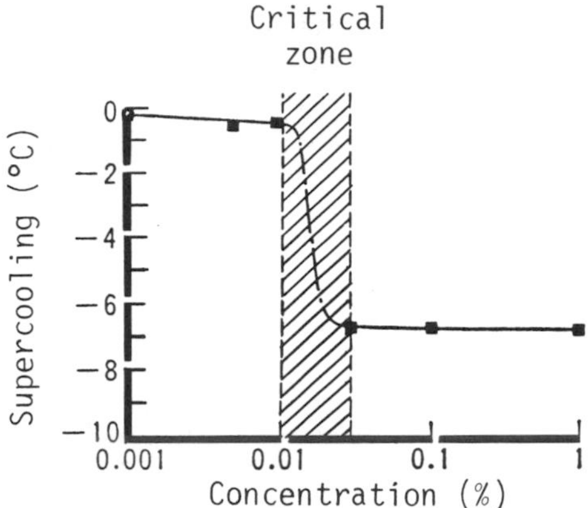

FIG. 8 *Relationship between the concentration of EMG-12 in water and its supercooling. Each plot is based on a mean of 10 independent measurements; their standard error was smaller than 0.5°C in every case.*

effect (Table 2). In our working hypothesis, the stability of supercooling depends on how efficiently the dispersed molecules adsorb onto the surface of the heterogeneous nuclei to prevent their nucleating action. To test this hypothesis we conducted an experiment to observe the adsorption process by differential thermal anal-

TABLE 2 *Degrees of supercooling* observed when aqueous dispersions of EMG-12, gelatin hydrolysate (EMG-0) and polyvinylpyrrolidone (PVP) were cooled in the presence of added silver iodide crystals***

Agent	Concentration (%)	Repetition	Degree of supercooling (°C)
EMG-12	*0.1*	*5*	*7.64 ± 0.24*
EMG-12	*0.01*	*5*	*0.44 ± 0.13*
EMG-0	*0.1*	*5*	*4.62 ± 0.28*
PVP	*0.1*	*5*	*7.92 ± 0.44*

* $T_m - T$ *(T_m: melting point; and T: temperature at which freezing is initiated)*
** *15 μg/3 ml dispersion*

lysis. Using a thermometer of special design, we observed that as soon as a silver iodide suspension in water was dropped into an 0.1 % EMG-12 dispersion, a distinct exothermic peak appeared, indicating that an adsorption process did proceed (Fig. 9-A). A similar process took place when PVP was used (Fig. 9-D). However, no clear peak resulted when EMG-0 was used at 0.1 % (Fig. 9-C) nor when EMG-12 was used at a level of lower than its critical micelle concentration (Fig. 9-B). It is probable that EMG-12 acts as an antifreeze agent only when it is dispersed in water at a sufficiently high concentration [26].

3. Comparison with a naturally occurring antifreeze protein

Extensive studies have been performed on antifreeze glycoproteins (AFGP) existing in the blood of winter polar fish including komai (Fig. 10) which literally means "fish under the ice" [7]. A great deal of information is available on AFGP-4 which is known to be one

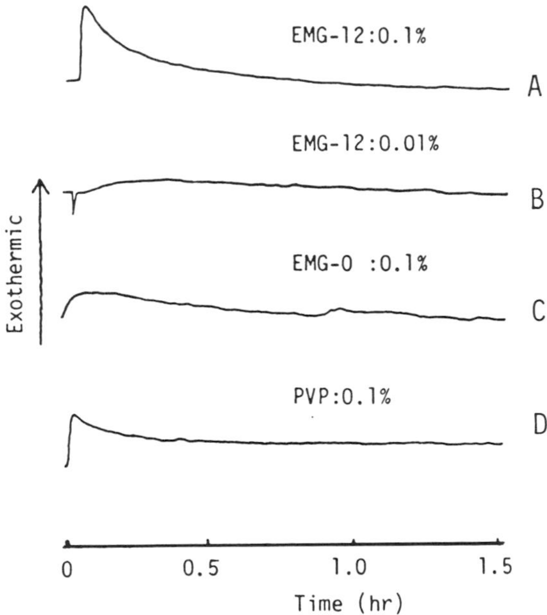

FIG. 9 *Time-course changes in exothermic reaction observed when a suspension of silver iodide crystals was dropped into each of surfactant dispersions.*

FIG. 10 *Komai (Eleginus gracilis), a type of fish harvested in the northernmost region of Japan.*

of the most potent AFGP homologues [8]. It would be interesting to compare the structure-function relationships of EMG-12 and this glycoprotein. Though both are involved in freezing retardation, this phenomenon occurs in the form of antinucleation in the case of EMG-12 and ice crystal growth inhibition in the case of AFGP-4 [7] [8]. The reader is referred to Fig. 1 for their hydrophilic-hydrophobic structures.

4. *Application* An attempt was made to apply EMG-12 as a cryoprotectant for use in freezer storage of human blood. We used systemic blood samples offered by our five healthy students. When the blood samples were stored at -5°C, all froze since the human blood serum freezes around -0.6°C. When EMG-12 was added at 0.5 % or higher prior to being stored at -5°C, all samples were unfrozen for at least 5 days (Table 3). Electron microscopic observation showed that a

TABLE 3 *Consequences of the storage of human systemic blood samples at -5°C for 5 days in the presence of added EMG-12*

Concentration of EMG-12 added (%)	Number of samples	Frequency	
		frozen	unfrozen
0	*5*	*5/5*	*0/5*
0.05	*5*	*5/5*	*0/5*
0.10	*5*	*2/5*	*3/5*
0.50	*5*	*0/5*	*5/5*
1.0	*5*	*0/5*	*5/5*

majority of erythrocytes looked normal in the unfrozen blood samples, whereas in the frozen blood samples there were a large number of erythrocytes that had suffered from crenation by hemolysis. Thus, the subzero temperature itself did not seem to be an unfavorable factor and it appears to be the freezing of the bulk water that causes such a hemolytic problem [27].

IV. CONCLUSION

Proteins are fundamentally important as nutrients, but now they are seen to be much more. Through the subtlety and versatility of their properties, they can become key ingredients that determine many parameters of food quality. However, most proteins still have scope for further improvement in their physical and functional properties.

The surfactancy of proteins has called a special attention of food scientists as well as of biochemists, since it determines an important part of physical and functional properties of proteins in food and biological systems.

The sophisticated methodology currently available to improve the surface properties of proteins emphasizes the potential of enzymatic modifications to their chemical structures. The covalent attachment of a highly hydrophobic amino acid ester to a hydrophilic protein with the aid of protease action could be a method for fulfilling the purpose.

Examples can be taken from our work on covalent attachment of L-leucine *n*-dodecyl ester to succinylated α_{s1}-casein to enhance its amphiphilicity. It is also possible to produce on an industrial scale an enzymatically modified gelatin to which leucine dodecyl ester is attached (EMG-12). This protein may be used as a potent surfactant, as suggested by data on its ability to decrease the surface tension of water, critical micelle concentration, phase characteristic, emulsifying activity and so forth.

EMG-12 has a particular function in retarding the freezing of water by supercooling. It stabilizes the supercooling of water even in the presence of silver iodide crystals added as ice nuclei.

This proteinaceous surfactant can be used as an agent for making an antifreeze emulsion and also for cryoprotection of sensitive food and biological systems under storage at subzero temperatures.

Thus, the significance of tailoring proteins into substances of increased added values by means of enzymatic modification is stressed. It will also be necessary to establish future research strategies for further development of this interesting methodology.

We are particularly grateful to Professor Toshio Kuroda of The Institute of Low Temperature Science, Hokkaido University, Japan, who made many pertinent suggestions for our work on cryophysical properties of EMG-12 and other enzymatically modified proteins. Our thanks are also given to Professor Teruyuki Fujita of The Institute of Applied Microbiology, The University of Tokyo, Japan, who had us join his group in thermal analysis.

REFERENCES

1. J. E. Kinsella and K. J. Shetty, *in "Functionality and Protein Structure" (ACS Symposium Series No. 92)*, A. Pour-El, ed., American Chemical Society, Washington, D.C., 1979, p. 37.
2. M. Shimizu, T. Takahashi, S. Kaminogawa, and K. Yamauchi, *J. Agric. Food Chem. 31:* 1214 (1983).
3. B. Ribadeau Dumas, G. Brignon, F. Grosclaude, and J.-C. Mercier, *Eur. J. Biochem. 25:* 505 (1972).
4. J.-C. Mercier, G. Brignon, and B. Ribadeau Dumas, *Eur. J. Biochem. 35:* 222 (1973).
5. D. H. Schlesinger and D. I. Hay, *J. Biol. Chem. 252:* 1689 (1977).
6. R. E. Feeney and Y. Yeh, *Adv. Protein Chem. 32:* 191 (1978).
7. R. E. Feeney, D. T. Osuga, D. S. Reid, and Y. Yeh, *Protein-Nucleic Acid-Enzyme 27:* 1645 (1972).
8. C. A. Bush, R. E. Feeney, D. T. Osuga, S. Ralapati, and Y. Yeh, *Int. J. Peptide Protein Res. 17:* 125 (1981).
9. M. Yamashita, S. Arai, Y. Imaizumi, Y. Amano, and M. Fujimaki, *J. Agric. Food Chem. 27:* 52 (1979).
10. M. Fujimaki, S. Arai, and M. Yamashita, *in "Food Proteins: Improvement through Chemical and Enzymatic Modification" (Advances in Chemistry Series No. 160)*, R. E. Feeney and J. R. Whitaker, eds., American Chemical Society, Washington, D.C., 1977, p. 156.

11. M. Watanabe and S. Arai, *in "Modification of Proteins: Food, Nutritional, and Pharmacological Aspects" (Advances in Chemistry Series No. 198)*, R. E. Feeney and J. R. Whitaker, eds., American Chemical Society, Washington, D.C., 1982, p. 199.

12. S. Arai, M. Yamashita, and M. Fujimaki, *in "Water Activity: Influences on Food Quality"* L. B. Rockland and G. F. Stewart, eds., Academic Press, New York-London-Toronto-Sydney-San Francisco, 1981, p. 489.

13. C. A. Zittle and J. H. Custer, *J. Dairy Sci. 46:* 1069 (1963).

14. P. Hoagland, *J. Dairy Sci. 49:* 783 (1966).

15. S. Arai, M. Watanabe, and S. Toiguchi, *Agric. Biol. Chem. 46:* 3085 (1982).

16. S. Toiguchi, S. Maeda, M. Watanabe, and S. Arai, *Agric. Biol. Chem. 46:* 2945 (1982).

17. M. Watanabe, H. Toyokawa, A. Shimada, and S. Arai, *J. Food Sci. 46:* 1467 (1981).

18. A. Shimada, E. Yazawa, and S. Arai, *Agric. Biol. Chem. 46:* 173 (1982).

19. A. Shimada, I. Yamamoto, H. Sase, Y. Yamazaki, M. Watanabe, and S. Arai, *Agric. Biol. Chem. 48:* 2681 (1984).

20. I. Yamamoto, K. Kusuhara, M. Matsumoto, A. Shimada, M. Watanabe, and S. Arai, *Agric. Biol. Chem. 48:* 2689 (1984).

21. M. Watanabe, A. Shimada, and S. Arai, *Agric. Biol. Chem. 45:* 1621 (1981).

22. M. Watanabe, N. Fujii, and S. Arai, *Agric. Biol. Chem. 46:* 1587 (1982).

23. S. Arai, M. Watanabe, and N. Fujii, *Agric. Biol. Chem. 48:* 1861 (1984).

24. S. Arai, M. Watanabe, and R. F. Tsuji, *Agric. Biol. Chem. 48:* 2173 (1984).

25. M. Watanabe, R. F. Tsuji, N. Hirao, and S. Arai, *Agric. Biol. Chem. 49:* 3291 (1985).

26. N. Hirao, M. Watanabe, S. Arai, and T. Fujita, *Agric. Biol. Chem.* under submission.

27. Unpublished.

4

Relationship of Structure to Taste of Peptides and Peptide Mixtures

Jens Adler-Nissen

Enzymes Research and Development
Novo Industri A/S
Bagsvaerd, Denmark

ABSTRACT

Peptides formed by enzymic hydrolysis of proteins are used today in certain foods as a source of soluble dietary nitrogen. The often reported bitter taste of such peptide mixtures is a phenomenon, which must be understood in the light of the amino acid composition of the protein and the hydrolysis kinetics. The bitter taste of the individual peptides is related to their average hydrophobicity and chain length, but in case of mixtures of peptides (i.e. protein hydrolyzates) the relationship is more complicated. Established ways of assessing the bitterness level of peptide mixtures are criticized and it is argued that the bitterness of protein hydrolyzates and the various ways of debittering them is a function of the hydrophobicity distribution of the peptide spectrum. Proteases having specificity for hydrophobic amino acids can be predicted to be preferable for producing protein hydrolyzates of low bitterness. Further reduction of the overall bitterness may be achieved by selective iso-electric precipitation of strongly bitter, hydrophobic peptides.

I. INTRODUCTION

The use of proteolytic enzymes for modification of the properties of food proteins has been studied extensively for the last two decades, and several more or less complete reviews of this field have appeared (1,2,3,4). The immediate effcts of the hydrolytic reaction are an increased solubility of the protein, in particular around the iso-electric point, a reduced viscosity and significant changes in the emulsifying, foaming and gelling properties of the protein. The molecular weight distribution of such food protein hydrolyzates is usually very broad, ranging from still unconverted protein to small peptides and free amino acids, depending on the specificity of the enzyme(s) and the nature and state of denaturation of the substrate (1,5).

The soluble peptide fraction of the protein hydrolyzate may be separated from the insoluble high-molecular-weight material to yield a 100% soluble mixture of peptides. Such peptide mixtures mainly derived from milk proteins, are produced commercially and have a long tradition for use in the clinical feeding of hospital patients (6,7). In the recent years other types of peptide mixtures have entered the food industry, notably iso-electric soluble soy protein hydrolyzate (ISSPH). ISSPH is produced by a controlled hydrolysis process (8,9) and finds application among other things for protein enrichment of fruit juices (10).

A prevailing problem in the food use of peptides is that most proteins yield bitter-tasting peptides during enzymic hydrolysis. These bitter peptides are characteristic by being rich in hydrophobic amino acids, and the objectionable taste which they impart on protein hydrolyzates has been acknowledged for long. Already in 1950 Cuthbertson (11) noted the offensive flavor of protein hydrolyzates meant for clinical feeding, and this has since been a major drawback to their use (12). In fact, the bitterness problem is generally lamented upon in the literature on the hydrolysis of protein for food use.

As a consequence of the practical importance of the bitterness problem a considerable number of publications describe ways of avoiding or reducing bitter taste in protein hydrolyzates. Many of these methods are based on the hydrophobic character of the bitter peptides; thus bitterness can be reduced by hydrophobic adsorption (13,14) or extraction with an organic solvent (15,16). Other works are concerned with the relationship between the appearance of bitterness and some other properties of the protein hydrolyzate, notably average hydrophobicity (17,18) and average peptide chain length (5,19,20). Bitter peptides as such have also been extensively studied, and much insight has been gained in the molecular properties responsible for inducing the sensation of bitterness (21).

The purpose of the present work is to review current theories on the bitter taste of peptides and protein hydrolyzates. Two characteristics, hydrophobicity and average peptide chain length, which are key elements in the existing theories, will be particularly focused upon. It will appear that the bitter taste of amino acids and single peptides are largely dependent of these two characteristics only, as generally acknowledged. This includes the well-known Q-rule which links the presence or absence of bitterness to the average hydrophobicity of the peptide (22). However, it will be demonstrated that the currently accepted extrapolation of the Q-rule to explain the presence or absence of bitterness in protein hydrolyzates (17) cannot be upheld. An alternative view on which factors contribute to the perceived bitterness of a protein hydrolyzate is therefore proposed.

II. GENERAL TRENDS IN BITTER PEPTIDE RESEARCH

The bitter taste of protein hydrolyzates was first investigated systematically by Murray and Baker in 1952 (6). They found that the bitter taste of a casein hydrolyzate could be lowered by a treatment with activated carbon. From the spent carbon a polypeptide-rich, strongly bitter fraction could be eluted. After acid hydrolysis of this fraction, the bitter taste changed to a meaty taste. These observations showed some important basic properties of the bitter taste: it is due to peptides rather than free amino acids, and these peptides can be adsorbed to a hydrophobic adsorbant.

In the following years bitter peptides were identified and characterized in protein hydrolyzates derived from several different food proteins, and around the mid-1970´es, Guigoz and Solms (23) could compile a list of some two hundred peptides for which the taste had been reported so far. The chain length of the bitter peptides varied, most were small or medium in length (2 - 12 amino acid residues) but three large bitter peptides containing 22, 24 and 27 amino acid residues were also included. Practically all the bitter-tasting peptides complied with the so-called Q-rule, which occupies a central position in the field of peptide bitterness.

A. THE Q-RULE

The Q-rule was formulated in 1971 by Ney (22), who surveyed the amino acid composition of around seventy bitter and non-bitter peptides, and suggested that the presence or absence of bitterness was determined by the hydrophobicity of the peptides. As a measure of the hydrophobicity Ney calculated what he called the Q-value, which is the average free energy for the transfer of the amino acid side chains from ethanol to water. These values were originally applied by Tanford (24) for assessing the relative hydrophobicity, denoted Δf_t, of amino acids in peptides and proteins, as will be discussed later. Ney found that all bitter peptides of known structure had Q-values above +1400 cal/mole, whereas all the non-bitter peptides had a Q-value below +1300 cal/mole. In between there was no correlation. This empirical correlation between the presence or absence of bitterness and the average hydrophobicity is called the Q-rule.

The Q-rule was further substantiated by observations on the taste of a stepwise synthesized heptapeptide (22). When a strongly hydrophobic amino acid (isoleucine) was added to a non-bitter tetrapeptide, the taste became bitter. The bitterness disappeared again when adding the two last, hydrophilic amino acids in accordance with the Q-rule.

In a subsequent work (17), Ney extrapolated the Q-rule by stating that the risk of a protein hydrolyzate being bitter could be predicted from the amino acid composition of the protein. Thus, hydrolyzates of rather hydrophobic proteins, such as casein, would tend to be bitter, while hydrolyzates of the hydrophilic gelatin would be non-bitter. The critical hydrophobicity range was the same as for individual peptides, i.e. 1300 - 1400 cal/mole.

Independent of Ney, Matoba and Hata (25) also proposed that hydrophobic amino acid side chains were responsible for the bitter taste, regardless of the amino acid sequence. Furthermore, they observed that a hydrophobic amino acid exerted the strongest bitterness when both its ends were blocked, e.g. by forming peptide linkages. The bitterness was comparatively weaker when the amino acid was in a terminal position and weakest when it was free. Consequently, solutions of peptides and peptide mixtures are more bitter than the equimolar mixture of amino acids, in agreement with the original observations of Murray and Baker (6).

In an extensive study, Wieser and Belitz (26,19) noted the (recognition) threshold values for amino acids, amino acid derivates and peptides. The importance of this work cannot be over-estimated for the understanding of peptide bitterness. In agreement with Matoba and Hata´s results these taste data quantitatively confirmed that dipeptides were more bitter than the corresponding free amino acids, and that the bitterness intensity was independent of the sequence of the two amino acids. Higher peptides tended to be even more bitter on a molar basis than dipeptides, for example the tetrapeptide, Phe-Gly-Phe-Gly was at least ten times more bitter than Phe-Gly (19).

B. THE BITTER-SWEET RECEPTOR MODEL

It is noteworthy that while the natural L-forms of the hydrophobic amino acids are bitter, the D-enantiomorphs are sweet (27). For both types of taste the presence of a hydrophobic area is essential, and the difference in taste sensation between the L- and D-enantiomorphs is due to the different spatial arrangement of $-NH_2$ and -COOH relative to the hydrophobic side chain (28). In continuation of these considerations Belitz and his group established a unified model for a bitter-sweet receptor (21). The receptor model consists of a hydrophobic pocket and two polar contact points, and the main difference between bitter and sweet compounds is if one or two of the polar contacts are stimulated. Accepting this model does not, of course, conclude anything with regard to the structure of real receptors, but the model allows a formal, uniform description of the molecular structure of bitter and sweet compounds (Fig. 1).

The hydrophobic region is important for the taste intensity of both L- and D-amino acids (21). This is illustrated by the threshold values of six hydrophobic amino acids shown in Fig. 2. It is apparent that a high hydrophobicity value, Δf_t, leads to a low threshold value, i.e. the more intense sweet or bitter is the amino acid. Tryptophan, being the most hydrophobic of all common amino acids, is top-ranked with respect to either bitterness (in case of the L-form) or sweetness (in case of the D-form).

C. THE WIESER-BELITZ CORRELATION

As mentioned the bitter taste of the hydrophobic amino acids is usually amplified when the amino acids form peptide linkages. In particular, combining with another hydrophobic amino acid leads to

<u>Bitter compounds:</u>

- Hydrophobic region } 3 Å distance
- A polar group }
- An apolar group, 'a'

<u>Sweet compounds:</u>

As bitter compounds but <u>two</u> polar groups involved. 'a' influences sweetness intensity

FIGURE 1 *Molecular architecture of bitter and sweet compounds. According to (21).*

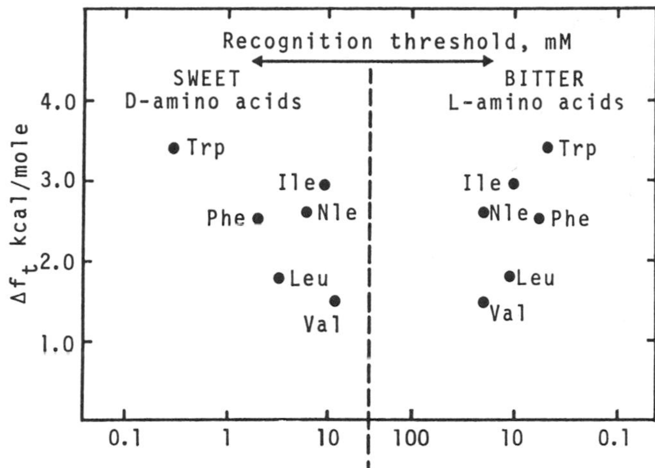

FIGURE 2 *Taste of hydrophobic amino acids. Nle = norleucine. Threshold data from (21).*

strongly bitter compounds, as shown in Table 1, which is adapted from the taste studies carried out by Wieser and Belitz (19). The Δf_t-values in the bottom row are those of the free amino acids as cited by Wieser and Belitz (19).

To account for the difference in taste intensity between dipeptides and free amino acids, Wieser and Belitz suggested the use of the total hydrophobicity, ΔF_t, of the two amino acids linked together (19). Thereby, it was possible to establish a correlation between threshold value and hydrophobicity value which could accommodate both free amino acids and dipeptides (Fig. 3). The calculation method was extended to cover tripeptides as well (the open circles on Fig. 3). Also for tripeptides a negative correlation between hydrophobicity and threshold value is observed, but the data are too few to assess whether or not tripeptides (and higher peptides) fit the correlation obtained for free amino acids and dipeptides.

The critical hydrophobicity range in Ney's Q-rule of +1.3 to +1.4 kcal/mole corresponds to $\Delta F_t \cong -4.6$ kcal/mole for dipeptides,

TABLE 1 Bitterness threshold values (nM) for combinations of hydrophobic amino acids.

	Trp	Phe	Ile	Leu	Val
none	5	6	11	12	21
Gly	13	16	20	21	75
Ala				20	70
Val			9	10	20
Leu	0.4	1.4	5.5	4.5	10
Ile	0.9		5.5	5.5	9
Phe		0.8		1.4	
Trp	0.25		0.9	0.4	
Δf_t	3.40	2.50	2.95	1.80	1.50

Source: (19).

which lies to the extreme right in Wieser and Belitz´s original correlation. This speaks for the general validity of the Q-rule for small peptides.

In their original work (19) Wieser and Belitz presented the same correlation as Fig. 3 except that their ΔF_t-values were derived

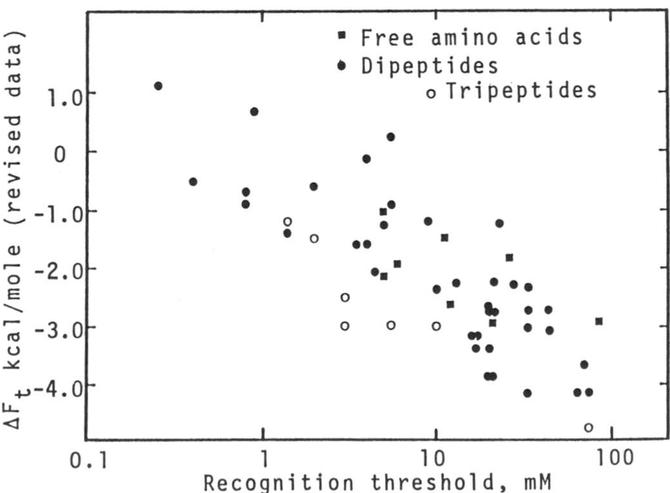

FIGURE 3 *Association of bitterness and total hydrophobicity. Threshold data from (19).*

from older sources (24). These source data have been revised later (29,30), but the gross impression of the correlation is the same, whatever set of data is used. However, the issue necessitates a discourse to be made on the meaning of the hydrophobicity values and how they are calculated.

III. HYDROPHOBICITY VALUES AND PEPTIDE BITTERNESS

A. *SIDE-CHAIN HYDROPHOBICITY,* Δf_t

The hydrophobicity values used by Ney (22) as well as by Wieser and Belitz were based on Tanford´s Δf_t-values (24), which were obtained from published data on the solubility of amino acids and related compounds in water and ethanol, respectively. From these data the change in free energy, ΔF_t, for the transfer of one mole amino acid from ethanol to water was calculated. As the contributions to the free energy from the amino group, the carboxyl group, and the side chain were taken to be roughly additive, the side chain contribution alone, Δf_t, could be calculated by subtracting ΔF_t for glycine (= -4630 cal/mole). Per definition, Δf_t for glycine is now zero, and the highest positive values are, of course, obtained for the aliphatic and aromatic side chains which reflect their poor solubility in water. The data were later re-evaluated by Nozaki and Tanford through new solubility measurements obtained in both ethanol and dioxane (29).

On the basis of Tanford´s original data Bigelow calculated the average hydrophobicity of more than 150 proteins and discussed these in relation to the structure, size, and function of these proteins (31). He assigned aspartic acid and glutamic acid the value zero on the likely assumption that these side chains would be charged and thus having no or negative Δf_t. The average hydrophobicity, $H\Phi_{ave}$, is simply the average of the individual Δf_t-values based on the molar composition.

After the publication of the revised Δf_t-values by Nozaki and Tanford, Bigelow and Channon recalculated Bigelow´s earlier values and compiled a list of 620 pure proteins with their hydrophobicity values, $H\Phi ave$ (30). The Δf_t-values listed by these authors are considered the current best estimate of the hydrophobicity values for evaluating peptide bitterness. The more recent data obtained by Fauchère and Pliška through two-phase partitioning presumably better reflect the conditions in globular proteins (32), but the hydrophobicities of for example tyrosine and proline is quite low in this scale and do not express the high bitterness potential of these two amino acids, cf. the data in (19). Furthermore, in the case of peptide bitterness the situation is that small, soluble molecules rather than large protein molecules are exerting the sensation of bitterness through hydrophobic interaction. Therefore, it can be argued that it is the most relevant to use the hydrophobicity values derived from solubility measurements of small peptides and amino acids.

In Table 2 the present hydrophobicity scale based on the rounded-off values of Bigelow and Channon (30) is given together with the two earlier sets of Δf_t-values. The use of Bigelow and Channon´s data is recommended, not the least because it is then not necessary to distinguish between acid and amide side chains. Since Bigelow and Channon (30) do not quote a value for hydroxyproline, it is estimated here to 1.8 kcal/mole knowing that an aliphatic hydroxy group decreases the hydrophobicity by about 800 cal/mole (29).

B. CHOOSING A SCALE - A MATTER OF TASTE ?

Ney (22) used the values given originally by Tanford (24). He later added histidine to the list with a value of 500 cal/mole, so he must have been acquainted with Nozaki and Tanford´s paper, but otherwise he kept the original values (18). The Q-value is calculated in the same way as $H\Phi ave$, i.e. as the average Δf_t on a molar basis (18,22). For proteins the value is calculated from the amino acid composition (17).

Wieser and Belitz, who also applied the original Δf_t-values, calculated the total hydrophobicity, ΔF_t, of the peptides rather

TABLE 2 Hydrophobicity values of amino acid side chains.

Amino acid	Δf_t in kcal/mole (25°C)		
	Present scale	Nozaki and Tanford, 1971	Tanford, 1962
Serine	-0.3	-0.3	0.04
Glycine	0.0	0.0	0.0
Aspartic acid	0.0	-	0.54
Asparagine	0.0	-	-0.01
Glutamic acid	0.0	-	0.55
Glutamine	0.0	-	-0.10
Threonine	0.4	0.4	0.44
Histidine	0.5	0.5	
Alanine	0.5	0.5	0.73
Arginine	0.75	-	0.73
Cysteine/cystine	1.0	-	-
Methionine	1.3	1.3	1.30
Lysine	1.5	-	1.50
Valine	1.5	1.5	1.69
Leucine	1.8	1.8	2.42
Hydroxyproline	1.8	-	-
Tyrosine	2.3	2.3	2.87
Phenylalanine	2.5	2.5	2.65
Proline	2.6	-	2.60
Isoleucine	2.95	-	2.97
Tryptophan	3.4	3.4	3.00

than the average hydrophobicity (19). To account for the effect of the peptide bond they subtracted ΔF_t of diglycine (= -5960 cal/mole) and triglycine (= -6780 cal/mole) in calculating the total hydrophobicity.

The difference between using the original Δf_t-values and the data recommended here is mainly a matter of principle and changing from one scale to another has few consequences. For example, for nine common food proteins the ranking with respect to hydrophobicity is the same whatever scale is used (Fig. 4). By the same token the correlation of Wieser and Belitz (19) did not appreciably improve by changing to the revised Δf_t-values (Fig. 3).

IV. FROM BITTER PEPTIDES TO PROTEIN HYDROLYZATES

As should be demonstrated from above presentation of theories concerning the bitter taste of individual peptides, this issue is quite well elucidated, both theoretically and empirically. However, it is a pertinent question to what extent these results can be applied to protein hydrolyzates which are complex mixtures of peptides.

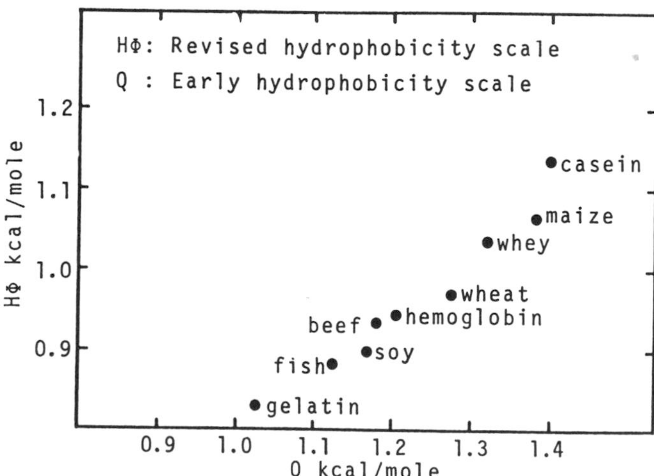

FIGURE 4 *Comparison of hydrophobicity calculations on common food proteins.*

A. THE INFLUENCE OF THE PEPTIDE CHAIN LENGTH

The formation of bitterness during hydrolysis of otherwise non-bitter protein molecules can be quantitatively described as follows (5). In the intact, globular protein molecule the majority of the hydrophobic side chains are concealed in the interior and cannot, therefore, interact with the taste buds. When the protein is degraded by proteolytic attack, peptides of varying size are formed. The largest peptides are still able to mask to some extent their hydrophobic side chains by hydrophobic interaction, whereby U-shaped peptides or clusters of peptides are formed. With further hydrolysis more and more hydrophobic side chains become exposed and bitterness increases, cf. Fig. 5. The bitterness reaches a maximum, however, because a hydrophobic amino acid exerts its strongest bitterness when both ends are blocked, e.g. by forming peptide linkages, as shown already by Hatoba and Hata (25). The bitterness is comparatively weaker when the amino acid is in a terminal position, and weakest when the amino acid is free. Consequently, an extensive hydrolysis usually results in a decreased overall bitterness which is illustrated by the successful use of exopeptidases for debittering casein hydrolyzates (33,34). Also the initial increase in bitterness has been confirmed experimentally (5,19).

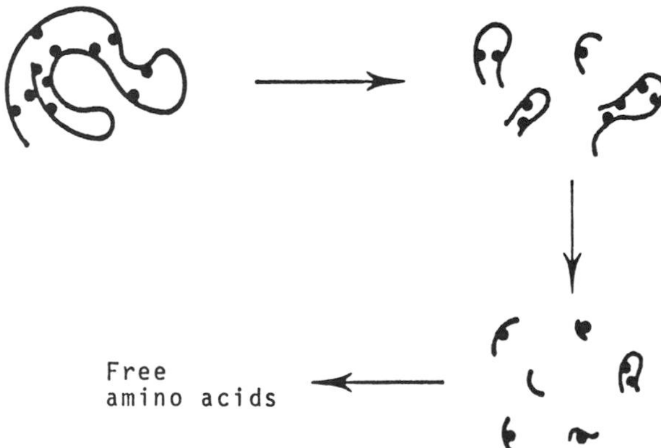

FIGURE 5 *Exposure of hydrophobic regions during enzymic degradation. (Black circles denote hydrophobic side chains).*

The critical peptide chain length, for which masking of hydrophobic side chains might be possible, is unfortunately open to conjecture. In peptides below five to seven amino acid residues masking appears unlikely to occur (5). However, the fact that even large peptides with 22 - 27 amino acid residues can be bitter, as mentioned in the previous, may push the critical chain length above 3000 Dalton. On the other hand, Ney found that hydrolyzates from common food proteins were bitter only if they contained appreciable amounts of peptide material below an apparent molecular weight of 6000 Dalton, as estimated by gel chromatography (20).

At present the question of the critical chain length is unresolved. A further complication arises when the hydrolyzate is separated, because strongly hydrophobic, bitter peptides are precipitated and thus removed, depending to a great extent on the pH during separation (9). This precipitation, at least in the case of ISSPH, has the consequence that the bitterness is relatively independent on the average peptide chain length (35). This conclusion does not imply a disregard of the influence of the peptide chain length on the bitter taste of individual peptides.

B. THE EXTRAPOLATED Q-RULE - THEORETICAL OBJECTIONS

In the above description the significance of the hydrophobic side chains for the formation of bitterness is clear, but the quantitative influence of the hydrophobicity on the bitterness is not dealt with. It appears that among all works concerned with bitterness of protein hydrolyzates only Ney´s extrapolated Q-rule links bitterness quantitatively to hydrophobicity (17). This extrapolation of the Q-rule has not been challenged since it was proposed, but it will be shown that there are so serious theoretical and empirical objections to it that the extrapolated rule must be rejected.

The existence of a qualitative association between amino acid composition of a protein and its proneness to bitter peptide formation can be deduced from the quantitative relationship between the bitterness of individual peptides and their hydrophobicity (Fig. 3): The more hydrophobic the protein the higher the statistical probability of forming very hydrophobic peptides. However, it is the presence and concentration of these very peptides and not the average hydrophobicity which are the cause of bitterness. Consequently, the bitterness depends on the distribution function of the hydrophobicity in the hydrolyzate and not just its average value, as the extrapolated Q-rule implies.

As shown in Fig. 3 for peptides the threshold value for bitterness decreases with increasing hydrophobicity, called H in the following. The bitterness intensity of a peptide at the concentration, C, will therefore be a monotoneously increasing function, B(H), of the hydrophobicity.

If the bitterness is measured as the quinine equivalent value (QEV) (5), then for a single peptide QEV will be the product of C and B(H). For all peptides in a protein hydrolyzate which have hydrophobicity values in the range, H to H + dH, the following expressions will hold:

$$dC = P(H) \times dH \tag{1}$$

$$dQEV = B(H) \times P(H) \times dH \tag{2}$$

where P(H) is the distribution function of the peptides according to hydrophobicity.

Assuming that the overall bitterness is additive with respect to the influence of the individual peptides, QEV is the integral of Eq. (2):

$$QEV = \int_{H(min)}^{H(max)} P(H) \times dH \tag{3}$$

Fig. 6 shows the hydrophobicity distribution P(H) of two hypothetical protein hydroly: tes, one with a broad distribution and another with a narrow dis ribution. The average hydrophobicity, HΦave, is the same for th two hydrolyzates. The broken line is the

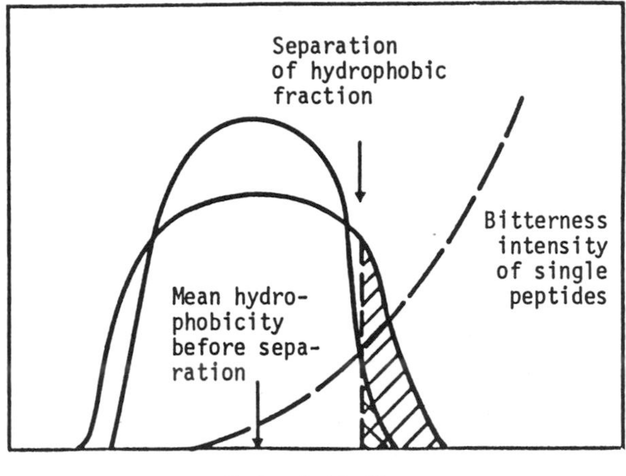

FIGURE 6 *The hypothetical distribution of two protein hydrolyzates according to hydrophobicity. The overall bitterness is the product of the distribution curve and the intensity curve.*

bitterness intensity function, B(H), and the shaded area illustrates the removal of highly hydrophobic peptides by a debittering process.

Fig. 7 shows the differential QEV distribution of the previous two hydrolyzates. Since QEV is the area of this distribution, it is evident that the hydrolyzate with the broadest distribution has the highest QEV. Also, the effect of debittering by hydrophobic peptide removal is quantitatively different for the two hydrolyzates.

Since QEV is lower in the hydrolyzate with the narrow distribution function, P(H), it follows that a single peptide with the same H_{ave} would have the lowest possible value of QEV. Consequently, for a constant C and H_{ave} the inequality holds.

$$\text{QEV (single pept.)} < \text{QEV (mixt.of pept.)} \tag{4}$$

This means again that if QEV (and C) should be the same in the two cases, the distribution function of the peptide mixture must be pushed to the left:

$$H_{ave} \text{ (mixt. of pept.)} < H_{ave} \text{ (single pept.)} \tag{5}$$

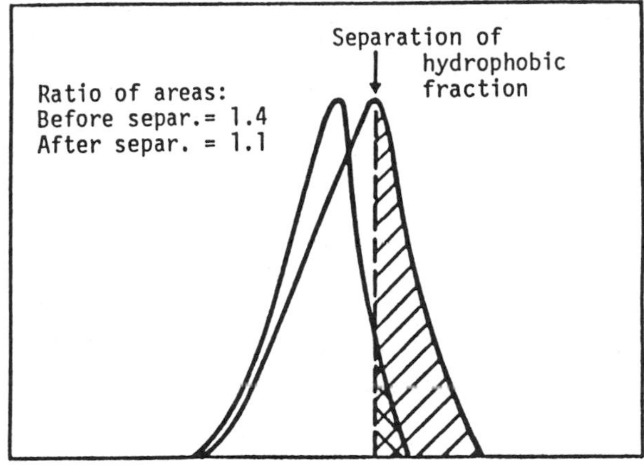

FIGURE 7 *The bitterness intensity distribution of two protein hydrolyzates.*

Eq. (5) shows that the critical hydrophobicity range for bitterness cannot be the same for peptides and protein hydrolyzates. Furthermore, if the distribution function is not known, the value of H_{ave} cannot be stated except that it is below H_{ave} for single peptides. The consequence of this is that not only is the extrapolated Q-rule refuted, but any quantitative rule based on the average hydrophobicity is likewise refuted.

C. THE EXTRAPOLATED Q-RULE - EMPIRICAL OBJECTIONS

The above rejection of the extrapolated Q-rule is purely theoretical, and the question is now if it matters in practice. When reviewing the Q-rule in 1979 Ney listed a number of proteins together with their Q-values and reported that observations on the bitterness of hydrolyzates of these proteins did comply with the Q-rule (18).

However, the Q-values which Ney refer to are considerably over-estimated. For example the high Q-value of whole casein (1605 cal/mole) is clearly at variance to the Q-values of the individual components of casein found elsewhere (16): α-S_1-casein: Q = 1343, κ-casein: Q = 1358, β-casein: Q = 1480 cal/mole. A recalculation of the Q-value for whole casein gives 1375 cal/mole using the same source of the amino acid composition as that referred to by Ney (17) and the Δf_t-values reported by Ney (18). Using established data for the amino acid composition (36,37) a value of Q = 1399 cal/mole is obtained, which is not much different from what Ney might have found.

Table 3 shows the Q-values (in kcal/mole) for six food proteins according to Ney (18) and the revised Q-values for comparison. The third columns shows $H\Phi ave$ according to Table 2. Ney stated that soy protein, wheat gluten, maize protein and casein all have been reported to yield bitter hydrolyzates (18). After revision of the Q-values soy protein and wheat gluten have Q-values below 1300 cal/mole. This is in contradiction to the extrapolated Q-rule, but in agreement with the above theoretical considerations which led to the fact that the average hydrophobicity of a protein hydrolyzate would be lower than that of a single peptide exerting the same bitterness at the same concentration.

The obvious conclusion to draw from the above is that the extrapolated Q-rule is unsubstantiated, both theoretically and empirically, and must be abandoned. The bitterness of protein hydrolyzates is simply a too complex affair to be meaningfully

TABLE 3 Revision of "standard" hydrophobicities of food proteins.

Protein	Q value kcal/mole		HΦ
	Ney(1979)	revised[a]	kcal/mole
Gelatin	1.28	1.03	0.83
Beef	1.30	1.18	0.94
Soy protein	1.54	1.17	0.90
Wheat gluten	1.42	1.28	0.97
Maize protein	1.48	1.38	1.06
Casein	1.60	1.40	1.14

[a]Adler-Nissen, 1985

characterized by the average hydrophobicity only, as also argued above. Table 4, which is based on the taste data of Ricks and co-workers (16), is quite illustrative in this respect. No clear relation between the hydrophobicity and the bitterness of these fifteen hydrolyzates can be seen; in fact the data suggest that the bitterness of a protein hydrolyzate is a result of that particular combination of enzyme and substrate.

D. BITTERNESS OF PROTEIN HYDROLYZATES - THE INFLUENCE OF THE ENZYME

The influence of the enzyme can be predicted to some extent from its specificity. As an example, the two well-known proteases, Alcalase (subtilisin Carlsberg) and trypsin can be taken. Both are endopeptidases with serine in the active site and their activity shows a maximum at slightly alkaline pH. However, trypsin cleaves at peptide bonds with lysine or arginine at the carbonyl side, while Alcalase (subtilisin Carlsberg) has a broad specificity with some preference for hydrophobic (in particular aromatic) amino acids in the same position (38). Hydrolyzates of a given substrate made with these two enzymes therefore differ with respect to the distribution and position of the hydrophobic amino acids. The hydrolyzate produced by the enzyme having hydrophobic specificity will consist of peptides with

TABLE 4 Bitterness threshold values (nM Leu-eq.) for fifteen different soluble protein hydrolyzates.

Substrate	Q kcal/mole	Thermolysin	Pepsin	Chymotrypsin
Pepsin	1184	>100	100	99
β-lactoglobulin	1343	15	50	100
α-S_1-casein	1347	>100	6.3	82
κ-casein	1358	25	11	8.4
β-casein	1488	82	4.2	3.1

Source: (16).

more hydrophobic amino acids in terminal position and fewer long sequences of hydrophobic amino acids in comparison with the other hydrolyzate. This should lead to a lower bitterness in the first hydrolyzate for two reasons: because the hydrophobicity distribution will be more narrow and because the terminal hydrophobic amino acids are less bitter than those in endo-position. This excess of terminal hydrophobic amino acids will have the further consequence that statistically seen the shorter the peptides are the more hydrophobic should they be. The reverse tendency should be observed in the other hydrolyzate.

However, in the case of the iso-electric soluble hydrolyzates the situation is more complicated than that above. At the iso-electric point a fraction of the peptides is precipitated and removed from the solution as mentioned previously. In particular, peptides with long sequences of hydropohobic amino acids will be sparingly soluble, and since the concentration of these is higher in the hydrolyzate made with the enzyme having non-hydrophobic specificity, a higher proportion of the hydrophobic amino acids will be precipitated. This means that the debittering effect of iso-electric precipitation should be larger.

Results from the production of a soluble hydrolyzate (ISSPH, cf. the introduction) and its subsequent separation by ultrafiltration support the above conclusions (Table 5). Soy protein concentrate was hydrolyzed to DH 10, which means that 10% of the peptide bonds were hydrolyzed during the reaction (For a detailed description see (9)). After separation, the amino acid composition (except Cys, Met, Trp) of the hydrolyzate was determined for calculation of mole% hydrophobic amino acids (Val + Leu + Ile + Phe + Tyr + Pro) and $H\Phi_{ave}$. The partition coefficient for nitrogen in aqueous 2-butanol was determined by Kjeldahl analyses, and the average peptide chain length, PCL, by analysis of the free amino groups in the hydrolyzate (39).

The partition coefficient is indicative of the concentration of small, hydrophobic, bitter peptides in protein hydrolyzates (16). All the data support the prediction that in a protein hydrolyzate

TABLE 5 Hydrophobicity of short versus long peptides.

Ultrafiltration of ISSPH DH 10	raw material	total ISSPH	permeate (75 %)	retentate (25 %)
Hydrophobe:				
- mole %	32.0	29.4	30.7	24.9
- HΦ	909	830	859	774
-Part. coeff[a]	~ 0	0.054	0.060	0.039
PCL	∞	7.0	6.6	12.5

[a]aqueous 2-butanol

produced with an enzyme having hydrophobic specificity (in casu Alcalase) the shorter peptides are the most hydrophobic.

In comparison with the raw material, the ISSPH is less hydrophobic, which can be ascribed to the precipitation of hydrophobic peptides (9). This precipitation is further illustrated in Table 6 which shows the hydrophobicity data of the ISSPH, of the solid residue at pH 4, and of the 2-butanol-extractable fraction of the

TABLE 6 Precipitation of hydrophobic peptides in ISSPH process.

Raw material:		Soy isolate		
Hydrolysis to DH 10:	protein	ISSPH	residue pH 4	extract from residue[a]
Hydrophobe:				
-mole %	32.2	28.7	37.4	47.3
-HΦ	883	820	1018	1169
-Part. coeff.[a]	~ 0	0.055	0.067	∞
PCL	∞	7.8	≥100	8.6

[a]aqueous 2-butanol

latter. The solubility curve of the extract (Fig. 8) shows that the peptides in this material are soluble at pH 8, which is the pH-value of the hydrolysis reaction.

If the soluble protein hydrolyzate is produced with an enzyme having non-hydrophobic specificity, it was predicted in the foregoing that the proportion of hydrophobic amino acids which can be removed by iso-electric precipitation is comparatively larger. This has been indirectly confirmed by amino acid analyses on other types of ISSPH produced by either Alcalase or trypsin to DH 5. The mole% hydrophobic amino acids of the two hydrolyzates were 27.9 and 24.8%, respectively.

V. CONCLUSIONS

This review of the theories on the bitterness of peptides and peptide mixtures should convince that the bitterness of single peptides is an issue rather well described and understood. Because peptide mixtures in the form of protein hydrolyzates are more relevant than single peptides as food ingredients, theories for assessing the relative bitterness of protein hydrolyzates are much needed. The

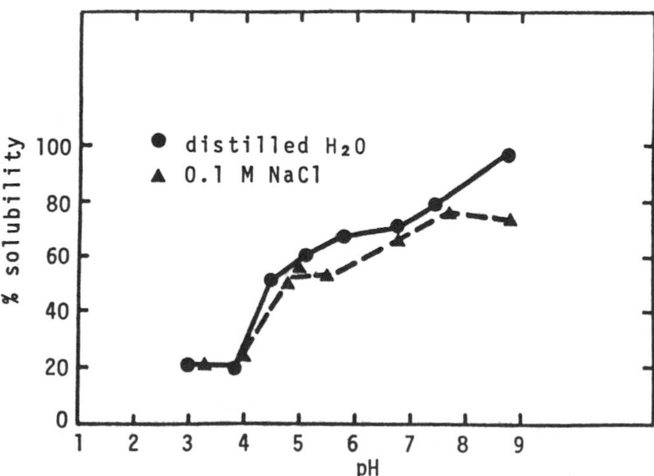

FIGURE 8 *Solubility curve for butanol-extracted peptides. From (9).*

most well-known of the existing theories, the Q-rule, is, however, based on false assumptions and cannot be upheld. A more satisfactory description of the bitterness of protein hydrolyzates has been proposed here by taking the hydrophobicity distribution of the peptides into account.

Enzyme specificity influences the hydrophobicity distribution significantly, and the bitterness of a protein hydrolyzate is therefore in general due to the combined effect of enzyme and substrate. For iso-electric soluble protein hydrolyzates, where a certain debittering takes place by precipitation of hydrophobic peptides, the enzyme specificity influences the relative bitterness in opposing directions. In reality, the bitterness of a particular soluble protein hydrolyzate cannot be immediately predicted, but a qualitative understanding of the various factors, which increase or decrease its bitterness, should pragmatically seen be a quite good substitute to a general theory.

REFERENCES

1. J. Adler-Nissen, Enzymatic hydrolysis of proteins for increased solubility, J. Agric. Food Chem., 24: 1090-1093 (1976).

2. T. Richardson, Functionality changes in proteins following action of enzymes, Adv. Chemistry Ser., 160: 185-243 (1977).

3. R. D. Phillips and L. R. Beuchat, Enzyme modification of proteins, ACS Symp. Ser., 147: 275-298 (1981).

4. S. Schwimmer, Source book of food enzymology, Avi, Westport. pp. 481-496 (1981).

5. J. Adler-Nissen and H. S. Olsen, The influence of peptide chain length on taste and functional properties of enzymatically modified soy protein, ACS Symp. Ser., 92: 125-146 (1979).

6. T. K. Murray and B. E. Baker, Studies on protein hydrolysis I - Preliminary observations on the taste of enzymic protein hydrolyzates, J. Sci. Food Agric., 3: 470-475 (1952).

7. R. L. Koretz and J. H. Meyer, Elemental diets - Facts and fantasies, Gastroenterology, 78: 393-410 (1980).

8. H. S. Olsen and J. Adler-Nissen, Industrial production and applications of a soluble enzymatic hydrolyzate of soya protein. Process Biochem., 14 (7): 6-11 (1979).

9. J. Adler-Nissen, Control of the proteolytic reaction and the level of bitterness in protein hydrolysis processes, J. Chem. Technol. Biotechnol., 34B: 215-222 (1984).

10. Anonymous, Soluble Rynkeby protein and protein drinks, Nordeur. mejeri-tidsskr., 9: 308 (1982).

11. D. P. Cuthbertson, Amino-acids and protein hydrolyzates in human and animal nutrition, J. Sci. Food Agric., 1: 35-41 (1950).

12. K. M. Clegg, Dietary enzymic hydrolyzates of protein, Biochemical aspects of new protein food (J. Adler-Nissen, B.O. Eggum, L. Munck, and H.S. Olsen, eds.), Pergamon Press, Oxford, pp. 109-117 (1978).

13. J. F. Roland, D. L. Mattis, S. Kiang and W. L. Alm, Hydrophobic chromatography: Debittering protein hydrolyzates, J. Food Sci., 43: 1491-1493 (1978).

14. N. B. Helbig, L. Ho, G. E. Christy, and S. Nakai, Debittering of skim milk hydrolyzates by adsorption for incorporation into acidic beverages, J. Food Sci., 45: 331-335 (1980).

15. G. Lalasidis and L.-B. Sjöberg, Two new methods of debittering protein hydrolyzates and a fraction of hydrolyzates with exceptionally high content of essential amino acids, J. Agric. Food Chem., 26: 742-749 (1978).

16. E. Ricks, B. Ridling, G. A. Iacobucci, and D. V. Myers, Approaches to analyse and optimize protein hydrolyzates, Biochemical aspects of new protein food (J. Adler-Nissen, B. O. Eggum, L. Munck and H. S. Olsen, eds.), Pergamon Press, Oxford. pp. 119-128 (1978).

17. K. H. Ney, Aminosäure-Zusammensetzung von Proteinen und die Bitterkeit ihrer Peptide, Z. Lebensm.-Untersuch. Forsch., 149: 321-323 (1972).

18. K. H. Ney, Bitterness of peptides: Amino acid composition and chain length, ACS Symp. Ser., 115: 149-173 (1979).

19. H. Wieser and H.-D. Belitz, Zusammenhänge zwischen Struktur und Bittergeschmack bei Aminosäuren und Peptiden. II. Peptide und Peptidderivate, Z. Lebensm. Untersuch.-Forsch., 160: 383-392 (1976).

20. K. H. Ney, Bitterkeit und Gelpermeationschromatographie von enzymatischen Proteinhydrolyzaten, Fette Seifen Anstrichm., 80: 323-325 (1978).

21. H.-D. Belitz, W. Chen, H. Jugel, R. Teleano, H. Wieser, J. Gasteiger, and M. Marsili, Sweet and bitter compounds: Structure and taste relationship, ACS Symp. Ser., 115: 93-131 (1979).

22. K. H. Ney, Voraussage der Bitterkeit von Peptiden aus deren Aminosäurezusammensetzung, Z. Lebensm.-Untersuch. Forsch., 147: 64-71 (1971).

23. Y. Guigoz and J. Solms, Bitter peptides, occurrence and structure, Chem. Senses Flavor, 2: 71-84 (1976).

24. C. Tanford, Contribution of hydrophobic interactions to the stability of the globular conformation of proteins, J. Am. Chem. Soc., 84: 4240-4247 (1962).

25. T. Matoba and T. Hata, Relationship between bitterness of peptides and their chemical structures, Agric. Biol. Chem., 36: 1423-1431 (1972).

26. H. Wieser and H.-D. Belitz, Zusammenhänge zwischen Struktur und Bittergeschmack bei Aminosäuren und Peptiden. I. Aminosäuren und verwandte Verbindungen, Z. Lebensm. Unters.-Forsch., 159: 65-72 (1975).

27. S. Eriksen and I. S. Fagerson, Flavours of amino acids and peptides, Internat. Flavours, Jan/Feb 1976, 13-16.

28. H.-D. Belitz and H. Wieser, Zur Konfigurationsabhängigkeit des süssen oder bitteren Geschmacks von Aminosäuren und Peptiden, Z. Lebensm. Untersuch. Forsch., 160: 251-253 (1976).

29. Y. Nozaki and C. Tanford, The solubility of amino acids and two glycine peptides in aqueous ethanol and dioxane solutions. Establishment of a hydrophobicity scale, J. Biol. Chem., 246: 2211-2217 (1971).

30. C. C. Bigelow and M. Channon, Hydrophobicities of amino acids and proteins, Handbook of biochemistry and molecular biology (G.D. Fasman, ed.), CRC Press, Cleveland, 3rd ed., Vol. 1, pp. 209-243 (1976).

31. C. C. Bigelow, On the average hydrophobicity of proteins and the relation between it and protein structure, J. Theoret. Biol., 16: 187-211 (1967).

32. J.-L. Fauchère and V. Pliška, Hydrophobic parameters of amino-acid side chains from the partitioning of N-acetyl-amino-acid amides, Eur. J. Med. Chem., 18: 369-375 (1983).

33. K. M. Clegg and A. D. McMillan, Dietary enzymic hydrolyzates of protein with reduced bitterness, J. Food Technol., 9: 21-29 (1974).

34. H. Umetsu, H. Matsuoka, and E. Ichishima, Debittering mechanism of bitter peptides from milk casein by wheat carboxypeptidase, J. Agric. Food Chem., 31: 50-53 (1983).

35. J. Adler-Nissen and H.S. Olsen, Taste and taste evaluation of soy protein hydrolyzates, Chemistry of foods and beverages - Recent developments (G. Charalambous and G. Inglett, eds.), Academic Press, New York, pp. 149-169 (1982).

36. Amino-acid content of foods and biological data on proteins, FAO, Rome. pp. 36-139 (1970).

37. J. R. Spies and D. C. Chambers, Chemical determination of tryptophan in proteins, Anal.Chem., 21: 1249-1266 (1949).

38. I. Svendsen, Chemical modifications of the subtilisins with special reference to the binding of large substrates. A review, Carlsberg Res. Commun., 41: 237-291 (1976).

39. J. Adler-Nissen, Determination of the degree of hydrolysis of food protein hydrolyzates by trinitrobenzenesulfonic acid, J. Agric. Food Chem., 27: 1256-1262 (1979).

5

Biologically Functional Peptides from Food Proteins: New Opioid Peptides from Milk Proteins

Hideo Chiba and Masaaki Yoshikawa

Department of Food Science and Technology
Kyoto University
Sakyo-ku, Kyoto, Japan

I. INTRODUCTION

The biological significance of proteins as food or feed has been taken merely to be providing a source of essential amino acids. However, examples have been shown in which oligopeptides derived from food proteins, especially from casein, had biological activities. From bovine casein hydrolysates, opioid peptides and peptide inhibitors of angiotensin I converting enzyme have been isolated (1-6). We have shown that fragments of human casein had opioid activity (7). An immunostimulating peptide was isolated from human casein hydrolysate (8). Opioid activity was also found in gluten hydrolysate (3). These results suggest that food proteins are potential precursors of biologically functional peptides. Milk proteins are biosynthesized to be ingested. Therefore, it is probable that the biological activities of oligopeptides derived from milk proteins have physiological importance. In this study, new opioid sequences were screened for in human and bovine milk proteins.

II. BASIC CONCEPT OF OPIOID PEPTIDES

In 1975, peptides having affinity for opiate receptors were isolated from brain and named enkephalins by Hughes *et al.* (9). Opioid peptides are defined as peptides like enkephalins that have both affinity for opiate receptors and opiate-like effects which are stereospecifically reversed by naloxone. Substances like naloxone that have both affinity for opiate receptors and anti-opiate effects are called opioid antagonists.

A. ENDOGENOUS OPIOID PEPTIDES

Various opioid peptides such as β-endorphin, α-neoendorphin and dynorphin have been found as endogenous ligands for opiate receptors, that is, opioid receptors (Table 1) (10-12). Many studies have been done on aspects such as isolation, characterization, biosynthesis, gene structure, physiological effects and structure-activity relationships of endogenous opioid peptides. All of the endogenous opioid peptides have a [Met]enkephalin or [Leu]enkephalin sequence, Tyr-Gly-Gly-Phe-Met/Leu, at their amino terminal regions. These peptides are biosynthesized as three different precursor proteins, proenkephalins A, B and proopiomelanocortin, and released by limited proteolyses (13). Some of the newly found opioid peptides are amidated at their carboxyl termini (14, 15). An example of opioid peptide having no enkephalin sequence is dermorphin, Tyr-D-Ala-Phe-Gly-Tyr-Pro-Ser-NH_2, which has been isolated from frog skin (16).

Besides analgesic activity, endogenous opioid peptides have various physiological effects such as regulations of respiration, body temperature, food intake and so on (Table 2) (17). Opioid peptides exert their activity by binding to specific receptors of the target cells (18). Opioid receptors are classified into the subtypes shown in Table 3. Individual receptors are responsible for specific physiological effects: for example, the μ-receptor for analgesia and suppression of intestinal motility, the δ-receptor for emotional behavior, and the κ-receptor for sedation and food intake. Certain ligands are specific for a receptor; dynorphin selectively

TABLE 1 *Structures of endogenous opioid peptides*

Peptides	Structures
Enkephalin family	
[Leu]enkephalin	Tyr-Gly-Gly-Phe-Leu
[Met]enkephalin	Tyr-Gly-Gly-Phe-Met
[Met]enkephalinyl-Arg-Phe	Tyr-Gly-Gly-Phe-Met-Arg-Phe
[Met]enkephalinyl-Arg-Gly-Leu	Tyr-Gly-Gly-Phe-Met-Arg-Gly-Leu
Peptide E	Tyr-Gly-Gly-Phe-Met-Arg-Arg-Val-Gly-Arg-Pro-Glu-Trp-Trp-Met-Asp-Tyr-Gln-Lys-Arg-Tyr-Gly-Gly-Phe-Leu
Adrenorphin	Tyr-Gly-Gly-Phe-Met-Arg-Arg-Val-NH_2
Amidorphin	Tyr-Gly-Gly-Phe-Met-Lys-Lys-Met-Asp-Glu-Leu-Tyr-Pro-Leu-Glu-Val-Glu-Glu-Glu-Ala-Asn-Gly-Gly-Glu-Val-Leu-NH_2
Neoendorphin/Dynorphin family	
α-Neoendorphin	Tyr-Gly-Gly-Phe-Leu-Arg-Lys-Tyr-Pro-Lys
Dynorphin	Tyr-Gly-Gly-Phe-Leu-Arg-Arg-Ile-Arg-Pro-Lys-Leu-Lys-Trp-Asp-Asn-Gln
Rimorphin	Tyr-Gly-Gly-Phe-Leu-Arg-Arg-Gln-Phe-Lys-Val-Val-Thr
Leumorphin	Tyr-Gly-Gly-Phe-Leu-Arg-Arg-Gln-Phe-Lys-Val-Val-Thr-Arg-Ser-Gln-Glu-Asp-Pro-Asn-Ala-Tyr-Tyr-Glu-Glu-Leu-Phe-Asp-Val
Endorphin family	
β-Endorphin	Tyr-Gly-Gly-Phe-Met-Thr-Ser-Glu-Lys-Ser-Gln-Thr-Pro-Leu-Val-Thr-Leu-Phe-Lys-Asn-Ala-Ile-Ile-Lys-Asn-Ala-Thr-Lys-Lys-Gly-Gln
α-Endorphin	Tyr-Gly-Gly-Phe-Met-Thr-Ser-Glu-Lys-Ser-Gln-Thr-Pro-Leu-Val-Thr

TABLE 2 *Physiological effects of opioid peptides*

Central effects

1. Analgesia
2. Catalepsy
3. Sedation and torpor
4. Respiratory depression
5. Hypotension
6. Regulation of body temperature
7. Regulation of food intake
8. Suppression of gastric secretion
9. Increase of GH, PRL and ADH levels
10. Decrease of LH, FSH, TSH and ACTH levels
11. Grooming
12. Regulation of sexual behavior

Peripheral effects

1. Suppression of intestinal motility
2. Potentiation of MSH activity

binds to the κ-receptor. However, some ligands have multiple binding abilities for different receptors: [Met]enkephalin for μ- and δ-receptors, and β-endorphin for μ-, δ- and ε-receptors.

B. *EXOGENOUS OPIOID PEPTIDES*

Brantl *et al.* found opioid activity in milk products and isolated opioid peptides from commercial peptone (1, 2, 19). The structures of the peptides, Tyr-Pro-Phe-Pro-Gly-Pro-Ile and Tyr-Pro-Phe-Pro-Gly, corresponded to the 60-66th and 60-64th residues of bovine β-casein. They named the peptides β-casomorphins 7 and 5, respectively. The latter was about 5 times as active as the former in the guinea pig ileum assay. Chang *et al.* reported that the synthetic β-casomorphin 4 amide, Tyr-Pro-Phe-Pro-NH_2, was about 5 times as active as β-casomorphin 5 in the guinea pig ileum assay (20). The

TABLE 3 *Sub-types of opioid receptors*

Receptor Sub-type	Specific Agonists	Localization		Physiological Effects
		Central	Peripheral	
μ	Morphine Morphiceptin [Met]enkephalin β-Endorphin Peptide E	Hypothalamus Thalamus Cerebral cortex-layers I and IV	Guinea pig ileum Mouse vas deferens	Analgesia Intestinal motility Hypothermia Bradycardia
δ	[Leu]enkephalin [Met]enkephalin β-Endorphin	Limbic system Cerebral cortex-layers II, III and V	Mouse vas deferens	Emotional behavior Hypotension
κ	Ketocyclazocine Dynorphin α-Neoendorphin Leumorphin	Hypothalamus Thalamus Guinea pig cerebellum	Rabbit vas deferens Mouse vas deferens Guinea pig ileum	Sedation Food intake
ε	β-Endorphin		Rat vas deferens	
σ	Allylnormetazocine			

tetrapeptide amide is specific for the μ-receptor and was named morphiceptin. Recently, morphiceptin and β-casomorphin 8, Tyr-Pro-Phe-Pro-Gly-Pro-Ile-Pro, were also found in an enzymatic digest of bovine casein (21). Because of the receptor specificity, morphiceptin has been used widely for studies of the μ-receptor. Potent derivatives of morphiceptin containing unnatural amino acids have been synthesized (22-24). Besides analgesic activity, β-casomorphin and its derivatives have various effects such as stimulation of insulin secretion and reduction of certain distress (25-30).

Zioudrou *et al.* found opioid activity in peptic digest of bovine casein and gluten (3). They named these exogenous opioid peptides exorphin. The structure of the opioid peptide they isolated from a peptic digest of α-casein fraction was Arg-Tyr-Leu-Gly-Tyr-Leu-Glu (4). The peptide corresponds to the 90-96th residues of α_{s1}-casein and is called α-casein exorphin. The peptide seems to be fairly specific for the δ-receptor.

We found a β-casomorphin-like sequence in human β-casein and synthesized the corresponding peptides (7). These peptides had affinities for opioid receptors.

The structures of these exogenous opioid peptides are different from those of endogenous opioid peptides. Only the common features among them are the presence of tyrosine residues at their amino termini and the presence of another aromatic residues, Phe or Tyr, in the third or fourth position. The structures of exogenous opioid peptides have added new concepts to the structure-activity relationships of opioid peptides.

III. INVESTIGATION STRATEGIES

Four kinds of techniques are required to find out novel sequences of biologically functional peptides in food proteins: first, establishment of assay system of biological activity, secondly, setting up of suitable isolation technique by which the active peptide should be obtained, thirdly, determination of chemical structure, and fourthly, chemical synthesis of peptides. Based on these four kinds of techniques, we searched for new opioid peptide sequences in milk proteins according to the following two strategies.

1) Amino acid sequences similar to the structure of known opioid peptides were looked for in the primary structures of milk proteins. The corresponding fragment peptides were synthesized and their opioid activities were evaluated.
2) Peptides having opioid activity were isolated from enzymatic digests of milk proteins by various chromatographic steps and their structures were determined.

A. *SYNTHESIS OF PEPTIDES*

The fragment peptides of milk proteins were synthesized essentially according to the solid-phase method of Merrifield (31). *t*-Butoxycarbonyl amino acid was bound to chloromethylated polystyrene resin and *t*-butoxycarbonyl group was removed by a treatment with trifluoroacetic acid (32). *t*-Butoxycarbonyl amino acids were bound successively to the resin in the presence of dicyclohexylcarbodiimide and 1-hyroxybenzotriazole (33). Peptide amides were synthesized on benzhydrylamine resin (34). The peptides were deprotected by liquid hydrogen fluoride and purified by reverse-phase liquid chromatography (35). Methoxylation of peptides was performed in methanol/hydrogen chloride (36).

B. *ASSAY SYSTEMS OF OPIOID ACTIVITY*

There are many ways to measure opioid activity (Table 4). In the present study, three assay systems were used. We adopted a radioreceptor assay for the first screening of opioid activity of peptides, because both agonist and antagonist activities can be detected by this method. Rat cerebrum membranes containing μ-, δ- and κ-receptors were incubated with opioid samples and 1 nM [^{3}H]naloxone (37). After washing off unbound naloxone on a glass filter disk, naloxone bound to the receptors was measured by a liquid scintillation counter. The amount of naloxone bound to receptors decreased as amounts of opioid sample increased (Fig. 1). The relative affin-

TABLE 4 *Assay methods of opioid activity*

1. Radioreceptor assay
2. Inhibition of electrically stimulated contraction of isolated organ preparations (guinea pig ileum, mouse or rabbit vas deferens)
3. Analgesic tests (tail pinch, tail flick, hot plate and writhing tests)
4. Inhibition of adenylate cyclase in neuroblastoma-glioma hybrid cells
5. Radioimmunoassay

ity of an opioid peptide for receptors is expressed by the IC_{50}, the concentration of a ligand to reduce the specific binding of $[^3H]$-naloxone by 50%.

As the second method, opioid activity was measured by the inhibition of electrically stimulated contraction of isolated organ

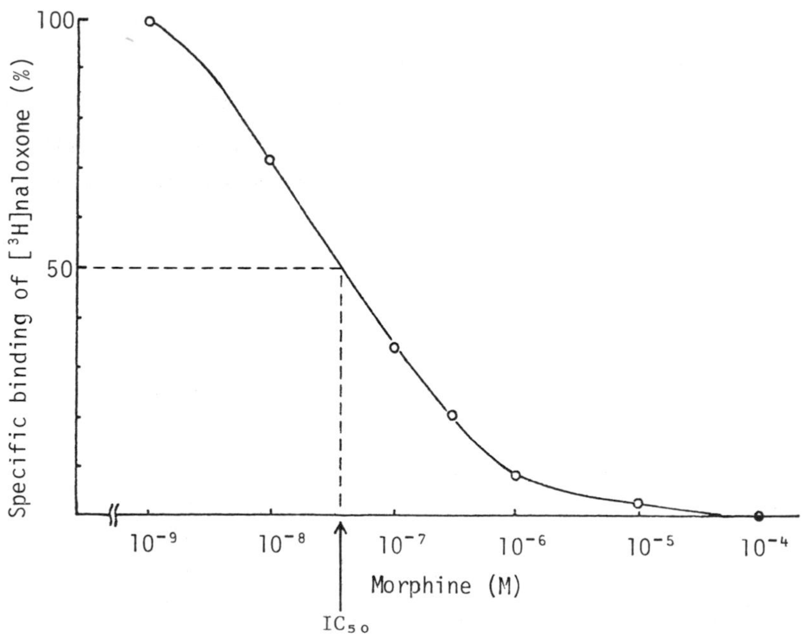

FIG. 1 *Binding of $[^3H]$naloxone to opioid receptors in the presence of morphine.*

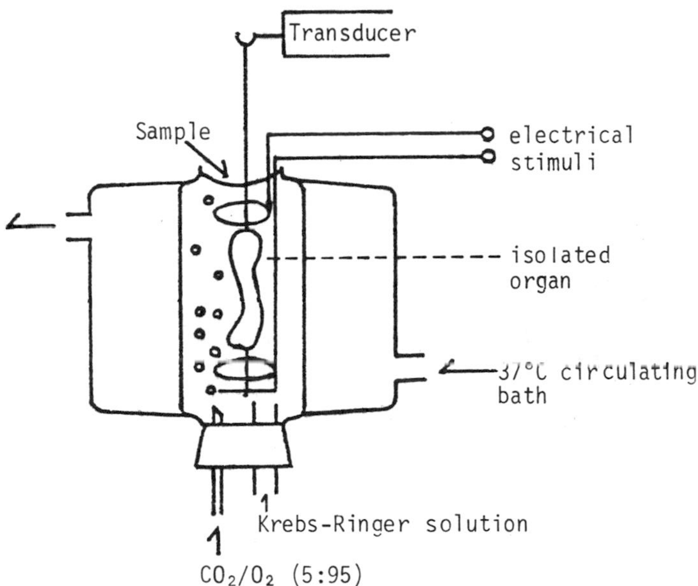

FIG. 2 *Apparatus for measurement of electrically stimulated contraction of isolated organs.*

preparations such as myenteric plexus of guinea pig ileum longitudinal muscle and mouse or rabbit vas deferens (38-40). An isolated organ was suspended in the Magnus tube filled with the Krebs-Ringer solution. Electrical pulses were given every ten seconds and the contraction of the organ was detected by an isometric transducer (Fig. 2). In guinea pig ileum, release of a neurotransmitter acetylcholine from nerve ending is suppressed by opioid peptides and the contraction is inhibited. The inhibition of contraction caused by opioid peptides is reversed by naloxone (See Fig. 3). Opioid activity in the isolated organ assays is also expressed by IC_{50}, the concentration of a ligand to reduce the contraction by 50%. Distribution of receptor sub-types among organs used for the measurement of opioid activity is shown in Table 5.

As the third method, analgesic activity was evaluated by a tail pinch test of mouse after an intracerebroventricular administration of peptides.

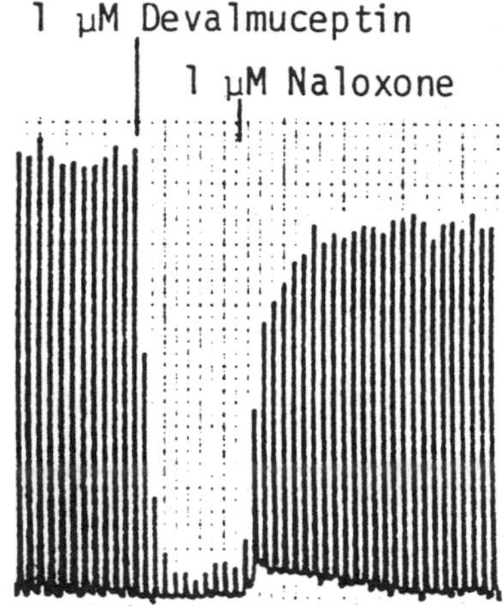

FIG. 3 *Inhibition of electrically stimulated contraction of guinea pig ileum longitudinal muscle-myenteric plexus by devalmuceptin.*

TABLE 5 *Distribution of opioid receptor sub-types among organs used for assays of opioid activity*

Organs	Receptor sub-types
Rat cerebrum	μ, κ, δ
guinea pig ileum	μ, κ
mouse vas deferens	μ, κ, δ
rabbit vas deferens	κ
rat vas deferens	ε

IV. OPIOID ACTIVITY OF CHEMICALLY SYNTHESIZED FRAGMENT PEPTIDES OF MILK PROTEINS

A. *OPIOID PEPTIDES SEQUENCE IN HUMAN β-CASEIN*

If β-casomorphin, which is derived from bovine β-casein, is physiologically important, similar peptides should be found in casein of other animal species. The β-casomorphin sequence is conserved in ovine β-casein (41). However, no β-casomorphin-like sequence was found in the primary structure of rat β-casein deduced from its cDNA sequence (42). We determined a part of the primary structure of human β-casein and found a β-casomorphin-like sequence, Tyr-Pro-Phe-Val-Glu-Pro-Ile-Pro, in it (7). According to the full primary structure of human β-casein determined by Greenberg *et al.*, this sequence corresponds to the 51-58th residues of the protein (43). Peptides of various length containing Tyr^{51} of human β-casein at their amino termini were synthesized and their affinities for opioid receptors were evaluated (7). Opioid activity of these peptides was measured in various assay systems and compared with that of bovine β-casomorphin in Table 6. Opioid activity of human β-casein[51-55], Tyr-Pro-Phe-Val-Glu, was smaller than that of β-casomorphin 5, Tyr-Pro-Phe-Pro-Gly. However, the activity of human β-casein[51-54], Tyr-Pro-Phe-Val, was comparable to that of β-casomorphin 4, Tyr-Pro-Phe-Pro. Substitution of the glycyl residue of β-casomorphin by the glutamyl residue seems to lessen the activity. The tetrapeptide amides, Tyr-Pro-Phe-Val-NH_2 and Tyr-Pro-Phe-D-Val-NH_2, were the most active among human β-casein fragments tested. Their potency is almost equal to that of morphiceptin, Tyr-Pro-Phe-Pro-NH_2, which is the most active opioid peptide among bovine β-casomorphin group (20). β-Casomorphin-like fragment peptides of human β-casein were more active in the guinea pig ileum assay than in the mouse vas deferens assay and almost inactive in the rabbit vas deferens assay. This means peptides tested have selectivity for the μ-receptor. Especially, two tetrapeptide amides were very specific ligands for the μ-receptor, like morphiceptin. Therefore, Tyr-Pro-Phe-Val-NH_2 and Tyr-Pro-Phe-D-Val-NH_2 were named valmuceptin and devalmuceptin, respectively. The tetrapeptide amide seems to be the essential structure for the activity.

Fig. 3 shows a typical results of the guinea pig ileum assay. The contraction of ileum was inhibited by devalmuceptin and the inhibitory effect was reversed by the addition of naloxone.

TABLE 6 *Opioid activity of β-casein fragments**

Peptides	IC_{50} (μM)			IC_{50} MVD / IC_{50} GPI
	RA**	GPI***	MVD****	
Human				
Tyr-Pro-Phe-Val (hβ[51-54])	*170*	*19*	*750*	*39.5*
Tyr-Pro-Phe-D-Val	*27*	*7.6*	*600*	*78.9*
Tyr-Pro-Phe-Val-NH_2 (valmuceptin)	*2.0*	*0.2*	*11.5*	*57.5*
Tyr-Pro-Phe-D-Val-NH_2 (devalmuceptin)	*0.58*	*0.17*	*28*	*165*
Tyr-Pro-Phe-Val-Glu (hβ[51-55])	*600*	*nd.*	*nd.*	-
Tyr-Pro-Phe-Val-Glu-Pro (hβ[51-56])	*540*	*25*	*350*	*14.0*
Tyr-Pro-Phe-Val-Glu-Pro-Ile-Pro (hβ[51-58])	*1300*	*25*	*540*	*21.6*
Tyr-Gly-Phe-Leu-Pro (hβ[59-63])	*300*	*270*	*>1000*	*>3.7*
Tyr-Pro-Ser-Phe-NH_2 (hβ[41-44]-NH_2])	*35*	*30*	*>1000*	*>33.3*
Bovine				
Tyr-Pro-Phe-Pro (β-casomorphin 4)	*30*	*30*	*nd.*	-
Tyr-Pro-Phe-Pro-NH_2 (morphiceptin)	*3.0*	*0.13*	*18.5*	*142*
Tyr-Pro-Phe-Pro-Gly (β-casomorphin 5)	*8.4*	*0.69*	*4.2*	*6.1*
Tyr-Pro-Phe-Pro-Gly-Pro-Ile (β-casomorphin 7)	*30*	*6.9*	*nd.*	-

* *Reference (49), ** Radioreceptor assay, *** Guinea pig ileum assay, **** Mouse vas deferens assay, nd.: not determined*

A mouse injected with 0.1 nmols of devalmuceptin intracerebroventricularly was insensitive to the tail pinch test for 15 min, that is analgesia.

A [Leu]enkephalin-like sequence, Tyr-Gly-Phe-Leu-Pro, is found in human β-casein. This human β-casein[59-63], which is adjacent to the β-casomorphin-like sequence, was also synthesized and its opioid activity was measured (Table 6). The peptide showed rather weak activity in assay systems tested.

In the primary structure of human β-casein, we found another possible opioid sequence (41-44th residues) and synthesized the corresponding tetrapeptide amide, Tyr-Pro-Ser-Phe-NH_2. This peptide amide showed opioid activity and named β-casorphin (Table 6). The corresponding sequence is not found in bovine β-casein (45).

Biologically active peptides derived from human and bovine β-caseins are illustrated in Fig. 4. Two kinds of biologically active peptides, opioid peptides and an immunostimulating peptide was found in the structure of human β-casein (7, 8). Bovine β-casein also contain two kinds of biologically active peptide sequence, opioid peptide and a peptide inhibitor for angiotensin I converting enzyme (2, 6). Thus, β-casein, which is present in milk of every mammalian species, seems to be a precursor of multiple biologically active peptides, like proopiomelanocortin.

B. *OPIOID PEPTIDES SEQUENCE IN WHEY PROTEINS*

We also found a possible opioid sequence, Tyr-Gly-Leu-Phe, in the primary structure of human and bovine α-lactalbumins (Fig. 5) (46, 47). Another possible sequence, Tyr-Leu-Leu-Phe, was found in bovine β-lactoglobulin (48). Tetrapeptide amides containing these sequences were synthesized and their opioid activity was evaluated. Both peptide amides showed rather weak opioid activity (Table 7). The tetrapeptide amides were named α-lactorphin and β-lactorphin, respectively. In the guinea pig ileum assay, inhibition of contraction of the organ by β-lactorphin was reversed partially by naloxone.

```
        1                                                                                       25
Bovine* Arg-Glu-Leu-Glu-Glu-Leu-Asn-Val-Pro-Gly-Glu-Ile-Val-Glu-Ser-Leu-Ser-Ser-Ser-Glu-Glu-Ser-Ile-Thr-Arg-
                                                                P       P   P   P
Human**                                     Arg-Glu-Thr-Ile-Glu-Ser-Leu-Ser-Ser-Ser-Glu-Glu-Ser-Ile-Pro-Glu-
                                            1       P           P       P   P   P               15

                                                                                                50
        Ile-Asn-Lys-Lys-Ile-Glu-Lys-Phe-Gln-Ser-Glu-Glu-Gln-Gln-Gln-Thr-Glu-Asp-Glu-Leu-Gln-Asp-Lys-Ile-His-
                                            P
        Tyr-Lys-Gln-Lys-Val-Glu-Lys-Val-Lys-His-Glu-Asp-Gln-Gln-Gln-Gly-Thr-Asp-Gln-His-Gln-Asp-Lys-Ile-Tyr-
                                                                                            40   β-

                                            60  β-casomorphin                                       75
        Pro-Phe-Ala-Gln-Thr-Gln-Ser-Leu-Val-Tyr-Pro-Phe-Pro-Gly-Pro-Ile-Pro-Asn-Ser----Leu-Pro-Gln-Asn-Ile-Pro-
        Pro-Ser-Phe-Gln-Pro-Gln-Pro-Leu-Ile-Tyr-Pro-Phe-Val-Glu-Pro-Ile-Pro-Tyr-Gly-Phe-Leu-Pro-Gln-Asn-Ile-Leu-
        casorphin                       50                                                  65

                                                                                                100
        Pro-Leu-Thr-Gln-Thr-Pro-Val-Val-Val-Pro-Pro-Phe-Leu-Gln-Pro-Glu-Val-Met-Gly-Val-Ser-Lys-Val-Lys-Glu-
        Pro-Leu-Ala-Gln-Pro-Ala-Val-Val-Leu-Pro-Val-Pro----Gln-Pro-Glu-Ile-Met-Glu-Val-Pro-Lys-Ala-Lys-Asp-
                                                                                            90

                                                                                                125
        Ala-Met-Ala-Pro-Lys-His-Lys-Glu-Met-Pro-Phe-Pro-Lys-Tyr-Pro-Val-Gln-Pro-Phe-Thr-Glu-Ser-Gln-Ser-Leu-
        Thr-Val-Tyr-Thr-Lys-Gly-Arg-Val-Met-Pro-Val-Leu-Lys-Gln-Pro-Thr-Ile-Pro-Phe-Phe-Asp-Pro-Gln-Ile-Pro-
                                                                                            115
```

```
                                                                                   150
Thr-Leu-Thr-Asp-Val-Glu-Asn-Leu-His-Leu-Pro-Pro-Leu-Leu-Leu-Gln-Ser-Trp-Met-His-Gln-Pro-His-Gln-Pro-

Lys-Leu-Thr-Asp-Leu-Glu-Asn-Leu-His-Leu-Pro-Leu-Pro-Leu-Leu-Gln-Pro-Ser-Met-Gln-Gln-Val-Pro-Gln-Pro-
                                                                                140

                                                                                   175
Leu-Pro-Pro-Thr-Val-Met-Phe-Pro-Pro-Gln-Ser-Val-Leu-Ser-Leu-Ser-Gln-Ser-Lys-Val-Leu-Pro-Val-Pro-Glu-

Ile-Pro-Gln-Thr-Leu-Ala-Leu-Pro-Pro-Gln-Pro-Leu-Trp-Ser-Val-Pro-Glu-Pro-Lys-Val-Leu-Pro-Ile-Pro-Gln-
                                                                                165

                                                                                   200
Lys-Ala-Val-Pro-Tyr-Pro-Gln-Arg-Asp-Met-Pro-Ile-Gln-Ala-Phe-Leu-Leu-Tyr-Gln-Gln-Pro-Val-Leu-Gly-Pro-

Glu-Val-Leu-Pro-Tyr-Pro-Val-Arg-Ala-Val-Pro-Val-Gln-Ala-Leu-Leu-Leu-Asn-Gln-Glu-Leu-Leu-Leu-Asn-Pro-
                                                                                190

                                    209
Val-Arg-Gly-Pro-Phe-Pro-Ile-Ile-Val

Pro-His-Gln-Ile-Tyr-Pro-Val-Pro-Glu-Pro-Ser-Thr-Thr-Glx-Ala-Asx-His-Pro-Ile-Ser-Val
                                                                    212
```

FIG. 4 *Biologically active peptides derived from bovine and human β-casein. Opioid sequences are boxed.* ═ *Phagocytosis peptide (8), --- Angiotensin converting enzyme inhibitor (6), * Bovine β-casein* A^2 *(45), ** The most highly phosphorylated form of human β-casein (43).*

α-Lactalbumin

	50	Reference
Human	-Thr-Glu-[Tyr-Gly-Leu-Phe]-Gln-Ile-Ser-Asn-Lys-	*(46)*
Bovine	-Thr-Glu-[Tyr-Gly-Leu-Phe]-Gln-Ile-Asn-Asn-Lys-	*(47)*

β-Lactoglobulin

	102	
Bovine	-Lys-Lys-[Tyr-Leu-Leu-Phe]-Cys-Met-Glu-Asn-Ser-	*(48)*

FIG. 5 *Opioid-like sequences in whey proteins.*

TABLE 7 *Opioid activity of whey protein fragments**

Peptides	IC_{50} (μM)	
	RA**	GPI***
Tyr-Gly-Leu-Phe-NH_2 (α-lactorphin)	*300*	*50*
Tyr-Leu-Leu-Phe-NH_2 (β-lactorphin)	*160*	*160*

** Reference (44), ** Radioreceptor assay,*
**** Guinea pig ileum assay*

V. OPIOID ANTAGONIST PEPTIDE FROM BOVINE κ-CASEIN

A. *ISOLATION AND STRUCTURAL ANALYSIS OF AN OPIOID ANTAGONIST FROM PEPTIDE DIGEST OF BOVINE κ-CASEIN*

As described above, many opioid agonist sequences were found in milk proteins by our synthetic approach. According to the second strategy, we tried to isolate opioid peptides from enzymatic digest of milk proteins. First, we purified an active peptide from a peptic digetst of bovine κ-casein, which showed high affinity for opioid receptors in the radioreceptor assay. κ-Casein was digested by pepsin at pH 1.4 (Fig. 6). After lyophilization, chloroform/methanol (65:35) soluble peptides were extracted and dissolved in water at pH 7. This extraction method gave high yield of activity. However, methoxylation of carboxyl groups was caused by methanol/hydrogen chlo-

κ-Casein solution (10 mg/ml)
- adjusted to pH 1.4 with HCl
- hydrolysed by pepsin (200 μg/ml) at 37°C for 5 hrs
- lyophilized
- extracted with $CHCl_3$/MeOH (65:35)

Organic extract
- evaporated
- dissolved in water at pH 7.4
- centrifuged

Crude extract
- reverse-phase liquid chromatography
- radioreceptor assay

Purified peptide

FIG. 6 *Purification scheme of active peptide from peptic digest of bovine κ-casein.*

ride, as will be described below. The crude extract was put on an octadecylsilane (ODS) column (Cosmosil $5C_{18}$, 10 × 250 mm, Nakarai Chemicals Inc.) and developed with a linear gradient of acetonitrile containing 0.1% trifluoroacetic acid. The activity was recovered in a fraction eluted at 27% acetonitrile. The active fraction was put on a cyanopropylsilane column (Cosmosil 5CN, 4.6 × 150 mm, Nakarai Chemicals Inc.) and developed with the same gradient of acetonitrile. The activity was eluted at 12% acetonitrile. The active fraction was purified again on an ODS column (Fig. 7). The purified peptide thus obtained was analysed with a gas-phase protein sequencer 470A (Applied Biosystems Inc). The sequence obtained was Ser-Arg-Tyr-Pro-Ser-Tyr (49). Recovery of the last tyrosine residue was very small. However, the result of amino acid analysis clearly suggested the presence of the second tyrosine residue. This peptide corresponds to the 33-38th residues of κ-casein (50).

The hexapeptide Ser-Arg-Tyr-Pro-Phe-Ser-Tyr was synthesized by the solid-phase method. However, the synthesized peptide was less

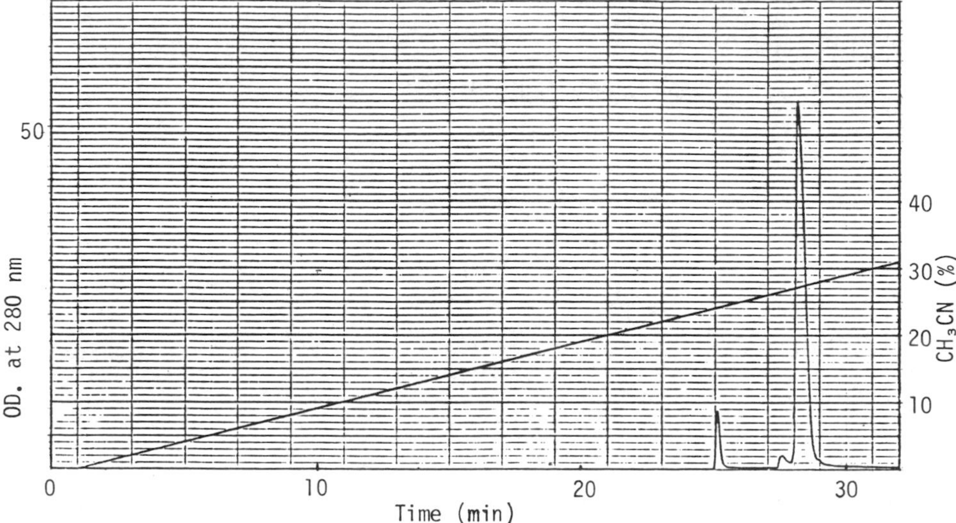

FIG. 7 *Purification of active peptide from bovine κ-casein on ODS column. Active fraction from cyanopropylsilane column was put on Cosmosil 5C$_{18}$ column (10 × 250mm) and the column was developed with a linear gradient of acetonitrile (0-40%) containing 0.1% trifluoroacetic acid. Flow rate was 4 ml/min. Binding activity to opioid receptors was recovered in the main peak.*

active than the isolated one in the radioreceptor assay (Table 8). Furthermore, the synthesized peptide was eluted at a lower concentration of acetonitrile from the ODS column than the isolated one. Upon the carboxypeptidase A treatment, only the synthesized peptide released a carboxyl terminal tyrosine. Thus modification of the carboxyl terminal in the isolated peptide was suggested. After the α-chymotrypsin treatment, the isolated peptide was eluted from the ODS column under the same conditions as the synthesized peptide, and released a carboxyl terminal tyrosine upon the carboxypeptidase A treatment (Table 8). Considering that α-chymotrypsin has esterase activity and that the peptide was treated with methanol in the presence of remaining hydrogen chloride during the extraction process, we thought that the α-carboxyl group of the isolated peptide was methoxylated. In fact, the synthetic methoxylated peptide had the activity comparable to that of the isolated peptide and was equal to

TABLE 8 *Identification of active peptide from κ-casein as Ser-Arg-Tyr-Pro-Ser-Tyr-OCH_3**

Peptides	IC_{50}** (μM)	Amino acid released by Cpase A	CH_3CN (%)*** Chymotrypsin**** (-)	(+)
Active peptide isolated from κ-casein digest	*15*	*no*	*27*	*24*
Ser-Arg-Tyr-Pro-Ser-Tyr (synthetic)	*250*	*Tyr*	*24*	*24*
Ser-Arg-Tyr-Pro-Ser-Tyr-OCH_3 (synthetic)	*20*	*no*	*27*	*24*

** Reference (49), ** Radioreceptor assay, *** Concentration of acetonitrile to elute the peptide from ODS column, **** Peptides were treated with α-chymotrypsin before HPLC analysis.*

it in many respects (Table 8). Thus, the structure of the active peptide isolated from a peptic digest of κ-casein was Ser-Arg-Tyr-Pro-Ser-Tyr-OCH_3.

B. *OPIOID ANTAGONIST ACTIVITY OF κ-CASEIN [33-38] METHYL ESTER, CASOXIN*

The κ-casein[33-38] methyl ester, Ser-Arg-Tyr-Pro-Ser-Tyr-OCH_3, did not have any opioid agonist activity in the guinea pig ileum assay, nor mouse and rabbit vas deferens assays. However, the peptide antagonized opioid activity of [Met]enkephalin in the guinea pig ileum assay and was an opioid antagonist as shown in Fig. 8. The peptide was named casoxin. The receptor specificity of casoxin was determined (Fig. 9). Casoxin counteracted morphiceptin in the guinea pig ileum assay and dynorphin[1-13] in the rabbit vas deferens assay but not [Leu]enkephalin in the mouse vas deferens assay. Therefore, casoxin is an opioid antagonist specific for the μ- and κ-receptors, but not for the δ-receptor. The antagonist activity of casoxin for the μ-receptor is about one hundredth that of naloxone.

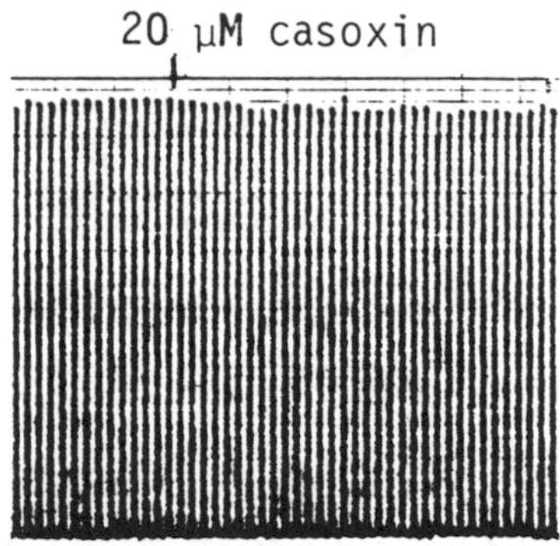

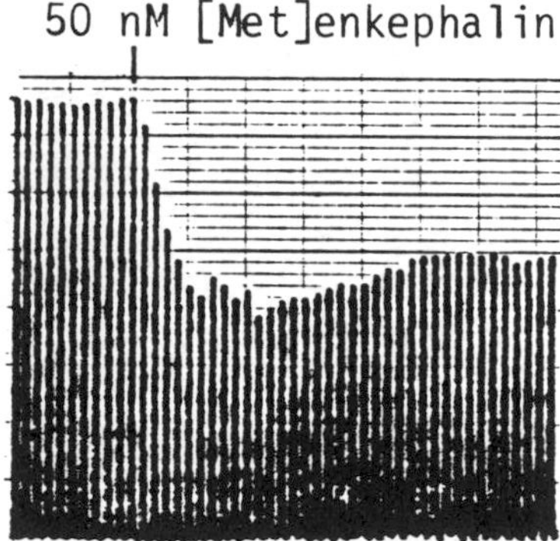

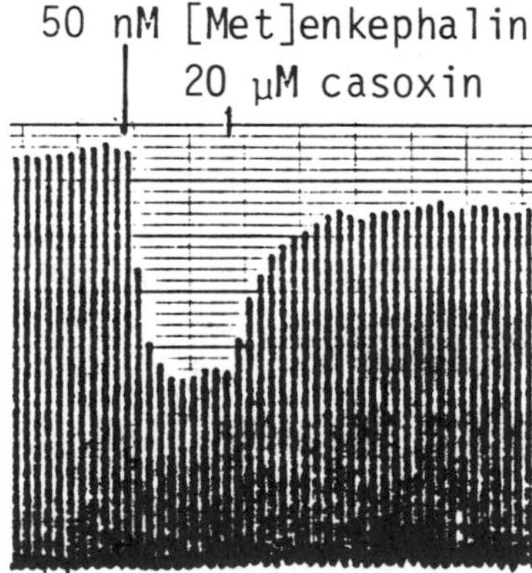

FIG. 8 *Opioid antagonist activity of casoxin 6 in guinea pig ileum longitudinal muscle-myenteric plexus.*

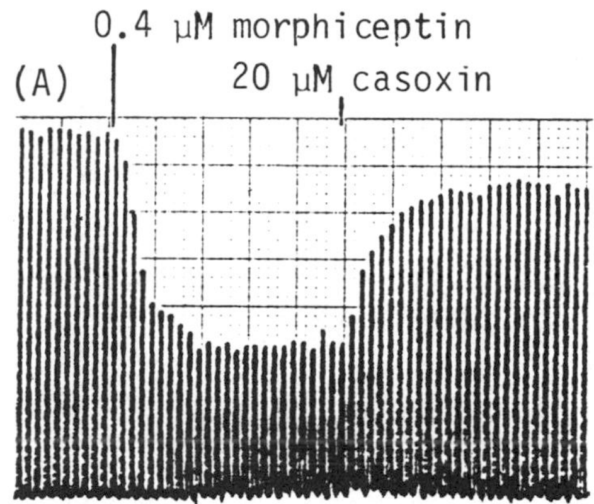

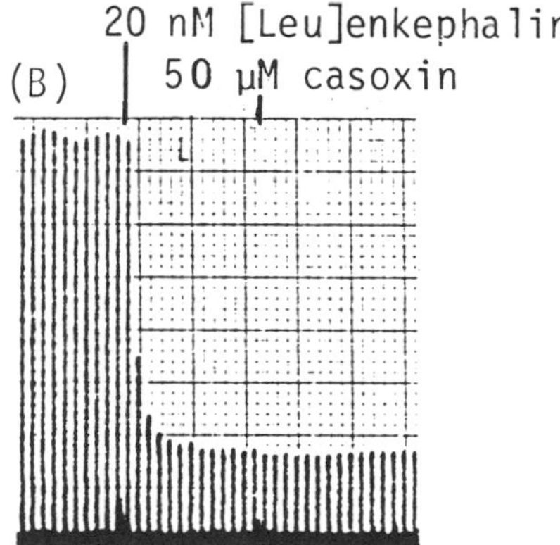

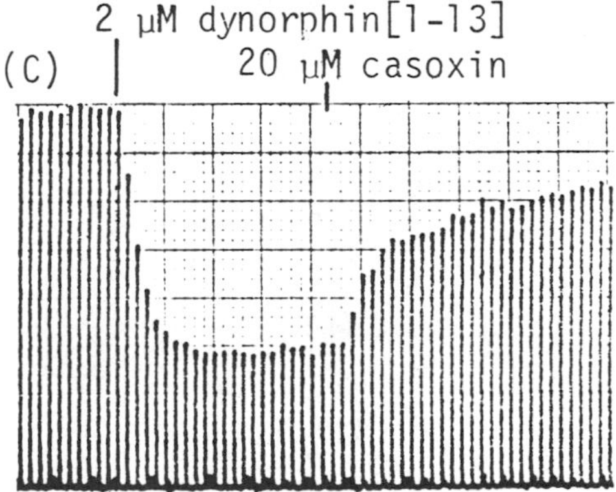

FIG. 9 *Opioid antagonist activity of casoxin 6 in various organ preparations. (A) Guinea pig ileum longitudinal muscle-myenteric plexus, (B) Mouse vas deferens, (C) Rabbit vas deferens.*

C. CHEMICAL SYNTHESIS OF κ-CASEIN FRAGMENTS

Various derivatives of casoxin were synthesized (Table 9). All of these peptides were opioid antagonists. Shorter peptide esters, Arg-Tyr-Pro-Ser-Tyr-OCH_3 (casoxin 5) and Tyr-Pro-Ser-Tyr-OCH_3 (casoxin 4), were more active antagonists than the hexapeptide ester (casoxin 6) in the guinea pig ileum assay. The antagonist activities were small when the α-carboxyl group of the second tyrosine was free. The activity was partly recovered when the second tyrosine residue was followed by glycine which is the next residue in κ-casein. This means dissociation of the α-carboxyl group of the second tyrosine is unfavorable for the antagonist activity.

Only a few peptidic opioid antagonists have been reported. β-Endorphin[1-27], a fragment of β-endorphin[1-31], is the only example of opioid antagonist of natural origin (51). Casoxins are the second example of opioid antagonist peptides with all natural amino acid residues. Especially, casoxin 4 is an opioid antagonist peptide with the minimal structure ever known. Examples of synthetic opioid antagonist are the peptides with N,N-bisallyl-tyrosine residue at their amino termini, which are specific for the δ-receptor (52, 53).

TABLE 9 *Affinities of κ-casein fragments to opioid receptors*

Peptides	IC_{50}* (μM)
Ser-Arg-Tyr-Pro-Ser-Tyr-OCH_3 (casoxin 6)	20
Arg-Tyr-Pro-Ser-Tyr-OCH_3 (casoxin 5)	8
Tyr-Pro-Ser-Tyr-OCH_3 (casoxin 4)	6
Ser-Arg-Tyr-Pro-Ser-Tyr	250
Arg-Tyr-Pro-Ser-Tyr	135
Tyr-Pro-Ser-Tyr	140
Tyr-Pro-Ser-Tyr-Gly	42
Tyr-Pro-Ser-Tyr-NH_2	30

* *Radioreceptor assay*

TABLE 10 *Opioid peptides from milk proteins*

Peptides		Origins	References
Agonists			
Tyr-Pro-Phe-Pro-Gly-Pro-Ile	(β-casomorphin 7)	bovine β-casein[60-66]	(2)
Tyr-Pro-Phe-Pro-Gly	(β-casomorphin 5)	bovine β-casein[60-64]	(2)
Tyr-Pro-Phe-Pro-NH_2	(morphiceptin)	bovine β-casein[60-63]-NH_2	(20)
Arg-Tyr-Leu-Gly-Tyr-Leu-Glu	(α-casein exorphin)	bovine α_{s1}-casein[90-96]	(4)
Tyr-Pro-Phe-Val-Glu-Pro-Ile-Pro		human β-casein[51-58]	present study
Tyr-Pro-Phe-Val-Glu-Pro		human β-casein[51-56]	present study
Tyr-Pro-Phe-Val-Glu		human β-casein[51-55]	present study
Tyr-Pro-Phe-Val		human β-casein[51-54]	present study
Tyr-Pro-Phe-Val-NH_2	(valmuceptin)	human β-casein[51-54]-NH_2	present study
Tyr-Pro-Phe-D-Val			present study
Tyr-Pro-Phe-D-Val-NH_2	(devalmuceptin)		present study
Tyr-Gly-Phe-Leu-Pro		human β-casein[59-63]	present study
Tyr-Pro-Ser-Phe-NH_2	(β-casorphin)	human β-casein[41-44]-NH_2	present study
Tyr-Gly-Leu-Phe-NH_2	(α-lactorphin)	human and bovine α-lactalbumin[50-53]-NH_2	present study
Tyr-Leu-Leu-Phe-NH_2	(β-lactorphin)	bovine β-lactoglobulin [102-105]-NH_2	present study
Antagonists			
Ser-Arg-Tyr-Pro-Ser-Tyr		bovine κ-casein[33-38]	present study
Arg-Tyr-Pro-Ser-Tyr		bovine κ-casein[34-38]	present study
Tyr-Pro-Ser-Tyr		bovine κ-casein[35-38]	present study
Ser-Arg-Tyr-Pro-Ser-Tyr-OCH_3	(casoxin 6)	bovine κ-casein[33-38]-OCH_3	present study
Arg-Tyr-Pro-Ser-Tyr-OCH_3	(casoxin 5)	bovine κ-casein[34-38]-OCH_3	present study
Tyr-Pro-Ser-Tyr-OCH_3	(casoxin 4)	bovine κ-casein[35-38]-OCH_3	present study
Tyr-Pro-Ser-Tyr-NH_2		bovine κ-casein[35-38]-NH_2	present study
Tyr-Pro-Ser-Tyr-Gly		bovine κ-casein[35-39]	present study

VI. DISCUSSION—FUTURE PROBLEMS

We have found that new opioid peptide sequences are widely distributed among milk proteins (Table 10). This fact suggests the physiological importance of these peptides. Although potency of opioid peptides found in milk proteins are smaller than those of endogenous opioid peptides, they may well have physiological effects because food proteins are usually ingested in fairly large amounts. Furthermore, infant animals are reported to be very sensitive to opioid peptides (54). Morphine itself was also found in milk (55). The physiological significance of these exogenous opiate and opioid peptides for infant and adult animals is a very interesting problem.

We found the opioid antagonist sequence in bovine κ-casein. The physiological importance of the coexistence of peptide sequences showing opposite biological effects, opioid and opioid antagonist ones, in milk proteins is not clear at present. Naloxone and naltrexone, synthetic opioid antagonists, have many physiological effects, as shown in Table 11. It is also an interesting problem whether casoxins, the opioid antagonist peptides obtained from κ-casein, will have some of the same effects or not.

TABLE 11 *Physiological effects of naloxone and naltrexone, synthetic opioid antagonists*

Effects	References
1. Anti-analgesic effect	
2. Improvement of various shocks: elevation of blood pressure	(56)
3. Improvement of disordered food intake, obesity and anorexia nervosa	(57, 58)
4. Modulation of growth rate	(59)
5. Increase of brain size	(60)
6. Enhancement of memory	(61)
7. Modulation of cancer	(62)
8. Improvement of constipation: enhancement of intestinal motility	(63)

Intact milk proteins have their specific functions such as casein micelle formation and regulation of lactose synthesis (Table 12) (64-66). Milk proteins exhibit various biological activities when they are partially digested as described above. Of course, they are finally a source of essential amino acids. This means milk proteins are highly functional substances.

Biologically active peptides are released by limited hydrolyses of well known proteins; an insulin stimulating peptide from serum albumin (67), vasoactive peptides from fibrinogen (68) and immunostimulating peptides (tuftsin and Fc fragment) from the constant region of immunoglobulin G, are reported (69, 70). Thus, there are many examples in which proteins are potential precursors of biologically functional peptides. Whitaker and Puigserver grouped proteolysis in vivo as either important for protein turnover or for expressing biological activity of proteins (71). The typical examples of the latter are the activation of zymogens and prohormones by limited proteolyses. Above results show that the hydrolyses of many proteins, including food proteins, which have been grouped as the former, can be grouped as the latter.

TABLE 12 *Biological functions of milk proteins*

	Biological functions		
	Intact proteins	Oligopeptides	Amino acids
Caseins	Casein micelle formation		Nutrition
α_{s1}-casein	Ca^{2+}-dependent aggregation	Opioid agonist	
β-casein	Ca^{2+}-dependent aggregation	Opioid agonist	
κ-casein	micelle stabilization	Opioid antagonist	
α-Lactalbumin	Lactose synthase subunit	Opioid agonist	Nutrition
β-Lactoglobulin	Retinol transport	Opioid agonist	Nutrition

There are many possible proteolytic reactions and these reactions prodice a number of peptides from food proteins. Instead of in vivo digestion, we have chosen peptic digestion and chemical synthesis as strategies to find out and isolate opioid peptides from food proteins. If we adopted in vivo digestion system to find out new biologically functional peptides in the structures of food proteins, the effective oligopeptides which should be produced in the middle course of protein hydrolyses by enzymes might be missed. Because varieties and amounts of peptides formed would be changed under the different conditions of in vivo digestion. From this reason, we adopted the peptic digestion system which has high reproducibility.

On the other hand, there is another aspect to be considered in relation to hydrolysis of food proteins. In fermented food, action of microbial proteases may be large. Proteases from contaminating microorganisms should be also considered. In fact, microbial proteases might be necessary for the release of β-casomorphin, which has originally been isolated from commercial peptone, because the peptide was not formed by the action of digestive proteases from β-casein (72). Moreover, proteases from food itself like milk protease would work on proteins during processing and storage of foods. Therefore, we synthesized fragments of milk proteins in order to cover opioid agonist and antagonist sequences.

To discuss possible physiological importance of opioid peptides found in this study, the information about formation, absorption and stability of these potential opioid peptides are of course necessary. We think that the evaluation of the physiological effects of biologically active peptides found in food proteins by a synthetic approach is important even if their release would be difficult under conditions tested. If a certain biological activity of a peptide would be truly desirable as a food constituent, we can add the synthetic peptide to foods. Furthermore, we can introduce the genes coding for the peptide or new susceptible sites to proteases into the existing proteins by genetic engineering techniques, so called protein engineering techniques, in the near future. If undesirable biological activity would be found, peptides responsible for the activity should be removed from the protein by a similar technique. In other words, information about "What kinds of biological activities should be desirable to be contained in food proteins ?" are very important to design ideal food proteins and their processing conditions.

VII. CONCLUSION

Oligopeptides derived from food proteins have been shown to have biological activities. To find new opioid sequences in milk proteins, we synthesized fragment peptides containing a tyrosine residue at their amino termini and evaluated their opioid activity. Various fragment peptides of human β-casein had opioid activity. Synthetic fragments of human and bovine whey proteins also showed opioid activity.

An opioid antagonist was obtained from chloroform/methanol extract of peptic digest of bovine κ-casein. The structure of the peptide was Ser-Arg-Tyr-Pro-Ser-Tyr-OCH_3, which corresponds to κ-casein[33-38] methyl ester. The antagonist was specific for μ- and κ-receptors and named casoxin. Various derivative peptides of casoxin were also opioid antagonists. This is the first example of opioid antagonist sequence found in food protein. Possible physiological importance of these biologically active peptides are discussed.

REFERENCES

1. V. Brantl, H. Teschemacher, A. Henschen and F. Lottspeich, *Hoppe-Seyler's Z. Physiol. Chem. 360:* 1211 (1979).
2. A. Henschen, F. Lottspeich, V. Brantl and H. Teschemacher, *Hoppe-Seyler's Z. Physiol. Chem. 360:* 1217 (1979).
3. C. Zioudrou, R. A. Streaty and W. A. Klee, *J. Biol. Chem. 254:* 2446 (1979).
4. S. Loukas, D. Varoucha, C. Zioudrou, R. A. Streaty and W. A. Klee, *Biochemistry 22:* 4567 (1983).
5. S. Maruyama and H. Suzuki, *Agric. Biol. Chem. 46:* 1393 (1982).

6. S. Maruyama, K. Nakagomi, N. Tomizuka and H. Suzuki, *Agric. Biol. Chem. 49:* 1405 (1985).

7. M. Yoshikawa, T. Yoshimura and H. Chiba, *Agric. Biol. Chem. 48:* 3185 (1984).

8. F. Parker, D. Migliore-Samour, F. Floc'h, A. Zerial, G. H. Werner, J. Jollès, M. Casaretto, H. Zahn and P. Jollès, *Eur. J. Biochem. 145:* 677 (1984).

9. J. Hughes, T. W. Smith, H. W. Kosterlitz, C. A. Fothergill, B. A. Morgan and H. R. Morris, *Nature 258:* 577 (1975).

10. C. H. Li and D. Chung, *Proc. Natl. Acad. Sci. USA 73:* 1145 (1976).

11. K. Kangawa, H. Matsuo, M. Igarashi, *Biochem. Biophys. Res. Commun. 86:* 153 (1979).

12. A. Goldstein, S. Tachibana, L. I. Lowney, M. Hunkapillar and L. Hood, *Proc. Natl. Acad. Sci. USA 76:* 6666 (1979).

13. S. Numa, *The Peptides: Analysis, Synthesis, Biology, vol. 6, Opioid Peptides: Biology, Chemistry, and Genetics* (S. Udenfriend and J. Meienhofer, eds.), Academic Press, Orlando, p. 1 (1984).

14. H. Matsuo, A. Miyata, K. Mizuno, *Nature 305:* 721 (1983).

15. B. R. Seizinger, D. C. Liebisch, C. Gramsch, A. Herz, E. Weber, C. J. Evans, F. S. Esch and P. Böhlen, *Nature 313:* 57 (1985).

16. P. C. Montecucchi, R. de Castiglione, S. Piani, L. Gozzini and V. Erspamer, *Int. J. Pept. Protein Res. 17:* 275 (1981).

17. V. Clement-Jones and G. M. Besser, *The Peptides: Analysis, Synthesis, Biology, Vol. 6, Opioid Peptides: Biology, Chemistry, and Genetics* (S. Udenfriend and J. Meienhofer, eds.), Academic Press, Orlando, p. 324 (1984).

18. S. J. Paterson, L. E. Robson and H. W. Kosterlitz, *The Peptides: Analysis, Synthesis, Biology, vol. 6, Opioid Peptides: Biology, Chemistry, and Genetics* (S. Udenfriend and J. Meienhofer, eds.), Academic Press, Orlando, p. 147 (1984).

19. V. Brantl and H. Teschemacher, *Naunyn-Schmiedeberg's Arch. Pharmacol. 306:* 301 (1974).

20. K.-J. Chang, A. Killian, E. Hazum and P. Cuatrecasas, *Science 212:* 75 (1981).

21. K.-J. Chang, Y. F. Su, D. A. Brent and J.-K. Chang, *J. Biol. Chem. 260:* 9706 (1985).

22. V. Brent, A. Pfeiffer, A. Herz, A. Henschen and F. Lottspeich, *Peptides 3:* 793 (1982).

23. K.-J. Chang, E. T. Wei, A. Killian and J.-K. Chang, *J. Pharmacol. Exp. Ther. 227:* 403 (1983).

24. H. Matthies, H. Stark, B. Hartrodt, H.-L. Ruethrich, H.-T. Spieler, A. Barth and K. Neubert, *Peptides 5:* 463 (1984).

25. V. Brantl, H. Teschemacher, J. Bläsig, A. Henschen ana F. Lottspeich, *Life Sci. 28:* 1903 (1981)

26. E. T. Wei, A. Lee and J. K. Chang, *Life Sci. 26:* 1517 (1980).

27. V. Schusdziarra, R. Schick, A. de la Fuente, J. Specht, M. Kller, V. Brantl and E. F. Pfeiffer, *Endocrinol. 112:* 885 (1983).

28. V. Schusdziarra, R. Schick, A. de la Fuente, A. Holland, V. Brantl and E. F. Pfeiffer, *Endocrinol. 112:* 1948 (1983).

29. V. Schusdziarra, R. Schick, A. Holland, A. de la Fuente, J. Specht, V. Maier, V. Brantl and E. F. Pfeiffer, *Peptides 4:* 205 (1983).

30. J. Panksepp, L. Normansell, S. Siviy, J. Rossi, III and A. J. Zolovick, *Peptides 5:* 829 (1984).

31. R. B. Merrifield, *J. Am. Chem. Soc. 85:* 2149 (1963).

32. W. L. Cosand and R. B. Merrifield, *Proc. Natl. Acad. Sci. USA, 74:* 2771 (1977).

33. W. König and R. Geiger, *Chem. Ber. 103:* 788 (1970).

34. G. L. Southard, G. S. Brooks and J. M. Peffee, *Tetrahedron Lett. 1969:* 3505.

35. S. Sakakibara, Y. Shimonishi, Y. Kishida, M. Okada and H. Sugihara, *Bull. Chem. Soc. Jap. 40:* 2164 (1967).

36. H. Fraenkel-Conrat and H. S. Olcott, *J. Biol. Chem. 161:* 259 (1945).

37. C. B. Pert and S. H. Snyder, *Proc. Natl. Acad. Sci. USA, 70:* 2243 (1970).

38. H. W. Kosterlitz, R. J. Lydon and A. J. Watt, *Brit. J. Pharmacol. 39:* 398 (1970).

39. G. Henderson, J. Hughes and H. W. Kosterlitz, *Brit. J. Pharmacol. 46:* 764 (1972).

40. T. Oka, K. Negishi, M. Suda, M. Matsumiya, T. Natsu and M. Ueki, *Eur. J. Pharmacol. 73:* 235 (1981).

41. B. C. Richardson and J.-C. Mercier, *Eur. J. Biochem. 99:* 285 (1979).

42. D. E. Blackburn, A. A. Hobbs and J. M. Rosen, *Nucleic Acid Res. 10:* 2295 (1982).

43. R. Greenberg, M. L. Groves and H. J. Dower, *J. Biol. Chem. 259:* 5132 (1984).

44. M. Yoshikawa, F. Tani, T. Yoshimura and H. Chiba, *Agric. Biol. Chem. submitted.*

45. B. Ribadeau-Dumas, G. Brignon, F. Grosclaude and J.-C. Mercier, *Eur. J. Biochem. 25:* 505 (1972).

46. J. B. C. Findlay and K. Brew, *Eur. J. Biochem. 27:* 65 (1972).

47. K. Brew, F. J. Castellino, T. C. Vanaman and R. J. Hill, *J. Biol. Chem. 245:* 4570 (1970).

48. G. Braunitzer, R. Chen, B. Schrank and A. Stangl, *Hoppe-Seyler's Z. Physiol. Chem. 354:* 867 (1973).

49. M. Yoshikawa, F. Tani, T. Yoshimura and H. Chiba, *in preparation.*

50. J. C. Mercier, G. Brignon and B. Ribadeau-Dumas, *Eur. J. Biochem. 35:* 222 (1973).

51. R. G. Hammond, Jr., P. Nicolas and C. H. Li, *Proc. Natl. Acad. Sci. USA 81:* 1389 (1984).

52. J. S. Shaw, L. Miller, M. L. Turnbull, J. J. Gromley and J. S. Morley, *Life Sci. 31:* 1259 (1982).

53. R. Cotton, M. L. Giles, L. Miller, J. S. Shaw and D. Timms, *Eur. J. Pharm. 97:* 331 (1984).

54. I. S. Zagon, P. J. McLaughlin, D. J. Weaver and E. Zagon, *Neurosci. Biobehav. Rev. 6:* 439 (1982).

55. E. Hazum, J. J. Sabatka, K.-J. Chang, D. A. Brent, J. W. A. Findlay and P. Cautrecasas, *Science 213:* 1010 (1981).

56. J. W. Holaday and A. I. Faden, *Nature 275:* 450 (1978).

57. R. L. Atkinson, *J. Clin. Endoc. Metab. 55:* 196 (1982).

58. R. Moore, I. H. Mills and A. Forster, *J. Royal. Soc. Med. 74:* 129 (1981).

59. I. S. Zagon and P. J. McLaughlin, *Life Sci. 33:* 2449 (1983).

60. I. S. Zagon and P. J. McLaughlin, *Science 221:* 1179 (1983).

61. M. Gallagher and B. S. Kapp, *Life Sci. 23:* 1973 (1978).

62. C. F. Aylsworth, C. A. Hodson and J. Meites, *Proc. Soc. Exp. Biol. Med. 161:* 18 (1979).

63. M.-J. Kreek, R. A. Schaefer, E. F. Hahn and J. Fishman, *Lancet 1983i:* 261.

64. D. F. Waugh and P. H. von Hippel, *J. Am. Chem. Soc. 78:* 4576 (1956).

65. U. Brodbeck, W. L. Denton, N. Tanahashi and K. E. Ebner, *J. Biol. Chem. 242:* 1391 (1967).

66. S. Pervaiz and K. Brew, *Science 228:* 335 (1985).

67. A. Ueno, Y.-M. Hong, N. Arakaki and Y. Takeda, *J. Biochem. 98:* 269 (1985).

68. T. Saldeen, *Ann. N. Y. Acad. Sci. 408:* 425 (1983).

69. K. Nishioka, A. Constantopoulos, P. S. Satoh, W. M. Mitchell and V. A. Najjar, *Biochem. Biophys. Acta 310:* 217 (1973).

70. M. A. Berman and W. O. Weigle, *J. Exp. Med. 146:* 241 (1977).

71. J. R. Whitaker and A. J. Puigserver, *Modifications of Proteins: Food, Nutritional, and Pharmacological Aspects* (R. E. Feeney and J. R. Whitaker, eds.), American Chemical Society, Washington D. C.) p. 57 (1982).

72. P. Petrilli, D. Picone, C. Caporale, F. Addeo, S. Auricchio and G. Marino, *FEBS lett. 169:* 53 (1984).

6

Genetic Engineering of Enzymes and Food Proteins

Rafael Jimenez-Flores, Young C. Kang, and Thomas Richardson

Department of Food Science and Technology
University of California
Davis, California

I. INTRODUCTION

One of the most important factors that determines whether or not a protein is usable in the fabrication of a food product is its functionality (1). The functionality of a food protein results from a combination of physico-chemical properties that define the behavior of the food protein in food systems. What is it about bovine milk caseins that promote coagulation to form cheese curd or about egg white proteins that favor whipping into a meringue?

At the molecular level much remains to be learned about the structural factors that determine the functional characteristics of food proteins (1). In Table 1 (1) are listed some structure-function relationship in food proteins that define their intrinsic molecular and functional properties.

It is readily evident that a detailed understanding of food protein functionality requires an intimate knowledge of the protein structure. Regretably, much of this information is

Table 1 *Structure-function relationships in food proteins (1)*

1. *Amino acid composition (major groups)*
2. *Amino acid sequence (segments/polypeptides)*
3. *Secondary/tertiary conformation (compact/coil)*
4. *Surface charge, hydrophobicity/polarity*
5. *Size, shape (topography)*
6. *Quaternary structures*
7. *Secondary interactions (intra- and inter-peptide)*
8. *Disulphide/sulphydryl content*
9. *Environmental conditions (pH, Redox, salts, temperature)*

lacking for most food proteins. However, some food proteins such as the bovine milk protein system have been studied in much detail. On the other hand, it is often possible to extract useful structure-function relationships at a less refined level without concern for detailed molecular structure. For example, the relative surface hydrophobicities of proteins can be determined from their interactions with hydrophobic probes that fluoresce in a hydrophobic environment (2,3). Thus, enchanced fluorescence of the probe in the presence proteins has been used as a measure of protein surface hydrophobicity. This in turn has been correlated with the interfacial activities of the proteins that govern emulsification and foaming properties of the proteins. As shown in Fig. 1 (2,3), examination of a series of proteins leads to relationships between the ability of a protein to reduce interfacial tension and its effectiveness as an emulsifying agent. Thus, it would appear possible to engineer the functionality of food proteins at a relatively gross level without necessary prior detailed knowledge of the molecular structure. Along these lines proteins have been chemically modified with a variety of reagents to prepare protein derivatives with altered functional properties (4,5).

Chemical modifications of proteins have provided much useful structure-function information on the functional behavior of food

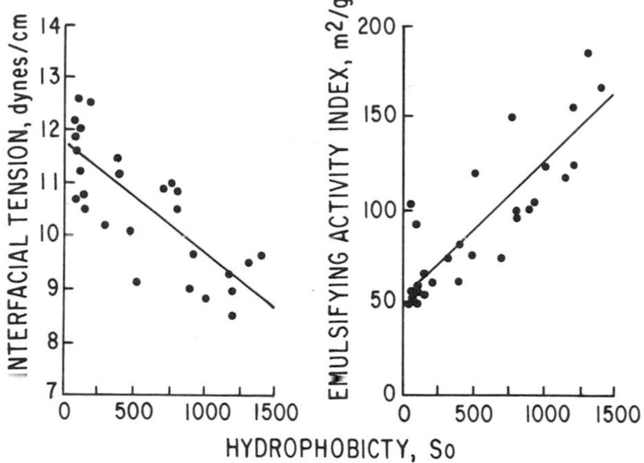

FIG. 1 *Correlations of hydrophobicity, S_o, with the interfacial tension and emulsifying activity index of proteins. Note, general trends with a decrease in interfacial tension being accompanied by an increase in emulsifying activity index. Proteins are identified in original article (from Ref. 3).*

proteins. For example, esterfication of side-chain carboxyl groups of β-lactoglobulin blocks potentially negative charges on the protein yielding protein derivatives with a higher pI (6-8). Not only does the protein have a greater net positive charge at neutral pH, but greater repulsive surface charges on the molecule disrupts the three dimensional structure of the protein thereby exposing internal hydrophobic patches. Thus, the positively charged, more hydrophobic β-lactoglobulin derivative can interact electrostatically and hydrophobically with negatively charged food proteins which can result in formation of potentially useful protein complexes. Although providing useful structure-function information on proteins, chemical derivatization often results in polydisperse protein products because of more-or-less random reactions with protein functional groups of varying reactivity. Thus, specificity of modification is lost which can hamper interpretation of structure-function information.

Recent developments in recombinant (6) DNA technology can be used to systematically alter single amino acids in the primary sequences of proteins (9-11). It is known that single changes in amino acids can markedly alter the functional characteristics of proteins. The substitution of a valine for a glutamate in hemoglobin is a classical example whereby the conformation and physical properties of the valine (sickle-cell) hemoglobin are totally altered compared to the normal glutamate hemoglobin (12). Consequently, it is possible to consider food protein functionality in the broader context of relatively large changes in protein structure or in the limited context of highly selective alterations in the primary sequence of the proteins using genetic engineering techniques.

In the genetic enigneering of enzymes, on the other hand, one probably requires the precision of modification inherent in recombinant DNA techniques to alter the K_m, V_{max}, specificity etc. of the enzyme. As we shall see, it is now feasible to genetically engineer enzymes to change their thermal inactivation, kinetic characteristics, and specificity to make them more suitable for use in food processing and analyses. On the other hand, manipulation of the functionality of food proteins for the most part will probably not require such precision and it becomes possible to alter their properties with more generalized changes. Nonetheless, this report will discuss the use of oligodeoxynucleotide directed mutagenesis for specific alterations in the primary sequence of enzymes to change their characteristics and give some hypothetical examples of how bovine caseins might be specifically modified to improve their functional behavior in cheese-making.

II. OLIGODEOXYNUCLEOTIDE SITE-DIRECTED MUTAGENESIS

Oligodeoxynucleotide site-directed mutagenesis has been used to make specific changes in the primary sequences of proteins in general and of enzymes in particular (9-11,13-16). Basically, there are seven general conditions required for oligodeoxynucleotide site-directed mutagenesis (9,10). A protein can be restructured by:

1. Cloning the relevant gene into an appropriate vector that is capable of autonomous replication (e.g. plasmids and viral genomes),
2. obtaining expression of the cloned gene,
3. determining the primary sequence of the DNA insert,
4. making the site to be mutagenized available in a single-stranded form,
5. using an appropriate, specifically-priming synthetic oligodeoxynucleotide containing base mismatches for altering the cloned sequence for specifically annealling to the single-stranded site,
6. performing site-directed mutagenesis wherein the sequences that are to remain wild-type must be regenerated with fidelity,
7. identifying mutant colonies with provisions for isolating and characterizating the mutant DNA.

Several strategies are available to achieve oligodeoxynucleotide site-directed mutagenesis (9-11). The strategy depicted in Fig. 2 (10) modifies a plasmid to yield a single stranded segment for site-directed mutagenesis. In this strategy, supercoiled plasmid circles are nicked in one strand by a restriction endonuclease in the presence of ethidium bromide. Exonuclease III is used to digest the nicked strand from the 3' end to yield a partial single stranded segment. The gapped circles are annealed or hybridized with a synthetic, homologous oligodeoxynucleotide (16 bases) carrying, by design, two mismatches. *In vitro* DNA synthesis is primed in part by the synthetic oligodeoxynucleotide heteroduplex plasmid circles. Molecular cloning and *in vivo* DNA replication generates homoduplexes, some of which have the sequence of the synthetic primer oligodeoxynucleotide. The plasmids can be purified and used to transform *Escherichia coli* a second time. Colony screening, with the same synthetic oligodeoxynucleotide, labeled with ^{32}P as a hydridization probe, allows identification of the desired mutant colony regardless of its phenotype.

This method was used to make a specific change in the β -lactamase (ampicillin resistance) gene of the plasmid pBR322 (10). The Ser-Thr (residues 70 and 71) dyad in the active site was inverted to Thr-Ser with two nucleotide mismatches as shown in Fig. 2. When the modified plasmid was used to transform *Escherichia coli*, the resultant mutant was ampicillin-sensitive indicating that the double-mismatch inactivated the β-lactamase. The foregoing method can be used as a general strategy because detection of mutants is at the level of DNA and involves only colony hybridization. Consequently, the procedure can be applied to any DNA sequence and does not depend on the phenotype of the mutant. Thus, a protein can be modified by inserting its cDNA into an appropriate expression vector and using oligodeoxy-nucleotide site-directed mutagenesis to alter the primary amino acid sequence. Site-directed mutagenic systems, such as those involving recombinant M-13 phage, that appear to be more efficient than the foregoing example have been developed (9). The above "gapped duplex" method suffers from the high degree of site specificity of the endonucleolytic nicking or cleavage reaction (which may prevent the targeting of single-stranded gaps to areas of interest), and frequently yield low efficiencies of mutagenesis (less than 10%) (9).

III. GENETIC ENGINEERING OF ENZYMES

The example alluded to in the previous section indicates it is possible to modify the active sites of enzymes using recombinant DNA technology. The manipulation of protein structure and function using recombinant DNA techniques has come to be known as "protein engineering" (15). In terms of the genetic engineering of enzymes, the list of properties that one would like to control in a predictable fashion would include (15):

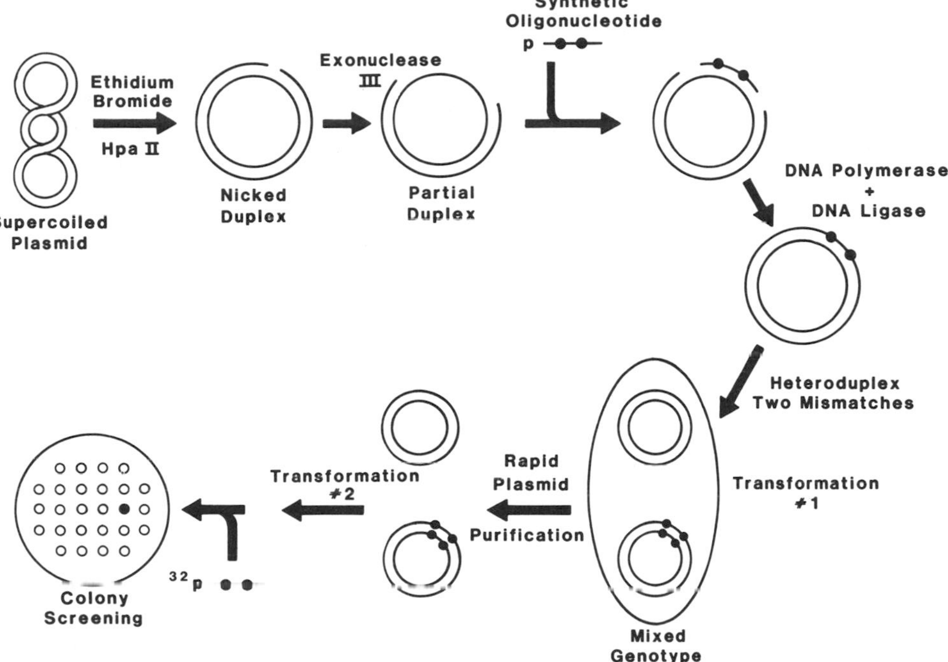

FIG. 2 *A strategy for oligodeoxynucleatide site-directed mutagenesis involving a plasmid (from Ref. 10).*

1. kinetic properties such as turnover number and Michaelis constant, K_m, for a given substrate,
2. thermostability and temperature optimum,
3. stability and activity in nonaqueous solvents,
4. substrate and reaction specificity,
5. co-factor requirements,
6. pH optimum,
7. protease resistance,
8. allosteric regulation,
9. molecular weight and subunit structure.

The following is a brief discussion illustrating how some of the foregoing properties of enzymes can be altered using recombinant DNA techniques. This emerging technology might prove

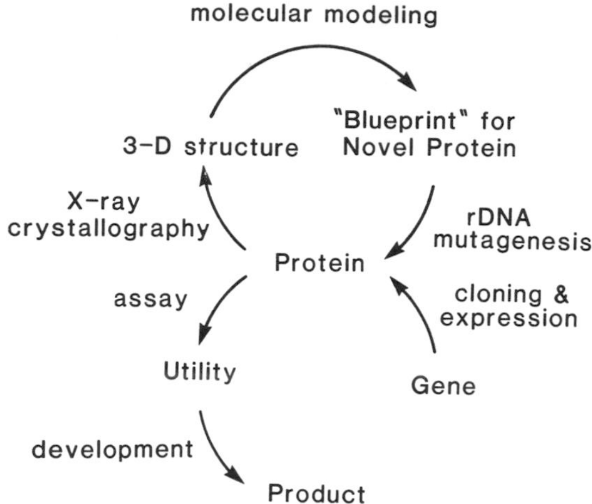

FIG. 3. *The iterative steps in protein engineering (from Ref. 14).*

useful in designing new catalysts for food processing and analyses.

As shown in Fig. 3 (14), the genetic engineering of an enzyme is ideally an iterative process. It involves oligodeoxynucleotide site-directed mutagenesis and molecular cloning to modify the enzyme coupled with the use of x-ray crystallography and molecular modeling to define the alterations in enzyme structure and to design new changes in enzyme characteristics. In many cases, the wild-type enzyme will be directly developed into a new product. Subsequently, however, new generations of enzymes can result for novel applications using the design loop depicted in Fig. 3. Once the parent systems are well understood, application of the iterative design process should facilitate the production and characterization of mutant proteins.

Rather than proceed with an exhaustive review of the genetic engineering of enzymes, it should prove instructive to give a few selected examples of how some of the basic properties of enzymes can be modified to possibly increase their utility in industrial processes.

A. *ENZYME THERMAL STABILITY (17)*

In globular proteins such as enzymes disulfide bonds provide conformational stability. Thus, the introduction of disulfide bridges into enzyme proteins might have the effect of enhancing the thermal stability of the enzyme.

Recently a disulfide bond was engineered into T4 lysozyme to stabilize the protein toward thermal inactivation (17). T4 Lysozyme is normally a disulfide-free protein. Its x-ray crystal structure is known and it has two unpaired cysteines and limited thermal stability. Theoretical calculations suggest that the degree of conformational stability conferred upon a protein by a cross link increases with the number of amino acids in the primary sequence that it spans. Upon examination of the x-ray crystallographic structure and with the aid of computer graphics, Perry and Wetzel (17) introduced a cysteine at position 3 which formed a disulfide bond with the natural cysteine at position 97 (Fig. 4).

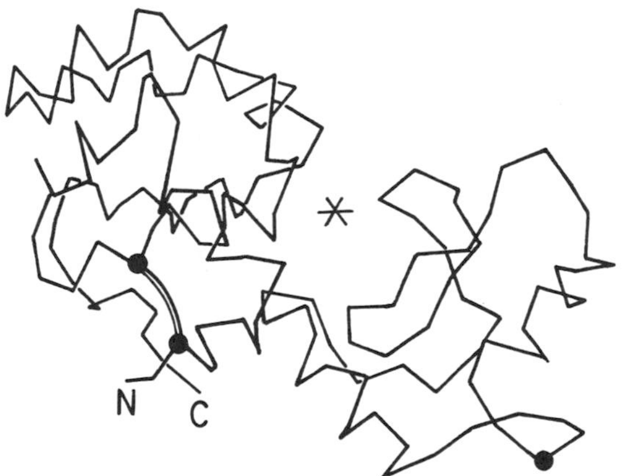

FIG. 4 *Computer graphics simulation of T4 lysozyme (Ile3 → Cys3) α-carbon chain showing the amino- and carboxyl-chain termini (N and C, respectively), the three cysteines (●) and the active site (star). Cys 3 and Cys 97 are connected by a schematic disulfide (from Ref. 17).*

Using synthetic oligodeoxynucleotide site-directed mutagenesis in an M13 phage system, these workers substituted Cys3 for Ile3 in the native enzyme. The new gene was expressed in Escherichia coli under control of the (trp-lac) hybrid tac II promoter, and the protein was purified for thermal inactivation studies.

Mild oxidation of modified enzymes with sodium tetrathionate generated a disulfide bond between the new Cys3 and Cys97, one of the two unpaired cysteines of the native molecule. The oxidized T4 lysozyme (Ile3 → Cys3) had a specific activity identical to that of the wild-type enzyme when measured at 20°C in a cell-clearing assay.

The cross-linked protein was more stable than the wild-type during incubation at elevated temperatures as determined by recovered enzymatic activity at 20°C. Wild-type and engineered T4 lysozymes were heated at 67°C and portions were removed at various times to determine residual activity (Fig. 5). The initial decay

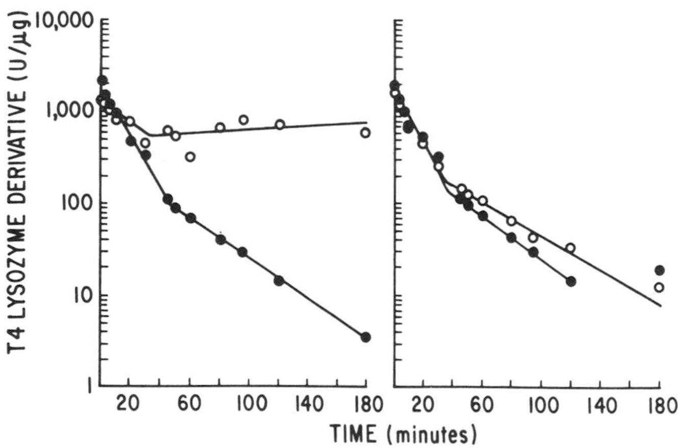

FIG. 5 *Kinetics of inactivation of lysozyme derivatives. At various times, samples were plunged into ice, diluted, and subjected to turbidity assay at 20°C. (a, left) Wild-type T4 lysozyme (●) and T4 lysozyme (Ile3 → Cys3) oxidized by incubation with sodium tetrathionate (O); (b, right) Wild-type T4 lysozyme (●) and reduced T4 lysozyme (Ile3 → Cys3) (O) in the presence of 10 mM β-mercaptoethanol (from Ref. 17).*

in activity in the wild-type lysozyme had a half-life of 11 minutes, whereas the disulfide-linked mutant decayed more slowly with an initial half life of 28 minutes. More dramatically, the activity of the disulfide form did not fall below 50% of the starting activity while the wild-type enzyme retained only 0.2% of its starting activity after 180 min. (Fig. 5a). On the other hand, in the presence of a reducing agent that freed SH groups, the mutant and wild-type enzymes behaved almost identical to thermal denaturation (Fig. 5b). This suggested that the cross-link and not the amino acid substitution *per se* provided the additional stability.

The foregoing studies were complicated by the existence of the additional cysteine at position 54. When Cys54 existed as a free thiol the modified enzyme showed a thermal stability similar to the wild-type enzyme at 67°C. However, when the free thiol of Cys54 was blocked (probably by thiosulfate from the tetrathionate oxidizing agent) the thermal stability of the mutant enzyme was enhanced as in Fig. 5a. Presumably, blocking of the free thiol group of Cys54 repressed thiol-disulfide exchange reactions important in the thermal inactivation of the enzyme.

Thus, the introduction of disulfide bridges into enzymes can enhance their thermal stabilities. This is probably true for proteins in general. Enhancing or decreasing (by removing disulfide bridges) the thermal stability of food proteins can have profound effects on their behavior toward thermal processing. Moreover, the presence of an odd free thiol group in food proteins can be involved in thiol-disulfide interchange reactions which can affect the course of thermal reactions in food proteins.

B. CHANGING ENZYME KINETIC PROPERTIES (16,18-22)

As noted earlier (Fig. 3), x-ray crystallography plays a key role in genetic engineering of enzymes because it provides an electron density map which together with the sequence of amino acid side chains can allow a three dimensional model of the enzyme to be built. Tyrosyl tRNA synthetase of Bacillus stearothermophilus has been crystallized and its structure solved. It thus provides a model to study protein engineering. The enzyme catalyzes the loading of tRNA in a two stage reaction. Initially, the tyrosine is activated by ATP to give an enzyme bound tyrosyl adenylate, and secondly the complex is attacked by the tRNA to give tyrosine-tRNA:

(1) Tyrosine + ATP   Tyrosyl-adenylate+Pyrophosphate

(2) Tyrosyladenylate + tRNA   Tyrosine-tRNA + AMP

Again, with the aid of x-ray crystallographic data and computer graphics, the active site was constructed to show possible hydrogen bond contacts between the enzyme and the tyrosyladenylate (Fig. 6) (16,16a). For example, His 48, Thr 51 and Cys 35 appear to form hydrogen bonds with the ribose moiety of tyrosyl adenylate.

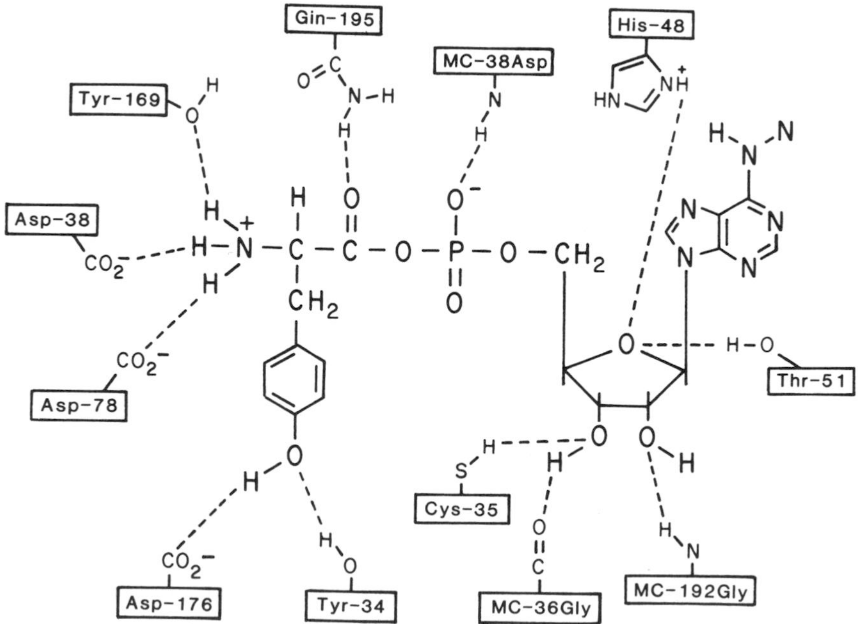

FIG. 6 *Side chains of the tyrosyl-tRNA synthetase that form hydrogen bonds with tyrosyl adenylate (from Ref. 16a).*

Table 2 *Protein Engineering to Alter Activity of the Tryosyl tRNA Synthetase in the Activation of Tyrosine (16,16a,18-22)*

Enzyme	k_{cat} (S^{-1})	K_m (mM ATP)
Wild type	*7.6*	*0.9*
Hrs 48 → Gly 48	*1.6*	*1.4*
Cys 35 → Gly 35	*2.8*	*2.6*
Thr 51 → Ala 51	*8.6*	*0.54*
Thr 51 → Pro 51	*12.0*	*0.058*

Mutations were constructed in the tyrosyl tRNA synthetase gene of B. stearothermophilus and the encoded enzymes were expressed in E. coli, purified and their kinetic and substrate-binding properties examined. Mutant enzymes have been prepared in which His 48 was altered to Gly, Thr 51 to Ala and Cys 35 to Gly, respectively (16,16a). This removes their hydrogen bonding capability to alter the catalytic rate constant (k_{cat}) and Michaelis constant (K_m) for ATP. As shown in Table 2, the His 48 → Gly 48 and Cys 35 → Gly 35 mutations resulted in decreases in k_{cat} and increases in K_m (ATP) compared to the wild-type enzyme with a major effect on k_{cat}. These mutations thus resulted in a deterioration in the activation of tyrosine by the tyrosyl-tRNA synthetase. On the other hand, with the loss of hydrogen bonding from the Thr 51 → Ala 51 mutation there appears to be an almost paradoxical increase in k_{cat} and decrease in K_m for ATP (16). This may reflect an unfavorable exchange of hydrogen bond energy, as the ATP substrate must displace a tightly bound water molecule from Thr 51 in order to form only a weak hydrogen bond with the enzyme (16,16a). Consequently, elimination of this unfavorable situation by the Thr 51 → Ala 51 mutation leads to more favorable binding energies.

The most remarkable changes in the catalytic properties of the enzyme occurred with a Thr 51 → Pro 51 mutation (Table 2). In the *E. coli* tyrosyl tRNA synthetase, compared to the *B. stearothermophilus* enzyme, Thr 51 was naturally replaced by proline. Since Pro is known to disrupt α-helix and other secondary structures the Pro 51 should destablize the polypeptide backbone in this region. To examine this prospect, the Thr 51 → Pro 51 mutant of the *B. stearothermophilus* enzyme was constructed (16,18,21). Surprisingly, the k_{cat} almost doubled and there was a remarkable 15 fold enhancement in the affinity for ATP in the activation reaction for tyrosine (21) (Table 2). This showed that enzyme affinities for substrates can be improved by *in vitro* manipulation - an observation that could have industrial potential. The reason for the improved affinity is not entirely clear. The Pro 51 might interact directly with the ribose, or distortion of the polypeptide backbone might strengthen an existing contact (e.g. His 48 or Cys 35). Current evidence, however, indicates that the improved affinity is due to an improved interaction with His 48 (16,16a,18,21).

The foregoing observations suggest the possibilities for modifying the kinetic properties of enzymes to prepare more effective industrial catalysts. Genetic engineering of acid proteases for use in the dairy industry while exploring their homologies in structure and differences in reactivity should prove instructive.

C. CHANGING ENZYME SPECIFICITY (23)

The function of trypsin an enzyme that cleaves proteins in specific locations has been selectively altered by changing one or two amino acid residues located near the enzyme's active site (23). Trypsin is a member of a large family of enzymes termed serine proteases that catalyze the hydrolysis of a wide variety of peptide bonds. However, trypsin is relatively specific and splits peptide bonds on the carboxyl side of lysine and arginine residues. Other serine proteases possess different specificities, but all are thought to use the same catalytic mechanism.

Again, the three dimensional structure of bovine trypsin has been determined by x-ray crystallography which proved very useful in designing structural alterations in the rat enzyme. Computer graphic analyses suggested that substitutions of the Ala for Gly residues at positions 216 and 226 in the binding cavity of rat pancreatic trypsin, homologous to bovine trypsin, might differentially affect Arg and Lys substrate binding of the enzyme. Thus, with the aid of oligodeoxynucleotide site-directed mutagenesis in a mammalian cell system, three mutant enzymes were prepared, the 216 and 226 single Ala mutants and the 216-226 double Ala mutant.

Comparison of rat and bovine trypsin indicates the catalytic residues His 57, Asp 102 and Ser 195 in the active site are conserved along with an Asp residue at the base of the substrate binding pocket which is thought to bind the basic Arg and Lys residues. Moreover, the Gly residues at positions 216 and 226 which seem to permit entry of large amino acid side chains into the hydrophobic active site are also conserved.

The ratio of k_{cat}/K_m is known as the specificity constant (18) and is a measure of changes in specificity of the mutant enzymes. Table 3 presents the kinetic measurements for the wild type enzyme and the three mutants. The revertant trypsin (Ala 226 → Gly 226) possessed kinetic constants similar to the wild-type enzyme, indicating that the activity effects were due to the single amino acid mutation. Each genetic modification produced an enzyme with distinctive kinetic characteristics. Although K_m values were higher and k_{cat} values were lower for the mutant enzymes compared to the wild-type trypsin there was a distinct change in specificity (k_{cat}/K_m) as a result of modification. Compared to the wild-type enzyme the specificity of Trypsin Gly 216 → Ala 216 definitely shifted toward Arg whereas in the Gly 226 → Ala 226 there was a definite increase in Lys over Arg. Note that the double mutant possessed little catalytic activity and the specificity for Arg was favored only slightly over Lys. The effects noted in Table 3 were not dominated by the K_m but by the

Table 3 *Kinetic Measurements of Wild-Type and Mutant Trypsins (23)*

Enzyme	Substrate	$K_m (\mu M)$[1]	k_{cat} (min^{-1})	k_{cat}/K_m $(min^{-1}\ \mu M^{-1})$	Arg/Lys[2]
Wild type	*Arg*	13.9 ± 0.2	1444 ± 11	104 ± 2	
	Lys	144.0 ± 5.0	1308 ± 20	9.1 ± 0.5	11.4 ± 0.9
226 Ala → Gly revertant	*Arg*	11.8 ± 0.5	1719 ± 42	146 ± 10	
	Lys	118 ± 3.0	1477 ± 18	12.5 ± 1.3	11.7 ± 2.2
216 Gly → Ala	*Arg*	393 ± 15	1017 ± 13	2.6 ± 0.2	
	Lys	3904 ± 427	337 ± 26	0.09 ± 0.01	28.9 ± 6.1
226 Gly → Ala	*Arg*	482 ± 26	13.0 ± 0.3	0.027 ± 0.002	
	Lys	3665 ± 248	172 ± 6.0	0.047 ± 0.005	0.57 ± 0.12
216,226 Gly → Ala	*Arg*	215 ± 20	1.10 ± 0.02	0.0051 ± 0.0006	
	Lys	331 ± 18	0.61 ± 0.01	0.0018 ± 0.0002	2.8 ± 0.8

1. *Substrates: D-Val-Leu-Arg-aminofluorocoumarin and Val-Leu-Lys aminofluoroumarin.*

2. *Ratio of k_{cat}/K_m values.*

k_{cat} values indicating a general decline in the rate of the enzyme reactions. Nonetheless, these data indicate that an understanding in the mode of binding of substrate in the active site can lead to structural alterations in the enzyme which can change specificities.

IV. GENETIC ENGINEERING OF FOOD PROTEINS

As we have seen for the genetic engineering of enzymes, x-ray crystallography and molecular modeling with computer graphics are important for the precise modifications that are essential. However, in the genetic engineering of food proteins to alter protein functionality, precise changes are probably not crucial to improvements in functional characteristics. Moreover, structural knowledge of most food proteins is sketchy at best; and one is forced to consider structural alterations at a less refined level than with enzymes. Although detailed structural knowledge of the food protein to be modified is highly desirable it may still be possible to manipulate the functionality of food proteins with only limited information.

The bovine milk proteins probably represent one of the best characterized group of food proteins (24,25). The caseins, in particular, are very functional milk proteins and serve as the basis for a major segment of the dairy industry, particularly in the manufacture of cheese. It is not possible to crystallize the caseins because of large domains of disordered structure; consequently, there exists no x-ray crystallographic data on their three dimensional structure. Nevertheless, the primary sequences of the bovine caseins are known (24,25) and much has been inferred about their three dimensional structure from various physico-chemical studies. As a result, it is possible to postulate alterations in the primary sequence of the caseins which may prove useful in enhancing their behavior in dairy processing and storage. It may eventually be possible to engineer the dairy cow or its mammary gland (26) to either overproduce selected normal caseins to modify the behavior of milk or to superimpose the production of modified, novel caseins on the biosynthesis of normal caseins to change the characteristics of dairy products, possibly in producing new foods.

Bovine caseins are a heterogeneous group of phosphoproteins precipitated at pH 4.6 and 20°C. The caseins are comprised of four major polypeptide families: α_{s1}-, α_{s2}-, β- and K-caseins (24). As phosphoproteins, they are phosphorylated, containing 8, 10-13, 5 and one seryl phosphates per monomer, respectively. In addition, some of the K-casein molecules are glycosylated, while others are not, resulting in electrophoretic heterogeneity. The caseins tend to associate via hydrophobic interactions to form small spherical submicellar complexes.

Seryl phosphate residues in α_{s1}-, α_{s2}-, and β-caseins at the surface of the sub-micelles interact with inorganic amorphous calcium phosphate complexes to form macromolecular protein aggregates termed "casein micelles", which are roughly spherical and stabilized in colloidal suspension by K-casein concentrated on their surfaces (27). The casein micelles can thus maximize the amount of calcium phosphate and protein in a limited volume for effective nutrition of the young.

The physical stability and behavior of colloidal casein micelles is important in determining the characteristics of dairy products. For example, physical association of micelles induced by thermal treatments (28) can lead to unwanted gelation problems in heated milk products. On the other hand, enzymatic action on the K-casein stabilizing the micelles can result in desired gelation in the preparation of cheese curd (29). These physical interactions of the milk proteins important in dairy processing are determined by the structural properties of the milk proteins. The primary amino acid sequences of the caseins largely dictate their secondary structures, their surface hydrophobicity and their sites of phosphorylation and glycosylation (24). Consequently, the primary sequence of amino acids defines to a great extent the properties of the caseins and their complexes.

We have seen that it is possible to alter the primary sequence of a protein using genetic engienering techniques (9-11). It thus follows that changes in the primary sequence of the caseins can alter their behavior toward enzymes and physical treatments such as heating. It is now within the realm of possibility to systematically change the primary sequences of food proteins to modify the physical properties or functionalities of those proteins. Recombinant DNA technology may eventually prove useful in designing not only novel enzymes but new and food proteins as well.

As we have seen, it is feasible to produce engineered enzymes and proteins in microorganisms (9-11,13-23). Systematic alterations in the primary sequences of proteins allows detailed basic studies of the structure-function relationships of isolated proteins. In the case of caseins, this basic information will furnish background needed to eventually engineer the dairy cow or its mammary gland (26) to produce milk proteins with more desirable properties. It may become possible to induce the mammary gland to overproduce certain of the caseins to modify the behavior of the milk protein system. In addition, biosynthesis of novel caseins with altered primary sequences might be superimposed on normal casein biosynthesis to engineer desirable properties into the milk protein complex. Although these possibilities are long term (perhaps 15-20 years), research on the genetic engineering of caseins from various species is progressing. The remainder of this report focuses on the genetic engineering of the caseins, gives some hypothetical examples of casein modifications and tries to anticipate how the food scientist can interact with the molecular biologist and geneticist to design more desirable food proteins.

A. CLONING CASEIN cDNAs

Casein cDNAs (DNAs complementary to the mRNA) derived from various species have been cloned into *Escherichia coli*. However, no one has yet inserted the casein cDNAs into expression vectors to achieve production of the caseins by microorganisms.

1. *Rat and other species* cDNAs coding for caseins from rat (30-33), from mouse (34-35), from rabbit (36), and from guinea pig (37) have been cloned.

DNA sequences coding for the rat caseins have been studied in greatest detail of any species. A major thrust of this research is to define DNA segments responsible for the hormonal regulation of casein biosynthesis.

Full length rat casein cDNA clones for α, β, and γ rat caseins have been prepared and characterized by restriction enzyme mapping and DNA sequencing (30-33). From derived amino acid sequence data the rat β-casein is about 38% homologous with bovine β-casein whereas the α-casein is aproximately 31% homologous with bovine α_{s1}-casein. However, the rat γ-casein appears to have a primary sequence unrelated to the bovine caseins.

More recently, the actual genomic structural genes for rat γ-casein (38) and β-casein (39) were isolated and characterized in detail from rat genomic libraries. They are both very large complex genes containing long intervening sequences between the exons.

2. *Bovine casein cDNAs* Bovine casein cDNAs coding for α_{s1}- (40-41) and K- (40) caseins have been cloned. Resultant plasmids were used to transform E. coli. In addition to the foregoing full length cDNAs, cDNA fragments coding for portions of the caseins have been cloned (42-45). Although none of the bovine cDNA sequences have been specifically inserted into expression vectors to produce proteins for further study, this opens the way for systematic structure-function studies of the caseins.

3. *Potential for engineering caseins* Initially it will be possible for the primary sequences of the caseins to be produced by microorganisms once the appropriate cDNA in an expression vector has been incorporated into a host cell such as E. coli.

One can then obtain sufficient protein to examine the effects of structural changes elicited using the aforementioned oligodeoxy-nucleotide site-directed mutagenesis.

In the longer term, reliance must be placed on the possibility of stable incorporation of casein structural genes into the bovine genome followed by appropriate development and expression of the gene. For example, it may be possible to inject the structural gene along with its controlling elements into the bovine embryo and have the gene stably integrated into the bovine genome for expression in the adult under appropriate circumstances (46-49). Once the gene is stably integrated into the genomic DNA, it is possible that it will be transmitted to the progeny via the germ cells (46,48,49). However, much research remains to be done before this potential becomes a reality. It is also unlikely that caseins will be produced by microorganisms for food use because of unfavorable economics and relatively low yields of proteins.

There are numerous modifications possible in altering the caseins to be more desirable from a functional point of view. Based on current knowledge of the caseinate system in milk, it is possible to propose changes in casein structure which may eventually prove useful.

For example, the ripening of cheese involves the proteolytic breakdown of the protein matrix to yield a more desirable texture. Conversion of α_{s1}-casein to α_{s1}-I-casein and an amino terminal peptide by cleavage of a Phe23-Phe24 or Phe24-Val25 bond by residual clotting enzyme (chymosin or other acid protease) is thought to be one of the primary changes in the caseins during maturation of cheddar cheese (50). It might be possible to develop caseins containing additional bonds with enhanced chymosin sensitivity to accelerate the rate of textural development in cheese. Conversion of Ile71 to Phe71 (by changing one nucleotide base) in α_{s1}-casein would generate an additional Phe71-Val72 bond in a region of α_{s1}-casein disordered by a series of seryl phosphate, other charged amino acid residues and an adjacent prolyl residue (24). The Val 72 can also be converted to a Phe (with one base change) or to Tyr or Trp with two base changes. This bond should be readily available for cleavage by residual acid proteases with a propensity for cleaving bonds adjacent to aromatic amino acids. Cleavage of a bond more toward the middle of the α_{s1}-casin (as at position 71) should maximize the rheological effects of proteolysis in promoting faster ripening of the cheese. More rapid textural development in cheese would be highly desirable from an economic, storage point of view.

In the clotting of milk during cheesemaking a Phe 105-Met 106 bond in K-casein is specifically cleaved by acid proteases such as chymosin (29). This releases a highly polar macropeptide from K-casein and destabilizes the casein micelle forcing the coagulation of the micelles to form cheese curd. The Phe-Met bond is not in itself especially susceptible to attack by chymosin. The Phe-Met bond in simple di-, tri-, and tetrapeptides is not readily hydrolyzed (29). Inclusion of the Phe-Met bond into longer peptides with structures analogous to those found normally in K-casein make the Phe-Met bond susceptible to attack by chymosin. It would appear that not only length but composition and sequence of the peptide segment containing the Phe-Met bond emerge as important determinants of enzyme-substrate interaction prior to hydrolysis (29). Norleucine used as an isosteric replacement for methionine increases the specificity constant (k_{cat}/K_m) by a factor of about 3 (29). Consequently, replacement of partners in the Phe-Met bond can enhance susceptibility towards hydrolysis. Moreover, the Phe 105-Met 106 bond in K-casein is probably in a particularly vulnerable region of the peptide sequence. In structures surrounding the chymosin-sensitive bond both α-helix and β-sheet structures have high probabilities. However, of potentially greater consequence is the presence of two predicted β-turns on either side of the susceptible sequence at residues 98 → 101 and 109 → 112. Moreover, the presence of another β-turn at 113 → 116 could cause the sensitive sequence to stand out on the molecular surface making it especially susceptible to attack by chymosin and other acid proteases (24).

Acid proteases used to clot milk are known to favor hydrolysis of peptide bonds flanked by aromatic residues such as Phe, Tyr and Trp (18,29). Thus, substitution of a Phe, Tyr, or Trp residue for Met in the susceptible sequence might enhance the rate of hydrolysis by acid proteases. Incorporation of such a K-casein analog into the micellar structure could increase the rate of K-casein hydrolyses thereby reducing the amount of enzyme required for coagulation. Using oligodeoxynucleotide-site-directed mutagenesis one could alter the ATG codon for Met with two base mismatches to derive the TGG codon for Trp.

V. CONCLUSIONS

We have seen how oligodeoxynucleotide site-directed mutagenesis can be used to engineer the primary sequences of proteins. Examples were given of how this powerful technique can be used to engineer enzymes and their protein substrates. Thus one can effect an alteration in the mode of action of a protease on its substrate by either modifying the enzyme or the susceptibility of bonds in the substrate.

In part, this has been a somewhat speculative review with an attempt to introduce the reader to the potential that genetic engineering holds forth upon successful application. Much additional research is necessary before successful genetic manipulation of plants and animals to tailor-make more functional proteins becomes a reality. As more is learned about the molecular basis for protein functionality in foods, it becomes increasingly possible to design proteins for particular uses. Consequently, the future holds the possibility of the food scientist working with the molecular biologist and geneticist to design food proteins with desired functionalities.

ACKNOWLEDGMENT

This review was made possible by support from the College of Agricultural and Environmental Sciences, University of California, Davis, California 95616

REFERENCES

1. J. E. Kinsella, Food Proteins (P.F. Fox and J. J. Condon, Eds.), *Applied Science Publ., New York*, pp. 51-103 (1982).

2. S. Nakai, , *J. Agric. Food Chem. 31*: 676 (1983).

3. A. Kato and S. Nakai, *Biochim. Biophys. Acta 624:* 13 (1980).

4. T. Richardson and J. J. Kester, *J. Chem. Ed. 61:* 325 (1984).

5. J. J. Kester and T. Richardson, *J. Dairy Sci. 67:* 2757 (1984).

6. N. L. Mattarella and T. Richardson, *J. Dairy Sci. 65:* 2253 (1983).

7. N. L. Mattarella, L. K. Creamer, and T. Richardson, *J. Agric. Food Chem. 31:* 968 (1983).

8. N. L. Mattarella and T. Richardson, *J. Agric. Food Chem. 31:* 972 (1983).

9. C. S. Craik, *Bio Techniques, 3:* 12 (1985).

10. G. Dalbadie-McFarland, L. W. Cohen, A. D. Riggs, C. Morin, K. Itakura, and J. H. Richards, *Proc. Natl. Acad. Sci. 79:* 6409 (1982).

11. M. J. Zoller and M. Smith, *Methods in Enzymol. 100:* 468 (1983).

12. L. Stryer, *Biochemistry, 2nd Ed., W. H. Freeman and Co., New York,* pp. 92,93 (1981).

13. T. H. Maugh II, *Science 223:* 269 (1984).

14. W. H. Rastetter, *Trends in Biotechnol. 1:* 80 (1983).

15. K. M. Ulmer, *Science 219:* 666 (1983).

16. G. Winter and A. R. Fersht, *Trends in Biotechnol. 2:* 115 (1984).

16a. A. R. Fersht, J-P. Shi, J. Knill-Jones, D. M. Lowe, A. J. Wilkinson, D. M. Blow, P. Brick, P. Carter, M. M. Y. Waye and G. Winter, *Nature, 314:* 235 (1985).

17. L. J. Perry and R. Wetzel, *Science 226:* 555 (1984).

18. A. R. Fersht, *Enzyme structure and mechanism, 2nd Ed., W. H. Freeman and Co., New York,* pp 105, 324, 386 (1985).

19. G. Winter, A. R. Fersht, A. J. Wilkinson, M. Zoller, and M. Smith, *Nature 299:* 756 (1982).

20. A. J. Wilkinson, A. R. Fersht, D. M. Blow, and G. Winter, *Biochemistry 22:* 3581 (1983).

21. A. J. Wilkinson, A. R. Fersht, D. M. Blow, P. Carter, and G. Winter, *Nature 307:* 187 (1984).

22. A. R. Fersht, J.-P. Shi, A. J. Wilkinson, D. M. Blow, P. Carter, M. M. Waye, and G. P. Winter, *Angew. Chem. Int. Ed. Engl. 23:* 467-538 (1984).

23. C. S. Craik, C. Largman, T. Fletcher, S. Roczniak, P. J. Barr, R. Fletterick, and W. J. Rutter, *Science 228:* 291 (1985).

24. H. E. Swaisgood, *Developments in Dairy Chemistry, Vol. 1 (P.F. Fox, ed.), Applied Science Publ., New York,* p. 1 (1982).

25. W. N. Eigel, J. E. Butler, C. A. Ernstom, H. M. Farrell Jr., V. R. Harwalker, R. Jenness, and R. Mcl. Whitney, *J. Dairy Sci., 67:* 1599 (1984).

26. S. Patton, U. Welsch, and S. Singh, *J. Dairy Sci. 67:* 1523 (1984).

27. D. G. Schmidt, *Developments in Dairy Chemistry, Vol. 1 (P.F. Fox, ed.), Applied Science Publ., New York,* p. 61 (1982).

28. P. F. Fox, *Developments in Dairy Chemistry, Vol. 1 (P.F. Fox, ed.), Applied Sci. Publ., New York,* p. 189 (1982).

29. D. G. Dalgleish, *Developments in Dairy Chemistry, Vol. 1, (P.F. Fox, ed.), Applied Sci. Publ., New York,* p. 157 (1982).

30. D. E. Blackburn, A. A. Hobbs and J. M. Rosen, *Nucl. Acids Res. 10:* 2295 (1982).

31. A. A. Hobbs and J. M. Rosen, *Nucl. Acid Res. 10:* 8079 (1982).

32. D. A. Richards, J. R. Rodgers, S. C. Supowit and J. M. Rosen, *J. Biol. Chem. 256:* 526 (1981).

33. D. A. Richards, D. E. Blackburn, and J. M. Rosen, *J. Biol. Chem. 256:* 533 (1981).

34. N. M. Mehta, M. Raafat El-Geuelly, J. Josh, R. B. Helling, and M. R. Bannerjee, *Gene 15:* 285 (1981).

35. L. G. Hennighausen and A. E. Sippel, *Eur. J. Biochem. 125:* 131 (1982).

36. Y. M. L. Suard, M. Tosi, J.-P. Kraehenbuhl, *Biochem. J. 201:* 81 (1982).

37. R. K. Craig, L. Hall, D. Parker, and P. N. Campbell, *Biochem. J. 194:* 989 (1981).

38. L.-Y. Yu-Lee and J. M. Rosen, *J. Biol. Chem. 258:* 10794 (1983).

39. W. K. Jones, L.-Y., Yu-Lee, S. M. Clift, T. L. Brown, and J. M. Rosen, *J. Biol. Chem. 260:* 7042 (1985).

40. A. F. Stewart, I. M. Willis, and A. G. Mackinlay, *Nucl. Acid Res. 12:* 3895 (1984).

41. M. Nagao, M. Maki, R. Sasaki, and H. Chiba, *Agric. Biol. Chem. 48:* 1663 (1984).

42. M. Maki, M. Nagao, M. Hirose, and H. Chiba, *Agric. Biol. Chem. 47:* 441 (1983).

43. I. M. Willis, A. F. Stewart, A. Caputo, A. R. Thompson, and A. G. Mackinlay, *DNA 1:* 375 (1982).

44. D. R. Kershulite, V. N. Ivanov, T. V. Kapelinskaya, A. S. Kaledin, and S. I. Gorodetsky, *Dokl. Akad. Nauk SSR 271:* 1502 (1983).

45. V. N. Ivanov, D. R. Kershulite, A. A. Bayer, A. A. Aklundova, G. E. Submova, E. S. Judinkova, and S. I. Gorodetsky, *Gene, 32:* 381 (1984).

46. J. D. Watson, J. Tooze, and D. T. Kurz, *Recombinant DNA; A Short Course, W. H. Freeman and Co., New York,* pp. 176, 200 (1983).

47. R. Kucherlapati and A. I. Skoultchi, *C.R.C. Crit. Rev., Biochem. 16:* 349 (1984).

48. R. D. Palmiter and R. L. Brinster, *Cell 41:* 343 (1985).

49. R. E. Hammer, V. G. Pursel, C. E. Rexroad, Jr., R. J. Wall, D. J. Bolt, K. M. Ebert, R. D. Palmiter, R. L. Brinster, *Nature 315:* 680 (1985).

50. L. K. Creamer and N. F. Olson, *J. Food Sci. 57:* 631 (1982).

7

Medical Applications of Protein Engineering

Ronald Wetzel

Biocatalysis Department
Genentech, Inc.
South San Francisco, California

I. INTRODUCTION

In only a few years the field of protein engineering has emerged as a very powerful approach for the study of protein structure/ function relationships (1). These new techniques already have been applied with significant success to fundamental problems in enzyme action and protein folding (2,3). In the more practical, and more stringent, area of medicine, however, results have been slow in coming. This chapter begins with a discussion of some problems and opportunities particular to the application of these methods to the design of pharmaceuticals. These points are illustrated with examples of protein engineering studies conducted both on proteins with potential medicinal applications, as well as proteins which are being used as model systems for the study of fundamental problems. While many examples are included, this is not intended to be comprehensive review of the literature of protein engineering.

Narrowly defined, protein engineering is the product of recent advances in molecular genetics, X-ray crystallography and computer graphics. Its real roots, however, lie in the classic approach of chemists to the elucidation of structure-function relationships, whether in reaction mechanisms or drug design: the synthesis and study of structural analogs. This approach has been vigorously applied to polypeptides in the past two decades, but progress has been inhibited by two dilemmas: Although chemical synthesis of most relatively short peptides is straightforward, there are significant problems in the determination of X-ray crystal structures of peptides; furthermore, those structures which have been determined cannot be assumed to hold for the solution state. In contrast, structures of proteins are more accessible and are probably more reliable approximations of solution structure, but proteins have been impossible to efficiently synthesize chemically.

There are a few examples of studies on polypeptides which defy the above generalizations, for example: insulin, a protein which is nonetheless chemically accessible, and somatostatin, a peptide with a solution conformation which can be modelled. Moreover, there are situations where primary sequence structure is a sufficient basis for the rational redesign of a protein. Because such chemical studies are part of the development of the ideas behind protein engineering, and because the chemical and molecular genetic approaches are unified by an appreciation for the primary importance of molecular mechanisms, several examples of chemical approaches are included in this chapter. This importance not withstanding, it is also clear that methods involving random mutagenesis, combined with genetic screening or selection procedures, can play an important role in protein engineering, especially in combination with a structure-based approach. The random mutagenesis approach will be briefly described and illustrated with an example.

II. PROTEIN ENGINEERING

Structure/function studies of peptides have been practiced for decades, but only recently has it become possible to contemplate studies unencumbered by the enormous technical difficulties in the chemical synthesis of longer polypeptides. This has been made possible through the use of recombinant DNA methods to clone and express genes into foreign host cells which allow overproduction of a single, genetically well-defined protein. Most commonly such "heterologous expression" is accomplished in E. coli (4,5), but expression is also possible in other systems such as bacillus (6), yeast (7) and mammalian tissue culture (8). The first application of recombinant DNA methods to produce a defined sequence variant of a protein, based on examination of an X-ray structure, was the synthesis of a "mini-C" proinsulin derivative reported in 1981 (9). The two fundamental approaches possible in utilizing recombinant DNA methods to generate variants are discussed below. Figure 1 shows an idealized scheme for the structural-mechanistic approach to protein engineering. If the protein of interest was not previously readily available, expression of a cloned gene allows the protein product to be studied for its biological properties and structural features. These, together with other information, form the basis of a working model for the means by which structure determines the function of interest. Based on this model, analogs can be designed - facilitated by computer graphics analysis if the 3-dimensional structure is known. These new structures can be produced by manipulating the structure of the cloned gene via directed mutagenesis. The properties of these variants are then determined to assess the functional consequences of structural change. At best, the process generates one or more variants of improved function. At worst, it generates variants which facilitate revision of the structure/function hypothesis. This in turn serves as the basis for the design and synthesis of further variants. The process is thus iterative, ideally requiring detailed evaluation of variants at each cycle.

This structural approach may be impacted by the random mutagenesis approach. Data from random mutagenesis studies, relating change in function to genetic lesions and amino acid changes, can contribute greatly to the development of a structure/function hypothesis. Further, molecules of altered

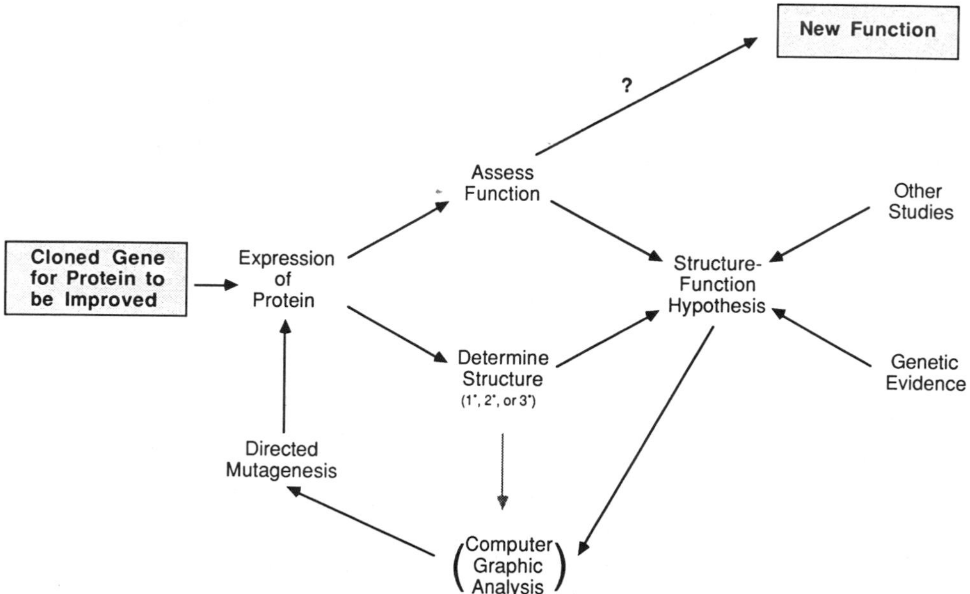

Figure 1. Scheme for protein engineering *via* a directed mutagenesis approach based on protein structure and mechanism.

function obtained by the random mutagenesis approach might be further refined by the structure-based approach.

Directed mutagenesis is normally thought of as involving single amino acid replacements. It can, however, include the synthesis of hybrids between homologous sequences (10,11), separate synthesis of domains (12), truncation of a polypeptide by the introduction of an early stop codon into its gene (13), removal of loops (9), and the synthesis of hybrid, mixed function proteins (14,15).

The above methods for generation of new variants of proteins would be much less powerful were it not for new developments in our abilities to determine and analyze three-dimensional structures of proteins. Although the rate-limiting step in most structure determinations, the production of highly diffracting crystals, remains a problem, advances in crystallization methodology have contributed significantly to reducing this obstacle. Data acquisition has become much more rapid with the availability of area detectors and the use of synchrotron radiation. These factors, in addition, allow structures to be determined at much higher resolutions leading to enhanced interpretability of the electron density maps. The availability of relatively inexpensive computing power has influenced the development of improved algorithms for structure refinement. This, when coupled to computer graphics, has led to greater efficiency in the refinement process and as a consequence led to improved accuracy of the resulting protein models.

Of course, computer graphics is also important as a tool for planning structural changes. One can replace specific amino acid side chains on the graphics screen and change side chain orientations to test for proper fits. It is, however, important to keep in mind that this is only a crude estimate of the effects a replacement may have on the structure and properties of a protein. Only recently have we begun to acquire structure/

function information on variants constructed by directed mutagenesis based on graphic analysis; over time, comparisons of predicted and observed effects will test the reliability of computer graphics simulations. It is clear we have a lot to learn: experiments in which many replacements have been engineered into specific sites in the enzymes subtilisin (16-18) and T4 lysozyme (19) have demonstrated that observed effects on solution properties can't always be rationalized on the basis of contemporary theory.

Figure 2 illustrates the random mutagenesis approach. Recombinant DNA methods are useful in extending this classical approach by (a) allowing for mutagenesis to be directed away from the host chromosome and to the gene of interest, (b) directing mutagenesis to a site or region of interest in the gene, (c) amplifying the ratio of singly-mutated/wild type genes, and (d) constructing the expression vector/host system to facilitate selection or screening (20-22).

In the selection process, the only surviving microbial colonies are those containing mutants of desired new function. Such selection schemes can be difficult to devise, however; this is especially true for hormones, lymphokines, and other mammalian proteins with no prokaryotic counterparts and no specific effects on microbial metabolism. An efficient screening system may be a reasonable alternative.

The scheme shows that this approach can be influenced by a structural approach. Based on a structure-function hypothesis, specific sites on the protein can be selected for "saturation mutagenesis", that is, the generation of up to 19 alternative amino acids at a position (16-19). Mutants obtained from directed mutagenesis may be further refined or characterized by the random mutagenesis approach. In a striking example of this, Ho *et al.* (23) recently used random mutagenesis and selection to obtain high activity single-site revertants from a truncated gene encoding a low activity ala-tRNA synthetase.

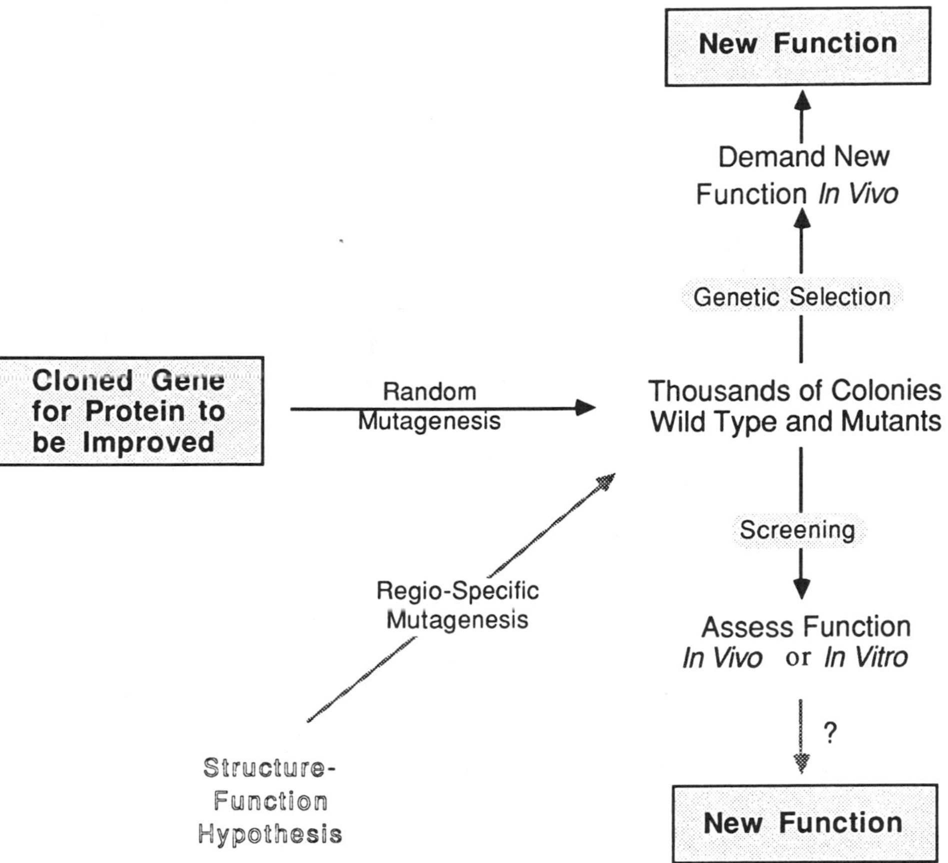

Figure 2. Scheme for protein engineering via random mutagenesis with genetic selection or screening for altered protein function.

III. PROTEINS AS PHARMACEUTICALS

Although only a few proteins have been used as drugs in the past, interest in the pharmaceutical potential of this class of molecules has vastly increased with the recent advances in high level expression of previously unavailable proteins (4-8). Proteins are subject to the same requirements as other pharmaceuticals; in some cases, however, their complex natures considerably increase the difficulty in meeting these requirements.

Proteins are relatively unstable structures. They are sensitive to chemical events such as air oxidation and pH- and enzyme-catalyzed hydrolysis. In addition, proteins have well-defined conformations which are often sensitive to extremes of pH and temperature. Two proteins can differ greatly in their sensitivities to these effects. Further, a structural change may have effects on some properties but not on others.

Stability is an issue during the synthesis of the polypeptide: a completed chain might suffer partial or essentially complete proteolysis during growth or extraction from the cells. Although not formally a stability issue, another important aspect is the integrity of the biosynthetic translation process and of post-translational modifications such as disulfide bond formation and glycosylation.

Stability is also an important issue in the formulation of the product for storage and administration. Thermal and oxidative damage are especially important. High temperatures can promote chemical damage (oxidation of Cys, Met, Trp, His and Tyr residues, deamidation of Gln and Asn residues) as well as conformational damage (folding changes often followed by aggregation or precipitation). Even at ambient temperatures, however, such decay processes are important, since only a few percent of decomposition may have deleterious effects while at the same time being difficult to detect analytically.

Proteins can also vary substantially in their serum lifetimes. These are influenced by the route of administration and by forces normally at work on the endogenous version of the protein, such as receptor number and clearance mechanisms. Unnatural elements of structure in the protein might also greatly influence serum lifetime. For example, it is well known that glycoproteins which have lost the sialic acid residues that often terminate their oligosaccharide chains have radically altered survival in the circulation due to the presence of carbohydrate-specific receptors in various tissues (24). A general feature of polypeptides is their extreme sensitivity to proteolytic conditions in the intestinal tract. This is one limitation to the oral administration of protein drugs; the other, more serious, is the barrier to transport of polypeptides by the mucosal membranes in the intestinal wall. This effectively limits the current use of proteins to parenteral routes such as intramuscular, subcutaneous, and intravenous injection.

As with any drug, side effects are of major importance to the use of proteins as pharmaceuticals. A potential side effect of major importance is immunogenicity, the ability of the injected protein to stimulate the synthesis of antibodies which can specifically recognize, clear, and possibly directly neutralize, the drug. Although the immune system of a developing organism learns to not respond to homologous proteins, many "self" proteins, injected in sufficient amounts or in a somewhat unnatural form, can generate such antibodies. The rules which govern the immunogenicity of a particular preparation of a protein are far from clear; however, it has been observed that preparations which differ only minutely in apparent purity and/or history, can differ considerably in immunogenicity. Finally, some protein hormones induce a number of effects *in vivo* which are not apparently related to the desirable activity; if these effects are not mechanistically associated with the desired activity, their sources may reside in distinctly different regions of the protein structure and thus may be amenable to surgical removal by protein engineering. At this point the question becomes, can one change the structure and function of one part of the molecule without affecting structure and functions elsewhere?

One can hope to isolate the desired therapeutic effect of the protein, as well as side effects, at the molecular level. The mechanisms by which protein hormones and lymphokines induce intracellular responses upon binding to cell surface receptors are not well understood. Is binding itself sufficient to induce the response, or are there secondary functions contained on the hormone which come into play after receptor binding? If the action of the protein is contained in a simple binding phenomenon, and if the binding surface can be identified, then it might be possible to design peptide analogs which mimic the required binding surface but which exhibit other, improved properties such as oral activity. Further, since non-peptide chemical structures can mimic the effective surface of a peptide ligand, as in the case of the interactions of opiates with enkephalin/endorphin receptors, it is possible that active non-peptide analogs of polypeptides might be designed and synthesized. Despite the attractiveness of this idea, it is sobering to realize that in this example of successful analog synthesis in nature, we do not as yet know the receptor-binding conformations of enkephalins and their relationship to the effective surface of the opiates (25).

A more immediate barrier to constructing such analogs is the identification of essential binding surfaces. The most significant problem is neither the elucidation of tertiary structure nor the generation of analogs. Rather, it is one of interpretation of structure/function results. The mechanisms by which a polypeptide sequence folds into a stable, unique solution structure are complex. An amino acid sequence change might produce only local, _direct_ effects on structure, due simply to the steric and chemical consequences of changes in side chain structures. Alternatively, sequence changes in one part of the molecule might cause _indirect_ effects on remote parts of the structure _via_ a series of conformational changes, as the folded polypeptide seeks a new free energy minimum. Distinguishing between direct and indirect effects is a continuing challenge in protein biochemistry. The best solution in general is probably an x-ray structure of the variant involved; short of this, one can build a case by investigating effects on solution structure using other techniques. One attractive possibility is the use of a bank of conformation-sensitive monoclonal antibodies generated against different epitopes of the protein. The number and locations of antigenic sites which are altered by a structural change should provide clues to the scope and nature of that change.

Some important protein pharmaceuticals, such as tissue plasminogen activator and factor VIII, are enzymes. Alteration of active site features of such enzymes in order to influence catalytic parameters may produce useful derivatives.

IV. ENGINEERING OF PHARMACEUTICALLY IMPORTANT POLYPEPTIDES

A. INSULIN

As the first protein to be used in a highly purified state as a major pharmaceutical, it is not surprising that insulin has been the subject of structure/function studies which might serve as a model for contemporary protein engineering experiments. With the determination of the three-dimensional structure of insulin (26) it became possible to try to deduce correlations in the sequences and properties of natural variants of the molecule isolated from different species; although many of these variants differ from each other at more than one position in the sequence, the locations in space (interior vs. exterior, solvated surface vs. subunit interface) of particular residues can be invoked in arguments for the relative importance of particular replacements (27). Many chemical analogs containing specific sequence changes have also been synthesized. Since insulin is composed of two relatively small polypeptide chains which can be reassembled via disulfide bond formation _in vitro_, it is possible to make derivatives by synthetic or semi-synthetic approaches (which are not possible with most larger polypeptides) (28). Enzymatic semi-synthesis (29) has also been used to generate modified insulins. A third approach is that of chemical modification of natural material (30). Recently it has become possible to characterize the insulin of diabetics; this has provided the amino acid sequences of naturally occurring mutants of insulin possessing reduced biological activity (31). Studies such as those listed above have helped to define a bioactive center of the molecule (30,32). The mechanisms by which replacements here and elsewhere affect receptor binding (direct vs. indirect, conformational effects), the conformational restrictions on the important residues, and any secondary functions (after receptor binding per se) on the molecule have not been elucidated.

Insulin is synthesized in the pancreas as preproinsulin. After the signal peptide is removed during secretion, proinsulin, in humans an 86-amino acid polypeptide, folds into a unique three-dimensional structure containing three disulfide bonds. Subsequent to this folding process, the C-peptide is enzymatically excised from the interior of the proinsulin sequence to generate insulin. Proinsulin is interesting in terms of protein folding as well as insulin structure/function relationships. The interior C-peptide must directly mask or indirectly affect formation of the insulin receptor binding site, since the presence of the C-peptide in proinsulin reduces biological activity (33). Despite the relatively long sequence of the C-peptide, much shorter peptides connecting the B and A insulin peptides allow proper folding. This was shown first by chemical crosslinking experiments of Busse and Carpenter (34) and later by genetically engineering a gene for a "mini-C proinsulin" (9). Molecules such as the latter, containing C peptides of different lengths and compositions, might be useful in helping to define the structures behind insulin's biological effects, and may also themselves be of some therapeutic value.

B. SOMATOSTATIN

Determination of X-ray crystal structures can be more difficult for peptides than for proteins. In addition, most peptides in solution tend to be more disordered than globular proteins. Thus, even if a structure is available, there is no guarantee that it will in any way resemble the receptor-binding conformation of the peptide. Although it is relatively straightforward to chemically synthesize simple analogs of a peptide hormone, the lack of information about biologically relevant structure(s) can limit interpretation of the biological effects of such analogs.

Because the 14-amino acid long peptide hormone somatostatin contains a disulfide bond between residues 3 and 14, the number of conformations it can achieve in solution is dramatically reduced compared to an uncrosslinked peptide of similar length. Given this head start, several groups have proposed models for the solution conformation of somatostatin (35,36). The model of Hirschmann and coworkers placed pairs of discontinuous residues (7 and 10; 6 and 11; 5 and 12) close in space in a ß-pleated sheet structure bounded on one side by the disulfide and on the other by a ß turn of residues 7-10. To test their model, the Merck group synthesized modified cyclic and bicyclic "carba" analogs of somatostatin. They found that residues 1-5 and 12-14 could be removed, and replaced by a short hydrocarbon chain, without impairing biological activity (36,37). Within this structure, they found they could replace phenylalanines 6 and 11 with a disulfide (generating a bicyclic derivative) and again retain activity. This suggested that (1) the two replaced phenylalanines are important in the natural structure for maintaining the tight 7-10 ß turn through a hydrophobic interaction, and are not involved directly in receptor binding, and (2) that only residues 7-10 are, at most, involved in the receptor interaction. Subsequently, the synthesis of many analogs with replacements at residues 7-10 confirmed the special importance of an aromatic side chain at position 7(Phe), and of Trp8 and Lys9 (38). These results suggest that even more abbreviated structures containing little more than the Phe, Trp and Lys side chains, held in the proper conformations, might also be biologically active.

These results are important for a number of reasons. They demonstrate the potential of the protein engineering approach, even when a crystal structure is unavailable. They clearly show that a large percentage of a polypeptide hormone's structure can be whittled away without jeopardizing biological activity. Some of these somatostatin analogs exhibit enhanced stability toward proteolytic damage *in vitro* and to metabolic clearance *in vivo* (37). One analog also exhibits oral activity (39), which native somatostatin lacks. These results illustrate how one might be able to improve the pharmaceutical properties of a natural polypeptide with analogs designed from structure/function data.

C. ALPHA AND BETA INTERFERONS

Only with the molecular cloning of interferon coding sequences was the primary amino acid sequence of any interferon elucidated; only with the expression of native alpha interferon protein in *E. coli* was the investigation of important structural features such as its disulfide arrangement and alpha helix content possible (40). Despite the availability of gram amounts of human beta interferon and several subtypes of human alpha interferons, no three-dimensional structures have been determined for these proteins. Nonetheless, several groups have attempted protein engineering of this system.

The human alpha interferons are a group of at least fifteen highly homologous polypeptides (40). Independently expressed in *E. coli*, many of these can be shown to exhibit unique profiles of antiviral activity when assessed in cell lines from different tissues (41). Early attempts on structure/function studies involved the construction of hybrid interferon proteins, by cutting and splicing the genes for two related interferons at common restriction sites, and comparison of the antiviral activities of the two parents and two daughters on a number of cell types and viruses (10,11). The results of a number of such efforts, involving the generation of many analogs, are quite difficult to interpret according to any simple models. One problem is the lack of a 3-dimensional structure. Another equally serious problem is that all of these analogs differ from all other interferons by multiple replacements; even with a structure, results on such a series might be difficult to interpret. It is likely that the wide variety of effects seen with the derivatives prepared so far are obtained from a mixture of direct (receptor binding site) and indirect (remote changes transduced by intervening structure to the receptor binding site) effects. In a related series of experiments, the Amgen group has designed and synthesized "consensus" interferons based on an analysis of sub-groups within the alpha interferon family of sequences (42). These derivatives also exhibit altered activities in various biological assays, and are equally difficult to interpret from a structural point of view.

In one case it has been possible to produce a substantial effect in the absence of detailed structural information. Human beta interferon derived from tissue culture is not particularly unstable, but the expression of the beta interferon gene in _E. coli_ leads to predominantly inactive, poorly behaved material, and what little active material can be extracted in such a system has very poor stability. In contrast to the alpha interferon series (where stability is by and large good), beta interferon possesses both a disulfide and a free cysteine residue. Since there was evidence for formation of incorrect disulfides in this system, and since thiol/disulfide interchange is known to be a cause of microheterogeneity among proteins possessing both disulfides and a free cysteine (43), Mark _et al_. (44) used directed mutagenesis to replace the lone cysteine in the cloned gene product with a serine; expression of this variant reportedly led to a higher yield of a more stable product.

D. IMMUNOGLOBULINS AND IMMUNOTOXINS

With the introduction of hybridoma techniques for isolation of monoclonal lines of antibody-producing cells, it has become possible to harness the power of the immune response to generate banks of antibodies with a range of affinities toward an antigen. Besides the importance of these methods for generating specific antibodies as research and commercial tools, they also make possible a powerful system for investigation of protein-protein interactions such as those which drive binding of ligands to receptors. Immunoglobulins are also of great potential in clinical diagnostics and therapeutics. Because of their compartmentalized, multi-domain structures, they offer an opportunity for the protein engineering approach to eliminate some functions as well as to add other functions at a site remote from the binding domain(s).

Both aspects of immunoglobulins have made them a target for heterologous expression and protein engineering. Because of their multi-chain structures which seem to demand special conditions *in vitro* and *in vivo* for proper folding, there has been only limited success in producing cloned immunoglobulins in *E. coli* (45,46) and yeast (47). Even so, the feasibility of producing fragments such as Fab in *E. coli* has been demonstrated (45). Since fragments like Fab and Fv (48) preserve the antigen binding site but eliminate the bulk of the molecule including other functions (such as binding to complement), such structures might be useful therapeutically both in their own right and as components of hybrids with additional functions. In addition, such hybrids may make important contributions to the development of clinically important immunodiagnostic techniques.

Immunotoxins, molecules which contain a cytotoxic element in addition to a cell-surface-directed antigen binding site, are one class of hybrids of potential therapeutic use (49; see articles elsewhere in this volume). Another is the recently described conjungate of urokinase and an anti-fibrin antibody (50). These molecules are now produced by chemically linking the immunoglobulin to the toxin or urokinase molecule. It may also be possible to synthesize such hybrid proteins by expression of fused genes or gene fragments.

The recent expression of cloned Ig genes in mammalian cell culture illustrates the feasibility of producing various modified immunoglobulins. In one demonstration project a hybrid protein was made combining active antibody Fab arms with active Staphylococcal nuclease (14). This suggests a route to immunotoxins possessing better defined, more reproducible structures.

Hybrid proteins composed of specific antibodies and some other active domain(s) are representative of a large group of potential pharmaceutical agents composed of a targeting domain and a domain capable of altering a biological structure. In principle, such proteins will be most conveniently and reproducibly manufactured by expression of hybrid genes designed by a protein engineering approach.

D. ALPHA-1 ANTITRYPSIN

This 394-amino-acid protein is a potent inhibitor of neutrophil elastase, a protease capable of catalyzing the breakdown of connective tissue (review, ref. 51). Individuals whose serum levels of alpha-1 antitrypsin are low are likely to develop early-onset emphysema. Because of this link, it seems possible that *in vivo* inactivation of normal levels of the protease inhibitor might also lead to this disease state. Since oxidants generated in cigarette smoke can inactivate alpha-1 antitrypsin *in vitro*, it further seems possible that the high incidence of emphysema among heavy smokers might be attributed to this effect.

It has been shown that oxidation of Met358, located at the P1 subsite for binding to serine proteases, leads to dramatic reduction in the protein's ability to inhibit elastase (52). Based on this data, Courtney *et al*. (53) and Rosenberg *et al*.(54) designed an oxidatively stable antitrypsin. Both groups replaced Met358 with a Val residue, making this choice because studies on the binding of peptides by elastases indicated a preference for Val over Met at the P1 site (51). The variants, produced in yeast (54) or *E. coli* (53), were both active inhibitors. Quantitative studies on the *E. coli* Met358 to Val variant showed it to be as good an inhibitor as the natural against human neutrophil elastase and somewhat better than natural against porcine pancreatic elastase. In addition, the variant was much less sensitive to oxidation by N-chlorosuccinimide than was the natural inhibitor. Similar results were obtained for the yeast material.

Recent experiments suggest that treatment of emphysemics with intravenous injections of alpha-1 antitrypsin may be a useful therapy (55). Expression systems such as those referred to here may make available enough material to support such therapy; furthermore, oxidatively stable variants like Val358 might make good substitutes or adjuncts to such therapy.

If a point mutation can convert antitrypsin to a more oxidatively stable molecule, why doesn't wild type antitrypsin contain some other residue than Met at position 358? Clearly, the Val358 protein, for example, retains good anti-elastase activity. It has been suggested that the oxidative sensitivity of natural alpha-1 antitrypsin may have a regulatory role (51), promoting tissue breakdown at sites of inflammation, where macrophages release locally high concentrations of oxygen radicals. If true, an individual producing only an oxidatively stable antitrypsin would lack this regulatory function. By the same token, therapies utilizing stable variants might be compromised by the same insensitivity to regulation. This potential pitfall is illustrative of the difficulties which may be encountered in attempts to impact highly evolved, highly complex mammalian humoural systems with protein engineered drugs.

F. SYNTHETIC VACCINES

In almost all cases, a primary criterion for the success of an engineered therapeutic will be the lack of significant immunogenicity (see above). One exception is the area of synthetic vaccines, where the efficient induction of neutralizing antibodies is the primary goal. Synthetic vaccines are an attractive alternative to traditional vaccines comprised of inactivated viruses. Not only will their production be less dangerous, not involving large scale culturing of the virus itself, but also the vaccine product should be more safe, since it cannot be contaminated by virulent particles or factors. These advantages are not gained easily, however. The principal difficulty in the preparation of synthetic vaccines is the construction of a surface which mimics the immunogenic regions of the natural virus.

Protein engineering can facilitate the correct presentation of the immunogenic region. Peptide components of viral coat proteins synthesized as part of hybrid proteins by recombinant DNA means have been shown to be antigenic, immunogenic, and to elicit a protective response against Foot-and-Mouth disease (56) and other viruses. Since these hybrids are new proteins whose folding patterns at present cannot be predicted, there is room for much improvement in the use of protein engineering in the design of vaccines. One recent approach is the use of a well-defined, regular particle as a matrix for presentation of an immunogenic polypeptide. Valenzuela _et al_. (57) constructed a gene encoding a single polypeptide composed of the hepatitis B surface antigen and the herpes simplex virus surface antigen. Expression of this gene in yeast generates particles reminiscent of Hepatitis B "Dane" particles but recognizable by anti-herpes surface antigen antibodies. There are no reports as to either the immunogenicity of the hybrid particles, or the neutralizing ability of any antibodies to Herpes Simplex virus which might be generated.

V. ENGINEERING OF OTHER MEDICALLY IMPORTANT PROTEINS

The elucidation of structure/function relationships of some medically important proteins might lead indirectly toward new pharmaceuticals. For example, in the future it may be possible to design synthetic ligands based on a knowledge of the three dimensional structure of a binding site such as a receptor surface. Such ligands might mimic, or interfere with, the normal interaction of a hormone with its receptor, or a substrate with an enzyme. For example, a knowledge of the structure of the aspartic proteinase renin might allow the design of improved inhibitors of this angiotensinogen processing enzyme (58). Knowing the binding surface, it should be possible to use computer graphics analysis to devise ligands with improved binding constants. Such an approach has proved successful in rationalizing and predicting the binding behaviour of various thyroxine derivatives to prealbumin (59).

While some progress has been made at elucidating the receptor binding site of insulin (see above), design of novel insulin analogs may not be possible until the structure of the receptor is known. This has recently become more feasible with the cloning and sequencing of the insulin receptor gene (60). Since the insulin receptor is a multidomain membrane bound protein, and since its efficient synthesis in a transformed cell line may prove difficult, it may be useful to turn to protein engineering methods to produce the extracellular insulin binding domain of the receptor. Compared to the intact receptor, such fragments may be simpler to produce in cell culture, as well as easier to crystallize.

Protein engineering studies on receptors (61) may prove valuable in other ways. Elucidation of signal transduction mechanisms may facilitate the design of drugs which bypass receptor binding to act more directly in inhibiting or triggering the intracellular response. In addition, since at least several oncogenes appear to be related to receptors, studies of mechanisms of signal transduction may lead to a better understanding of molecular mechanisms of oncogenesis.

In fact, protein engineering has been used more directly to study the mutagenic activation of proto-oncogenes. Levinson and colleagues produced a series of c-Ha-rasl oncogenes containing each of the 20 possible amino acids at position 12 of the gene product. They found that 18 of the 20 produce transformed phenotypes, while only those containing glycine (the amino acid in the normal cellular gene) and proline are inactive. This trend was interpreted as suggesting that the transforming character of the protein *ras* product was dependant on the existence of an alpha helix containing residue 12 (62).

VI. MODEL STUDIES

The complexity of biological systems will always make the use of protein engineering techniques in the design of pharmaceuticals an especially challenging undertaking. This field suffers at present from a more immediate problem: the lack of information on the proteins involved. For many of these, it has only been a few years since we have had amino acid sequence data (provided, in most cases, by cDNA cloning). Heterologous expression of cloned cDNA has made some of these proteins available in reasonable amounts and purity. For protein engineering efforts to be efficient, we should like to have available a three-dimensional structure, detailed characterization of solution properties, bioassays, receptor assays, immunochemical criteria for the native structure, thermostability data, selection or screening methods, etc. [For protein engineering efforts to be properly focused, a clinical experience of some depth would also be desirable.] While fascinating insights into structure/function relationships have come from studies based on little more than a knowledge of the amino acid sequence, too often such efforts have led to many variants, much data, and no progress. Until some of these systems are further developed, important general features of protein structure/function relationships can be gleaned from protein engineering studies on well-studied, relatively well-understood model systems which may have little or no medicinal importance in themselves. This section is a selective review of such studies.

A. THERMOSTABILITY

The measurement of the thermodynamic stability of a protein, that is, the analysis of the energy requirements for the reversible unfolding of a protein, is of fundamental importance for a number of reasons. Such analyses, and the kinetic experiments that grow out of them, are important tools for elucidating how amino acid sequence determines structure (protein folding). In addition, this information is useful in understanding how proteins achieve stability toward various environmental stresses, in which inactivation often involves protein unfolding followed by the operation of irreversible processes. Finally, information on the energy contributions of amino acid contacts within a single folded polypeptide will be of immense value in better understanding the energetics of protein-protein interactions such as receptor-ligand and antibody-antigen binding.

Several groups have taken a random mutagenesis approach to the analysis of how sequence influences thermostability. Using the power of classical microbial genetics coupled with equally powerful new methods for generating random mutations, these groups have quickly collected substantial new information on Staphylococcal nuclease (63), T4 lysozyme (19,64,65) and lambda repressor (66). These methods rely on genetic screens or selections designed to probe for increased thermostability.

Using a plate assay for lytic competance of T4 phage mutants, a series of T4 lysozyme temperature sensitive mutants has been isolated and characterized (64). More recently, Alber and Wozniak devised a scheme for screening for lysozyme variants of increased stability, after mutagenizing the lysozyme gene in an expression vector (65). Such methods have generated a series of single site variants of altered thermostability. The 3-dimensional structures (19) and detailed folding thermodynamics (64) of some of these have been studied in efforts to elucidate the molecular basis of thermostability. At present, it appears that the resolution of X-ray crystallography, and/or our ideas about the quantitation of binding energies, are not sophisticated enough to account for many observed effects. One approach to better understanding the effects of some mutations is to investigate the properties of a series of variants containing amino acid replacements at the site of interest (19).

B. ROLES OF DISULFIDE BONDS

Many pharmaceutically important proteins contain disulfide bonds. These bonds are felt to provide extra stability to proteins which must function in an extracellular milieu which might be more stressful than the cytoplasm. Although it is well-documented that intramolecular crosslinks can supply a small amount of free energy to the global stabilization energy of a protein, this energy could in most cases easily be supplied by other sorts of stabilizing interactions. In fact many globular proteins function well without the benefit of disulfides. The question arises as to whether these bonds might not provide other benefits to some proteins.

This possibility is being addressed in a series of experiments involving the introduction of disulfides by directed mutagenesis into a protein which is not crosslinked in nature. The subject of these studies, T4 lysozyme, was chosen for a number of reasons (67), including its relatively low stability toward irreversible thermal inactivation. Replacement of Ile3 with Cys makes possible disulfide formation with the wild type residue Cys97; this crosslinked derivative is essentially fully active and is significantly more stable than the wild type against irreversible thermal inactivation (67). Removal of an unpaired Cys at position 54 further stabilizes disulfide-crosslinked T4 lysozyme (68). Studies of such crosslinked, double variants indicate that the disulfide radically changes the mechanism by which T4 lysozyme undergoes thermal inactivation, in a way which is not consistent with a simple thermodynamic contribution as discussed above (69). This suggests that disulfides can stabilize proteins by mechanisms involving changes in folding and unfolding pathways, and that the effects can be substantial.

While the 3-97 disulfide stabilizes T4 lysozyme against incubation in the range of 60-85°C, the wild type and the variant are both stable in the 5-60°C range. Do such studies then have any bearing on the role of disulfides under physiological conditions? This is a question being addressed by further experiments in this series, which involve the introduction of temperature sensitive loci into disulfide-crosslinked variants (69).

C. REPLACEMENT GENERALIZATIONS

Amino acid replacements designed into proteins may be intended to alter some properties of the protein while preserving others. Ideally the design will rely on an x-ray crystal structure and a refined understanding of the relationship between the local structure and the properties of interest. Since we cannot yet rationalize some effects even with the help of a refined crystal structure, there may be some value in an empirical approach. Are there particular amino acid residues which, in general, can be used to replace other residues with the same end result on the properties of proteins? One can look to nature for some guidance here. But while there are clear trends in the observed replacement frequencies within homologous families (70), the trends are almost certainly biased by differences in mutational frequencies between codons that differ in one, two or three nucleotides.

As an example of the problem, one can ask what residue might best replace cysteine; cysteine replacements in the absence of structural information can be useful diagnostic devices since it may be possible to uncover disulfide bonding relationships in this way (44,61,71). The residue replacement that is routinely used in such studies is serine for cysteine, since their side chains differ in only a sulfur to oxygen substitution. Yet the properties of these two side chains may be significantly different. Sulfur is much less electronegative than oxygen, and as such will be more comfortable than oxygen in a hydrophobic core. For the same reason thiols are less likely to be found in hydrogen bonds than hydroxyls. The fact that many protein thiols are resistant to modification suggests that they are in fact often buried in hydrophobic pockets. The point here is that it may be dangerous to interpret the loss of activity of a protein variant harboring a Cys to Ser replacement as evidence for the involvement of the wild type Cys in a disulfide. Replacement of an unpaired Cys might inactivate a protein by other mechanisms, either because the Cys is specifically required for structure and function, or because it is involved in structure which is incompatible with the more electronegative, hydrogen bonding side chain of Ser.

The other sulfur-containing side chain, methionine, may also be misunderstood. The replacement of Met has become increasingly important because of the oxidative stability it may impart (16,53,54). The substitutions chosen, whether or not guided by structural information, tend to be large hydrophobic side chains. Given the repertoire of replacements available in biosynthesis, these may be the best one can do; but sulfur is considerably different from carbon. In one set of experiments on subtilisin, Estell, Wells and co-workers carried out a systematic replacement of Met222 with all possible amino acids (16,10). Although they found many replacements which exhibited good retention of specific activity and improved oxidative stability, the best replacement for retention of activity, at least with some substrates, was Cys. At the same time, hydrophobic, non-aromatic residues such as Ile and Leu were among the poorest replacements. The availability of methods for the saturation mutagenesis of particular sites should facilitate the generation of data bases which, with or without further structural information, may provide some general rules for making substitutions.

D. ENZYME STRUCTURE/ACTIVITY RELATIONSHIPS

Some potential pharmaceuticals, notably those involved in blood clotting and fibrinolysis, are enzymes; in other cases enzymes such as angiotensin-converting enzyme are of interest as potential targets of activity-modulating agents. While the most straightforward path to the synthesis of a better tissue derived plasminogen activator (t-PA) is the study of t-PA structure/function relationships, progress with particular enzymes will proceed more rapidly the more we know about enzymes in general. Perhaps of most relevance to both classes mentioned above are other proteases. Because of considerable interest in subtilisin as a bulk industrial enzyme, studies on this molecule have been solidly supported in industry and many interesting results are emerging regarding protease specificity, transition state stabilization, and stability (19). Pancreatic proteases like trypsin (72) are also being studied in terms of molecular mechanisms of protease specificity.

The most extensively studied enzyme by protein-engineering methods is *B. stearothermophilis* Tyr-tRNA synthetase. Examination of the structure of the enzyme, co-crystallized with the aminoacyl adenylate intermediate bound, reveals a hydrogen bonding network holding the intermediate in place. Many of these H-bonds can be broken by relatively minor replacements of amino acid side chains. By systematically breaking these H-bonds in a series of variants, Fersht, Winter and colleagues have explored the roles of H bonds in providing binding energy for substrate specificity (73) as well as for transition state stabilization and catalysis (74). Other important studies on this enzyme currently underway include an investigation of subunit interactions as they relate to stability of the symmetric dimer and "half-of-the-sites" reactivity, and a characterization of the tRNA binding site (which is transparent in the crystal structure because of temporal or spatial disorder in the protein fold).

VII. CONCLUSION

Like all proteins, medically important molecules such as hormones, serum proteinases and receptors have evolved to exist in complex environments and to serve complex functions. It is as if they have been constructed by a committee forced to strike manageable compromises between retention of structure and demands for tissue specificity, receptor binding, signal transduction, serum lifetime, suppressed immunogenicity and other requirements. In some therapies the complete loss or modulation of a subset of these functions may be acceptable, when weighed against the benefits from enhancement of other functions. In some cases, the selective elimination of activities, such as those responsible for side effects, may actually be desirable. Protein engineering offers the possibility of editing nature's molecular compromises for improved therapeutic value. Protein pharmaceuticals tailored for altered pharmacokinetics, alternate routes of administration, elimination of side effects, or addition of new properties may be important improvements over natural structures.

The complexities of protein molecules may also create problems for protein engineers. Even though the structure responsible for a particular side effect may be identified, it may still be difficult to neutralize the effect without either losing desirable functions, such as serum lifetime or receptor binding, or generating new undesirable functions, such as immunogenicity or conformational instability.

On the other hand, it may not prove so difficult to engineer the properties of globular proteins in a controlled manner. Even if indirect effects occur, they may be tolerable in some engineered drugs if the pharmaceutical provides a unique therapy against a serious disease. Whether or not the limited redesign and engineering of nature's structures proves a viable approach, protein engineering can facilitate drug design in another way: by helping to elucidate structure-function relationships behind the action of nature's molecules, it may be possible to radically redesign them to yield novel, chemically synthesized structures which mimic natural interactions while possessing improved pharmacologic properties.

ACKNOWLEDGMENTS

I gratefully acknowledge Wayne Anstine for preparation of the figures and manuscript, and thank Tony Kossiakoff, J. Ramachandran, Jim Wells, Rodney Pearlman and Hugh Niall for critical reading of the manuscript.

REFERENCES

1. W. H. Rastetter, Enzyme engineering: applications and promise, Trends Biotech. 1: 80 (1983).

2. M. Inouye and R. Sarma, eds.: Protein Engineering New York, Academic Press, in press.

3. D. Oxender, ed.: Protein Modification and Design. New York Alan R. Liss, Inc., in press.

4. R. Wetzel and D. V. Goeddel, Synthesis of polypeptides by recombinant DNA methods, The Peptides (E. Gross and J. Meienhofer, eds.). Academic Press, New York, p. 1 (1983).

5. Harris, T. J. R., Expression of eucaryotic genes in E. coli, Genetic Engineering (R. Williamson, ed.), Vol. 4. Academic Press, New York, p. 127 (1983).

6. L. Band and D. J. Henner, Bacillus subtilis requires a "stringent" Shine-Dalgarno region for gene expression, DNA 3: 17 (1984)

7. R. A. Hitzeman, C. Y. Chen, F. E. Hagie, J. M. Lugovoy, and A. Singh, Yeast: An alternative organism for foreign protein production, in: Recombinant DNA Products: Insulin, Interferon and Growth Hormone (A. P. Bollen, ed.). CRC Press, Baca Raton, FL, p.47 (1984).

8. R. Kucherlpati and A. I. Skoultchi, Introduction of purified genes into mammalian cells, CRC Critical Reviews in Biochemistry, Vol. 16, #4:349 (1985).

9. R. Wetzel, D. G. Kleid, R. Crea, H. L. Heyneker, D. G. Yansura, T. Hirose, A. Kraszewski, A. D. Riggs, K. Itakura, and D. V. Goeddel, Expression in Escherichia coli of a chemically synthesized gene for a "mini-C" analog of human proinsulin, Gene 16: 63 (1981).

10. M. Streuli, A. Hall, W. Boll, W. E. Stewart II, S. Nagata, and C. Weissman, Target cell specificity of two species of human interferon-alpha produced in Escherichia coli and of hybrid molecules derived from them, Proc. Natl. Acad. Sci. (USA) 78: 2848 (1981).

11. P. K. Weck, S. Apperson, N. Stebbing, P. W. Gray, D. Leung, H. M. Shepard, and D. V. Goeddel, Antiviral activities of hybrids of two major human leukocyte interferons, Nucleic Acids Res. 9: 6153 (1981).

12. M. M. Y. Waye, G. Winter, A. J. Wilkinson, and A. R. Fersht, Deletion mutagenesis using an "M13 splint": the N-terminal structural domain of tyrosyl-tRNA synthetase (B. stearothermophilis) catalyses the formation of tyrosyl adenylate, EMBO J. 2: 1827 (1983).

13. A. E. Franke, H. M. Shepard, C. M., Houck, D. Leung, D. V. Goeddel, and R. M. Lawn, Carboxyterminal region of hybrid leukocyte interferons affects antiviral specificity, DNA 1: 223 (1982).

14. M. S. Neuberger, G. T., Williams, and R. O. Fox, Recombinant antibodies possessing novel effector functions, Nature 312: 604 (1984).

15. L. Bulow, P. Ljungcrantz, and K. Mosbach, Preparation of a soluble bifunctional enzyme by gene fusion, Bio/technology 3: 821 (1985).

16. D. A. Estell, T. P. Graycar, and J. A. Wells, Engineering an enzyme by site-directed mutagenesis to be resistant to chemical oxidation, J. Biol. Chem. 260: 6518 (1985).

17. J. A. Wells, M. Vasser, and D. B. Powers, Cassette mutagenesis: An efficient method for generation of multiple mutations at defined sites, Gene 34: 315 (1985).

18. J. A. Wells, D. B. Powers, R. R. Bott, B. A. Katz, M. H. Ultsch, A. A. Kossiakoff, S. D. Power, R. M. Adams, H. H. Heyneker, B. C. Cunningham, J. V. Miller, T. P. Graycar, and D. A. Estell, Protein engineering of subtilisin in Protein Modification and Design (D. Oxender, ed.). Alan R. Liss, New York (1986). In press.

19. T. Alber, M. G. Gruetter, T. M. Gray, J. A. Wozniak, L. H. Weaver, B.-L. Chen, E. N. Baker, and B. W. Matthews, Structure and stability of mutant lysozymes from bacteriophage T4, Protein Modification and Design (D. Oxender, ed.). Alan R. Liss, New York (1986), in press.

20. D. Botstein and D. Shortle, Strategies and applications of in vitro mutagenesis, Science 229: 1193 (1985).

21. R. M. Myers, L. S. Lerman, and T. Maniatis, A general method for saturation mutagenesis of cloned DNA fragments, Science 229: 242 (1985).

22. H. Liao, T. McKenzie, and R. Hageman, Isolation of a thermostable enzyme variant by cloning and selection in a thermophile, Proc. Natl. Acad. Sci. (USA) 82 (1985), in press.

23. C. Ho, M. Jasin, and P. Schimmel, Amino acid replacements that compensate for a large polypeptide deletion in an enzyme, Science 229: 389 (1985).

24. G. Ashwell and J. Harford, Carbohydrate-specific receptors of the liver, Ann. Revs. Biochem. 51: 531 (1982).

25. P. W. Schiller, Conformational analysis of eukephalin and conformation-activity relationships, The Peptides, Vol. 6 (S. Underfriend and J. Meienhofer, eds.), Academic Press, New York, p.219 (1984).

26. T. L. Blundell, J. F. Cutfield, S. M. Cutfield, E. J. Dodson, G. G. Dodson, D. C. Hodgkin, D. A. Mercola, and M. Vijayan, Atomic positions in rhombohedral 2-zinc insulin crystals, Nature 231: 506 (1971).

27. T. L. Blundell and S. P. Wood, Is the evolution of insulin Darwinian or due to selectively neutral mutation?, Nature 257: 197 (1975).

28. G. Schwartz and P. G. Katsoyannis, Synthesis of des (tetrapeptide B^{1-4}) and des (pentapeptide B^{1-5}) human

insulins. Two biologically active analogues, Biochemistry 17: 4550 (1978).

29. J. S. Fruton, Proteinase - catalyzed synthesis of peptide bonds, Advs. Enzymol. 53: 239 (1982).

30. R. A. Pullen, D. G. Lindsay, S. P. Wood, I. J. Tickle, T. L. Blundell, A. Wollmer, G. Krail, D. Brandenburg, H. Zahn, J. Glieman, and S. Gammeltoft, Receptor-binding region of insulin, Nature 259: 369 (1976).

31. M. Haneda, S. J. Chan, S. C. M. Kwok, A. H. Rubenstein, and D. F. Steiner, Studies on mutant human insulin genes: Identification and sequence analysis of a gene encoding [Ser-B24]insulin, Proc. Natl. Acad. Sci. (USA) 80: 6366 (1983).

32. C. R. Kahn, What is the molecular basis for the action of insulin?, Trends Biochem. Sci. 4: 263 (1979).

33. S. S. Yu and A. E. Kitabchi, Biological activity of proinsulin and related polypeptides in the fat tissue, J. Biol. Chem. 248: 3753 (1973).

34. W. -D. Busse, S. R. Hansen, and F. H. Carpenter, Carbonylbis(L-methionine p-nitrophenyl ester). A new reagent for the reversible intramolecular cross-linking of insulin, J. Amer. Chem. Soc. 96: 5947 (1974).

35. L. A. Holladay, and D. Puett, Somatostatin conformation: Evidence for a stable intramolecular structure from circular dichroism, diffusion, and sedimentation equilibrium, Proc. Natl. Acad. Sci. (USA) 73: 1199 (1976).

36. D. F. Veber, F. W. Holly, W. J. Paleveda, R. F. Nutt, S. J. Bergstrand, M. Torchiana, M. S. Glitzer, R. Saperstein, and R. Hirschmann, Conformationally restricted bicyclic analogs of somatostatin, Proc. Natl. Acad. Sci. (USA) 75: 2636 (1978).

37. D. F. Veber, F. W. Holly, R. F. Nutt, S. J. Bergstrand, S. F. Brady, R. Hirschmann, M. S. Glitzer, and R. Saperstein, Highly active cyclic and bicyclic somatostatin analogues of reduced ring size, Nature 280: 512 (1979).

38. S. F. Brady, R. R. Nutt, F. W. Holly, W. J. Paleveda, R. G. Strachan, S. J. Bergstrand, D. F. Veber, and R. Saperstein, Synthesis and biological activity of somatostatin analogs of reduced ring size, Proceedings of the Seventh American Symposium (D. H. Rich and E. Gross, eds.), Pierce Chemical Co., Rockville, Ill, p. 653 (1981).

39. D. F. Veber, R. M. Freidinger, D. S. Perlow, W. J. Paleveda, F. W. Holly, R. G. Strachan, R. F. Nutt, B. H. Arison, C. Homnick, W. C. Randall, M. S. Clitzer, R. Saperstein, and R. Hirschmann, A potent cyclic hexapeptide analogue of somatostatin, Nature 292: 55 (1981).

40. K. C. Zoon and R. Wetzel, Comparative structures of mammalian interferons, Handbook of Experimental Pharmacology Vol. 71 (Came, P. E. and Carter, W. A., eds.). Springer-Verlag, Heidelberg, p. 79 (1984).

41. P. K. Weck, S. Apperson, L. May, and N. Stebbing, Comparison of the antiviral activities of various cloned human interferon-alpha subtypes in mammalian cell cultures, J. Gen. Virol. 57: 233 (1981).

42. K. Alton, Y. Stabinsky, R. Richards, B. Ferguson, L. Goldstein, B. Altrock, L. Miller, and N. Stebbing, Production, characterization and biological effects of recombinant DNA derived human IFN-alpha and IFN-gamma analogs, The Biology of the Interferon System 1983 (E. De Maeyer and H Schellekens, eds.). Elsevier, Amsterdam, p. 119 (1983).

43. H. A. McKenzie, G. B. Ralston, and D. C. Shaw, Location of sulfhydryl and disulfide groups in bovine beta-lactoglobulins and effects of urea, Biochemistry 11: 4539 (1972).

44. D. F. Mark, S. D. Lu, A. A. Creasey, R. Yamamoto, and L. S. Lin, Site-specific mutagenesis of the human fibroblast interferon gene, Proc. Natl. Acad. Sci. (USA) 81: 5662 (1984).

45. S. Cabilly, A. D. Riggs, H. Pande, J. E. Shively, W. E. Holmes, M. Rey, L. J. Perry, R. Wetzel, and H. L. Heyneker Generation of antibody activity from immunoglobulin polypeptide chains produced in Escherichia coli, Proc. Natl. Acad. Sci. (USA) 81: 3273 (1984).

46. M. A. Boss, J. H. Kenten, C. R. Wood, and J. S. Emtage, Assembly of functional antibodies from immunoglobulin heavy and light chains synthesized in E. coli, Nucleic Acids Res. 12: 3791 (1984).

47. C. R. Wood, M. A. Boss, J. H. Kenten, J. E. Calvert, N. A. Roberts, and J. S. Emtage, The synthesis and in vivo assembly of functional antibodies in yeast, Nature 314: 446 (1985).

48. J. Hochman, D. Inbar and D. Givol, An active antibody fragment (Fv) composed of the variable portions of heavy and light chains, Biochemistry 12: 1130 (1973).

49. J. M. Lord, L. M. Roberts, P. E. Thorpe and E. S. Vitetta, Immunotoxins, Trends in Biotech. 3: 175 (1985).

50. C. Bode, G. R. Matsueda, K. Y. Hui, and E. Haber, Antibody-directed urokinase: A specific fibrinolytic agent, Science 229: 765 (1985).

51. R. W. Carrell, J.-O. Jeppsson, C.-B. Laurell, S. O. Brennan, M. C. Owen, L. Vaughan, and D. R. Boswell, Structure and variation of human alpha-one antitrypsin, Nature 298: 329 (1982).

52. K. Beatty, J. Bieth, and J. Travis, Kinetics and association of serine proteinases with native and oxidized alpha-one proteinase inhibitor and alpha-one antichymotrypsin, J. Biol. Chem. 255: 3931 (1980).

53. M. Courtney, S. Jallat, L.-H. Tessier, A. Benavente, R. G. Crystal, and J. -P. Lecocq, Synthesis in E. coli of alpha-one antitrypsin variants of therapeutic potential for emphysema and thrombosis, Nature 313: 149 (1985).

54. S. Rosenberg, P. J. Barr, R. C. Najarian, and R. A. Hallewell, Synthesis in yeast of a functional oxidation-resistant mutant of human alpha-one antitrypsin, Nature 312: 77 (1984).

55. J. E. Gadek, H. G. Klein, P. V. Holland, and R. G. Crystal, Replacement therapy of alpha-one antitrypsin deficiency, J. Clin. Invest. 68: 1158 (1981).

56. D. G. Kleid, D. Yansura, B. Small, D. Dowbenko, D. M. Moore, M. J. Grubman, P. D. McKercher, D. O. Morgan, B. H. Robertson, and H. L. Bachrach, Cloned viral protein vaccine for foot-and-mouth disease: responses in cattle and swine, Science 214: 1125 (1981).

57. P. Valenzuela, D. Coit, M. A. Medina-Selby, C. H. Kuo, G. Van Nest, R. L. Burke, P. Bull, M. S. Urdea, and P. V. Graves, Antigen engineering in yeast: Synthesis and assembly of hybrid hepatitis B surface antigen-herpes simplex 1 gD particles, Bio/technology 3: 323 (1985).

58. B. L. Sibanda, T. Blundell, P. M. Hobart, M. Fogliano, J. S. Bindra, B. W. Dominy, and J. M. Chirgwin, Computer graphics modelling of human renin, FEBS Letts. 174: 102 (1984).

59. J. M. Blaney, E. C. Jorgensen, M. L. Connolly, T. E. Ferrin, Langridge, S. J. Oatley, J. M. Burridge, and C. C. F. Blake, Computer graphics in drug design: Molecular modeling of thyroid hormone-prealbumin interactions, J. Med. Chem. 25: 785 (1982).

60. A. Ullrich, J. R. Bell, E. Y. Chen, R. Herrera, L. M. Petruzzelli, T. J. Dull, A. Gray, L. Coussens, Y.-C. Liao, M. Tsubokawa, A. Mason, P. H. Seeburg, C. Grunfeld, O. M. Rosen, and J. Ramachandran, Human insulin receptor and its relationship to the tyrosine kinase family of oncogenes, Nature 313: 756 (1985).

61. M. Mishina, T. Tobimatsu, K. Imoto, K.-i. Tanaka, Y. Fujita, K. Fukuda, M. Kurasaki, H. Takahashi, Y. Morimoto, T. Hirose, S. Inayama, T. Takahashi, M. Kuno, and S. Numa, Location of functional regions of acetylcholone receptor alpha-subunit by site-directed mutagenesis, Nature 313, 364 (1985).

62. P. H. Seeburg, W. W. Colby, D. J. Capon, D. V. Goeddel, and A. D. Levinson, Biological properties of human c-Ha-ras1 genes mutated at codon 12, Nature 312, 71 (1984).

63. D. Shortle and B. Lin, Genetic analysis of Staphylococcal nuclease: Identification of three intragenic "global" suppressors of nuclease-minus mutations, Genetics 110: 539 (1985).

64. R. Hawkes, M. G. Gruetter, and J. Schellman, Thermodynamic stability and point mutations of bacteriophage T4 lysozyme, J. Mol. Biol. 175: 195 (1984).

65. T. Alber and J. A. Wozniak, A genetic screen for mutations that increase the thermal stability of phage T4 lysozyme, Proc. Natl. Acad. Sci. (USA) 82: 747 (1985).

66. M. H. Hecht, J. M. Sturtevant, and R. T. Sauer, Effect of single amino acid replacements on the thermal stability of the amino terminal domain of phage lambda repressor, Proc. Natl. Acad. Sci. (USA) 81: 5685 (1984).

67. L. J. Perry, and R. Wetzel, Disulfide bond engineered into T4 lysozyme: Stabilization of the protein toward thermal inactivation, Science 226: 555 (1984).

68. L. J. Perry, and R. Wetzel, Unpaired Cys54 interferes with the ability of an engineered disulfide to stabilize T4 lysozyme, Biochemistry, in press.

69. R. Wetzel, Investigation of the structural roles of disulfides by protein engineering; a study with T4 lysozyme, Protein Engineering (M. Inouye and R. Sarma, eds). Academic Press, New York (1986), in press.

70. Atlas of Protein Sequence and Structure (M. O. Dayhoff, ed.) National Biomedical Research Foundation, Silver Spring, MD, p. D-2, (1972).

71. T. Shiroishi, G. A. Evans, E. Appella, and K. Ozato, Role of a disulfide bridge in the immune function of major histocompatibility class I antigen as studied by in vitro mutagenesis, Proc. Natl. Acad. Sci. (USA) 81: 7544 (1984).

72. C. S. Craik, C. Largman, T. Fletcher, T. Roczniak, P. J. Barr, R. Fletterick, and W. J. Rutter, Redesigning trypsin: Alteration of substrate specificity, Science 228: 291 (1985).

73. A. R. Fersht, J. -P. Shi, J. Knill-Jones, D. M. Lowe, A. J. Wilkinson, D. M. Blow, P. Brick, P. Carter, M. M. Y. Waye, and G. Winter, Hydrogen bonding and biological specificity analysed by protein engineering, Nature 314: 235.

74. T. N. C. Wells and A. R. Fersht, Hydrogen bonding in enzymatic catalysis analyzed by protein engineering, Nature 316, 356 (1985).

8

Mechanism-Based Enzyme Inactivators for Medical Uses

Richard B. Silverman

Department of Chemistry and
Department of Biochemistry, Molecular Biology, and Cell Biology
Northwestern University
Evanston, Illinois

I. ENZYME INHIBITORS IN DRUG DESIGN

Many drugs in clinical use today are specific enzyme inhibitors. Enzyme inhibition is an important approach to drug design for several reasons. Once an enzyme is inhibited, its substrates cannot be converted to products, and this would be useful if a disease state results from a deficiency of one of the substrates, an excess of a product, or even from the formation of small amounts of the product which trigger particular physiological events. In a foreign organism, such as a bacterium or virus (or in a tumor cell), specific enzyme inhibition can shut down vital metabolic growth processes which may result in death of the undesirable organisms or tumor cell (chemotherapy).

A. *REVERSIBLE ENZYME INHIBITORS*

Most enzyme inhibitor drugs are reversible inhibitors, that is, their interactions with an enzyme are non-covalent and therefore,

$$\begin{array}{ccc} E + I & \rightleftharpoons & E\cdot I \\ {\scriptstyle +S}\downarrow\uparrow{\scriptstyle -S} & & \\ E\cdot S \rightleftharpoons E\cdot P \rightleftharpoons E + P & & \end{array}$$

FIG. 1 *Competition of substrate and inhibitor for an enzyme active site.*

LeChatelier's Principle is important to the stability of the enzyme-inhibitor complex. In order for these drugs to be effective, then, a steady state concentration of the drug must be present at all times. As the concentration of the drug diminishes, the equilibrium shifts to the left and free enzyme becomes available to catalyze the reaction with substrate (Fig. 1). In terms of drug administration, dosages must be taken periodically over an extended time interval to maintain an appropriate concentration of the drug inside the enzyme.

B. *IRREVERSIBLE ENZYME INHIBITORS*

An improved approach, theoretically at least, would be the use of an irreversible enzyme inhibitor (also termed inactivator) drug, one which forms a covalent bond to the target enzyme. This approach no longer would be governed by LeChatelier's Principle and, therefore, once the inactivator reacts with the enzyme, an event which may require only one inactivator molecule per enzyme molecule, it would not be necessary to maintain a concentration of the drug in the body. Of course, the gene encoding for the inactivated enzyme will produce new enzyme, so additional drug will be needed, but this process may take several hours or even days. So, why not always use an irreversible inactivator? Generally, irreversible inactivators are reactive compounds. Therefore, not only will they react with the target enzyme, but they also can react with other enzymes and macromolecules. These non-specific reactions lead to toxic side effects, which may limit the use of the drug. Irreversible inactivators containing reactive functional groups are termed affinity labeling agents; many of these types of compounds are cancer chemotherapeutic agents.

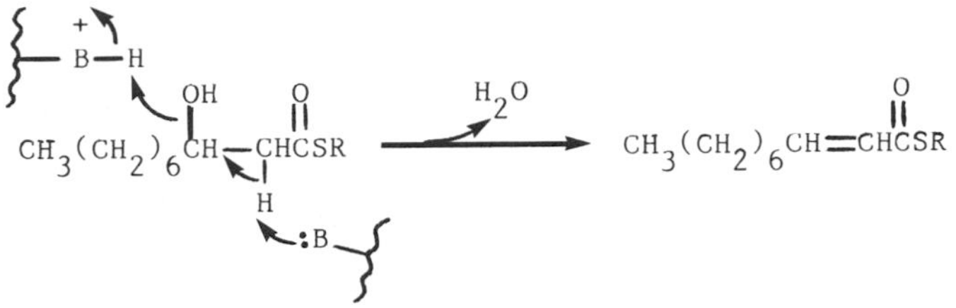

$$CH_3(CH_2)_6CH(OH)-CH(H)C(O)SR \xrightarrow{-H_2O} CH_3(CH_2)_6CH{=}CHC(O)SR$$

FIG. 2 *Mechanism for β-hydroxydecanoyl thioester dehydrase.*

C. MECHANISM-BASED ENZYME INACTIVATORS

In 1970 an alternative approach to irreversible enzyme inactivation was described by Bloch and coworkers (1). An unreactive compound, whose structure is similar to that of the substrate of a target enzyme, and which is converted into a reactive compound by the mechanism of action of that enzyme, was shown to be an effective irreversible inactivator. The compound first described in these terms was 3-decynoyl-N-acetylcysteamine and the enzyme was β-hydroxydecanoyl thioester dehydrase. Because the basis for the inactivation by this compound (Fig. 3) depends upon the mechanism of the enzyme (Fig. 2), these types of irreversible inactivators are known as mechanism-based enzyme inactivators. Other names have been used to describe these inactivators, e.g.,

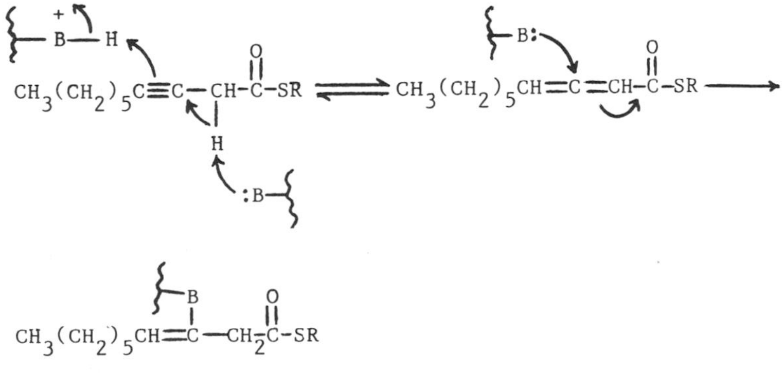

$$CH_3(CH_2)_5C{\equiv}C-CH(H)-C(O)-SR \rightleftharpoons CH_3(CH_2)_5CH{=}C{=}CH-C(O)-SR \longrightarrow$$

$$CH_3(CH_2)_5CH{=}C(B)-CH_2C(O)-SR$$

FIG. 3 *Proposed mechanism for inactivation of β-hydroxydecanoyl thioester dehydrase by 3-decynoyl thioesters (1).*

enzyme-activated irreversible inhibitors (2) and suicide inactivators (or substrates) (3). Except for the diazoketone inactivators (4), this is the first example of a mechanism-based inactivator whose mechanism of inactivation was described. However, this is not the first mechanism-based inactivator. In fact, several pharmaceuticals, especially monoamine oxidase inhibitors used in the treatment of depression and hypertension, were used clinically long before their mechanism of action was elucidated. The example of Bloch and coworkers, however, set the wheels in motion for the rational design of target enzyme inactivators. The major advantageous property of the mechanism-based inactivators is their chemical unreactivity. This minimizes nonspecific reactions with other macromolecules and, therefore, potentially lowers their toxicity. One of the drugs that will be discussed here (α-difluoromethylornithine) is so non-toxic that 30 g a day for several weeks have been given to patients with only minor side effects (5). In addition to their low intrinsic reactivity, these compounds are activated, ideally, only by the mechanism of the target enzyme. This gives an additional factor of specificity, especially if the structure of the compound can be tailored to coincide with the active site structure of the target enzyme. Of course, also in their favor is the fact that these compounds are irreversible inactivators, so fewer doses would need to be administered.

II. EXAMPLES OF MECHANISM-BASED ENZYME INACTIVATORS AS DRUGS

In this section eight clinically-used drugs have been selected as specific examples of mechanism-based enzyme inactivators, and their molecular mechanisms of inactivation are described. Five of these compounds are currently on the drug market (tranylcypromine, 5-fluorouracil, clavulanic acid, chloramphenicol, and allopurinol), two are in the latter stages of clinical trials (α-difluoromethylornithine and γ-vinyl GABA), and one is in an early phase of clinical trials ((E)-2-(3',4'-dimethoxyphenyl)-3-fluoroallylamine). Two of the drugs were designed by Mother

Nature (chloramphenicol and clauvulanic acid) and three were designed as mechanism-based inactivators for specific enzymes (α-difluoromethylornithine, γ-vinyl GABA, and (E)-2-(3',4'-dimethoxyphenyl)-3-fluoroallylamine). The other three compounds were synthesized as inhibitors of enzymes, but with the intention that they would be competitive reversible inhibitors, and only years after their discovery were the mechanisms of inactivation of these compounds elucidated. Merrell-Dow Pharmaceuticals appears to be the pioneering drug company in the design of specific mechanism-based enzyme inactivators for medical uses (their program began in 1973) and the fruits of their early labors are now being reaped (γ-vinyl GABA, α-difluoromethylornithine, and (E)-2-(3',4'-dimethoxyphenyl)-3-fluoroallylamine). The compounds are discussed according to the type of disease that they treat: two are antidepressant agents (tranylcypromine and (E)-2-(3',4'-dimethoxyphenyl)-3-fluoroallylamine), two are antitumor agents (5-fluorouracil and α-difluoromethylornithine, which also is an antiprotozoan agent), two are antibiotics (clavulanic acid and chloramphenicol), one is an anticonvulsant agent (γ-vinyl GABA), and one is a uricosuric agent (allopurinol).

A. *trans-2-PHENYLCYLOPROPYLAMINE (TRANYLCYPROMINE)*

In the early 1950's it was observed in clinical studies with an antituberculosis drug that a mood-elevating effect resulted. This compound was shown to be a potent inhibitor of monoamine oxidase (MAO)(6). By the late 1950's, it was found to have an antidepressant effect and a search for lower toxic inhibitors of MAO was underway. One of the less toxic compounds was trans-2-phenylcyclopropylamine (tranylcypromine, **1**), which was prepared as a conformationally rigid analogue of amphetamine (**2**) in an

1 *tranylcypromine*  **2** *amphetamine*

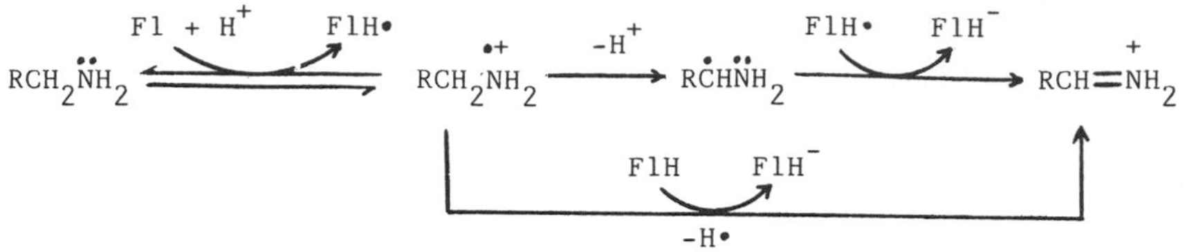

FIG. 4 *Proposed mechanism for monoamine oxidase-catalyzed amine oxidation (9-15).*

attempt to enhance central stimulant properties (7). It turned out to be a potent MAO inactivator (8), and by the 1960's it was prescribed for the treatment of depression.

Mitchondrial monoamine oxidase is one of the enzymes responsible for the catabolism of biogenic amines and it is believed that inhibition of this enzyme increases the intracellular concentration of certain pressor amines which leads to the elevation of mood effect. The mechanism of amine oxidation catalyzed by MAO is unknown, but various studies in my labs with mechanism-based inactivators (9-15) have pointed to a mechanism involving one-electron transfers from the amine to the flavin cofactor (Fig. 4). In 1980 it was reported that tranylcypromine was a mechanism-based inactivator of MAO (16). The mechanism proposed involved a two-electron oxidation to 2-phenylcyclopropaniminium ion (Fig. 5) followed by nucleophilic attack of an active site amino acid residue. Acid precipitation in the presence of 2,4-dinitrophenylhydrazine of the enzyme inactivated by [^{14}C]tranylcypromine gave a radioactive 2,4-dinitrophenylhydrazone that was identified incorrectly as that of 2-phenylcyclopropanone. In

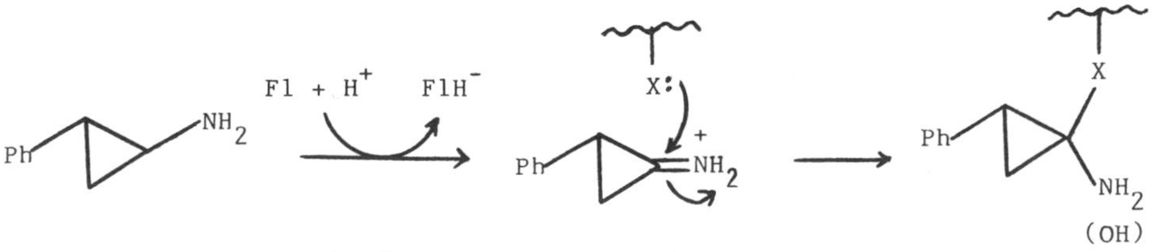

FIG. 5 *Singer mechanism (16) for inactivation of monoamine oxidase by tranylcypromine.*

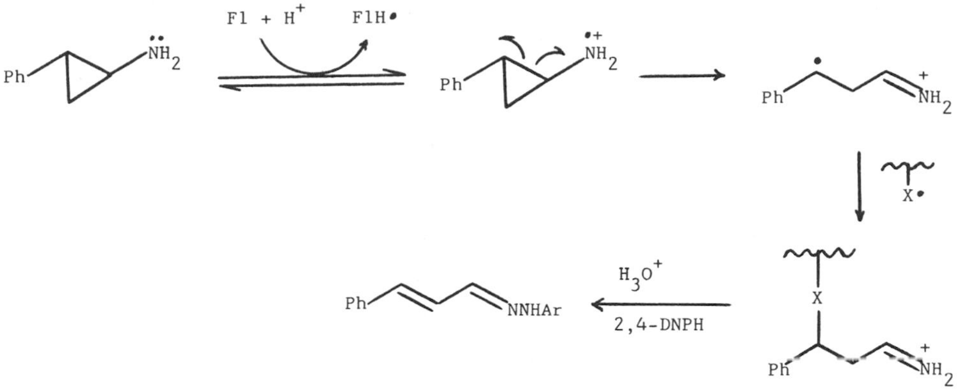

FIG. 6 *Silverman mechanism (11) for inactivation of monoamine oxidase by tranylcypromine.*

1983 this inactivation mechanism was reinvestigated (11) and the compound released from the labeled enzyme by acid in the presence of 2,4-dinitrophenylhydrazine was identified definitively as the 2,4-dinitrophenylhydrazone of cinnamaldehyde. This is consistent with radical mechanism of inactivation as shown in Fig. 6.

B. (*E*)-2-(3',4'-*DIMETHOXYPHENYL*)-3-*FLUOROALLYLAMINE*

Although tranylcypromine is in current medical use, it, and other MAO inhibitors, are not the drugs of choice for most cases of depression (tricyclic antidepressants are prescribed much more frequently). This is not because of ineffectiveness, but, rather, because of toxicity. In fact, for several months in 1964, tranylcypromine was withdrawn from the market because of an increasing number of deaths resulting from its use. Death, which was caused by hypertensive crisis, was soon found to be related to the concurrent consumption of foods (e.g., cheeses) high in pressor amine content, particularly tyramine. This, so-called, "cheese effect" was traced to release of norepinephrine upon ingestion of tyramine. Norepinephrine, a vasoconstrictor, raises the blood pressure, and since MAO has been inactivated, the excess norepinephrine cannot be degraded. The blood pressure, therefore, rises uncontrollably. In the late 1960's it was discovered that there are two forms of MAO in humans, termed MAO A and MAO B (17). The difference in these forms is related to which biogenic amines are more readily oxidized; presumably the oxidation mechanism for the two forms are the same, but binding properties of the substrates differ. It was thought that if compounds were prepared that were specific inactivators of only one of the two forms (by structural modification), then the increased concentration of biogenic amines, and mood elevation effect, still could be accomplished. However, the MAO isoenzyme not inactivated would be available to degrade excess pressor amines. This inactivator specificity has been achieved, and it has been shown that specific inactivation of MAO B in humans is devoid of a cheese effect (18). A specific MAO B mechanism-based inactivator was designed recently for the treatment of depression without dietary restrictions, and currently is in clinical trials. This compound, prepared by Merrell-Dow Pharmaceuticals (19) is (E)-2-(3'-4'-dimethoxyphenyl)-3-fluoroallylamine (**3**).

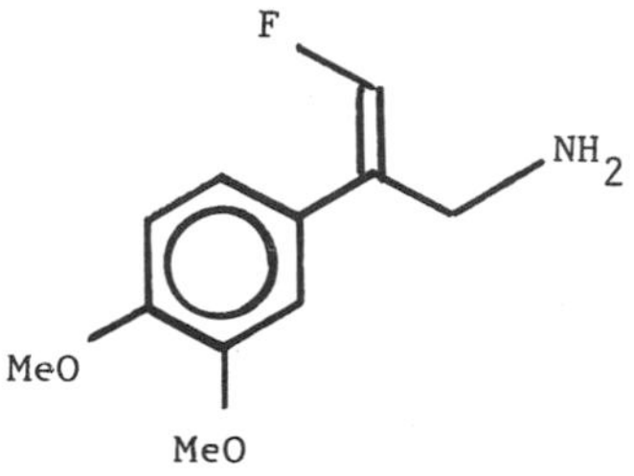

F

NH_2

MeO

MeO

3 *(E)-2-(3',4'-dimethoxyphenyl)-3-fluoroallylamine*

In early clinical trials, administration of this compound with tyramine showed no increase in the cardiovascular effects of tyramine alone and no impairment of the central nervous system (20). Little in the way of mechanistic work has been reported for the compound. As an allylamine derivative, it has been proposed that it is oxidized by MAO B to a Michael acceptor (Fig. 7) (21). No experimental evidence for this mechanism has been presented yet, nor has the identity of the X group attached to the

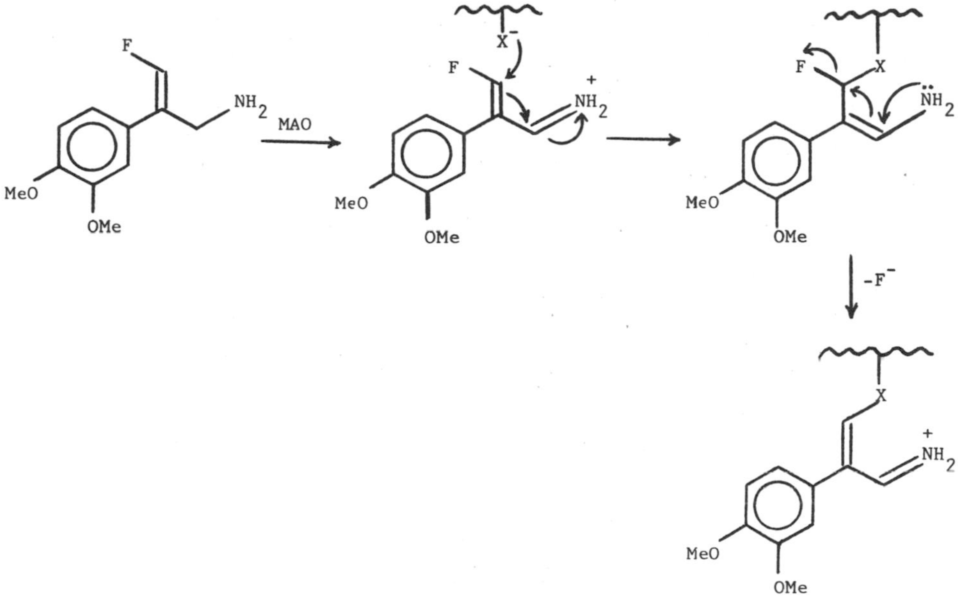

FIG. 7 *Proposed mechanism for inactivation of monoamine oxidase by (E)-2-(3',4'-dimethoxyphenyl)-3-fluoroallylamine (21).*

enzyme been revealed. The oxidation of the amine to the immonium ion presumably involves a one-electron transfer.

C. 5-*FLUORO*-2'-*DEOXYURIDYLATE*

5-Fluoro-2'-deoxyuridylate (**4**) is not the form of the antitumor agent used in cancer chemotherapy. The two compounds prescribed, 5-fluorouracil and 5-fluoro-2'-deoxyridine (floxuridine), are actually prodrugs, metabolic precursors, to the mechanism-based

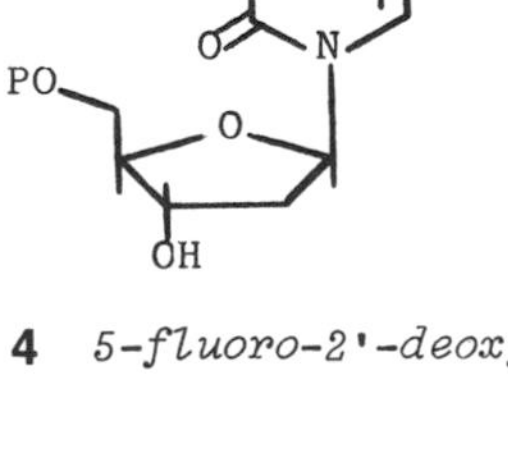

4 *5-fluoro-2'-deoxyuridylate*

inactivator, 5-fluoro-2'-deoxyuridylate. Because the van der Waals radius of fluorine (1.35 Å) is similar to that of hydrogen (1.20 Å), 5-fluorouracil can be metabolized to 5-fluorouridine (by uridine phosphorylase) then to 5-fluorouridylate (by uridine kinase), or it may be converted directly to 5-fluorouridylate with 5-phosphoribosyl-1-pyrophosphate (by orotate phosphoribosyl transferase). 5-Fluorouridylate is then reduced to 5-fluoro-2'-deoxyuridylate (by ribonucleotide reductase). Floxuridine is converted directly to 5-fluoro-2'-deoxyuridylate in one step catalyzed by thymidine kinase, and, therefore, has a therapeutic advantange over 5-fluorouracil. However, floxuridine also is a

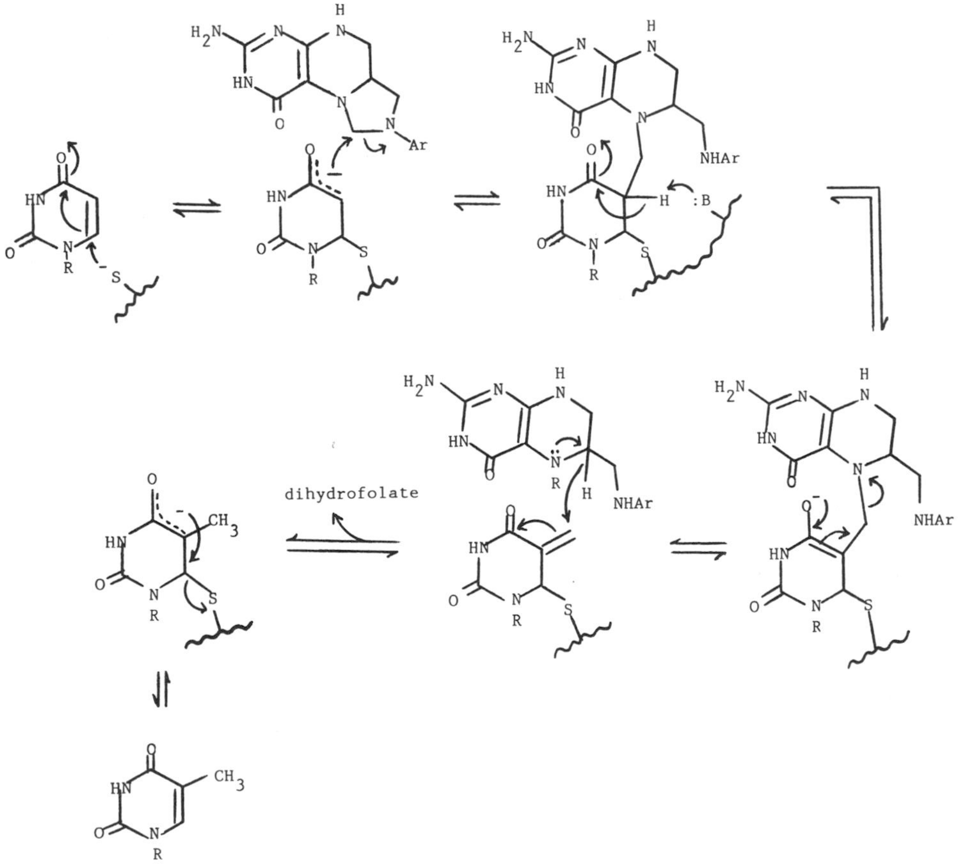

FIG. 8 *Proposed mechanism for thymidylate synthetase (22-25).*

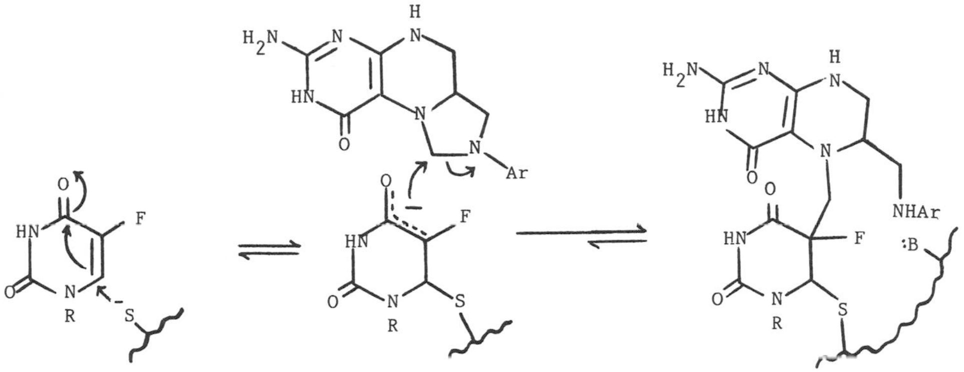

FIG. 9 *Proposed mechanism for inactivation of thymidylate synthetase by 5-fluoro-2'-deoxyuridylate* (22-25).

good substrate for thymidine- and deoxyuridine phosphorylases and, therefore, is rapidly degraded to 5 fluorouracil.

The principal site of action of 5-fluoro-2'-deoxyuridylate is thymidylate synthetase, the enzyme that catalyzes the transfer of a methyl group from N^5, N^{10}-methylenetetrahydrofolate to 2'-deoxyuridylate. The mechanism for this reaction is shown in Fig. 8. The steps up to ternary complex (enzyme-nucleotide-methylenetetrahydrofolate) formation are well documented (22-25), but breakdown to product is mostly hypothetical. Since thymidylate synthetase is the only de novo source of thymidylate, it is an essential enzyme for the biosynthesis of DNA (but not of RNA or proteins). The activity of this enzyme is substantially elevated in rapidly proliferating cells relative to normal cells. Also, since uracil is one of the precursors of thymidylate, it and 5-fluorouracil are taken up in tumor cells more rapidly than in normal ones. Inactivation of thymidylate synthetase, which results in the blockage of thymidylate biosynthesis, produces a phenomenon known as "thymineless death" and this can explain the cytotoxicity of 5-fluorouracil and its derivatives. Several research groups (Santi (22), Heidelberger (23), Dunlap (24), and Maley (25)) have examined the mechanism of inactivation of thymidylate synthetase by 5-fluoro-2'-deoxyuridylate in detail. The generally accepted mechanism is shown in Fig. 9. Note that this

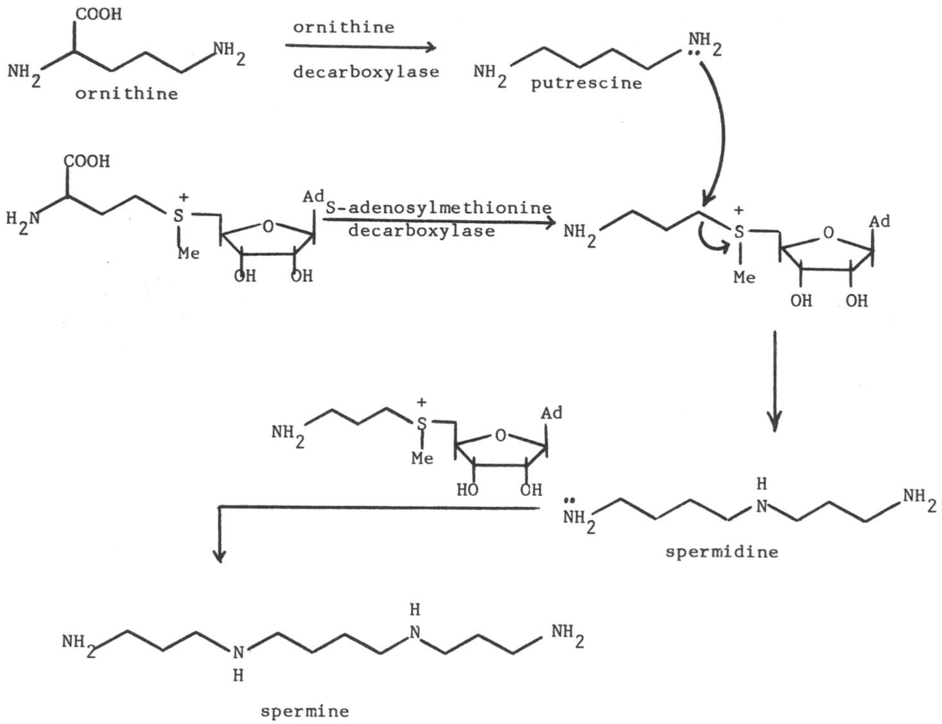

ornithine

ornithine decarboxylase

putrescine

S-adenosylmethionine decarboxylase

spermidine

spermine

FIG. 10 *Biosynthesis of putrescine and polyamines.*

mechanism mimics that of the substrate, but the ternary complex cannot break down readily since the proton at C-5 of the pyrimidine is replaced by fluorine and a F^+ cannot be abstracted. This mechanism was elucidated long after 5-fluorouracil was designed as a competitive inhibitor (26).

D. α-*DIFLUOROMETHYLORNITHINE* (α-*DFMO*)

The polyamines, spermidine and spermine, and their diamine precursor, putrescine, are implicated in the initiation and maintenance of rapid cell growth and differentiation and have important, but not yet well defined, regulatory functions in proliferation of malignant cells. The biosynthesis of these amines originates with ornithine and is shown in Fig. 10. Induction of ornithine decarboxylase, the rate-determining step in polyamine

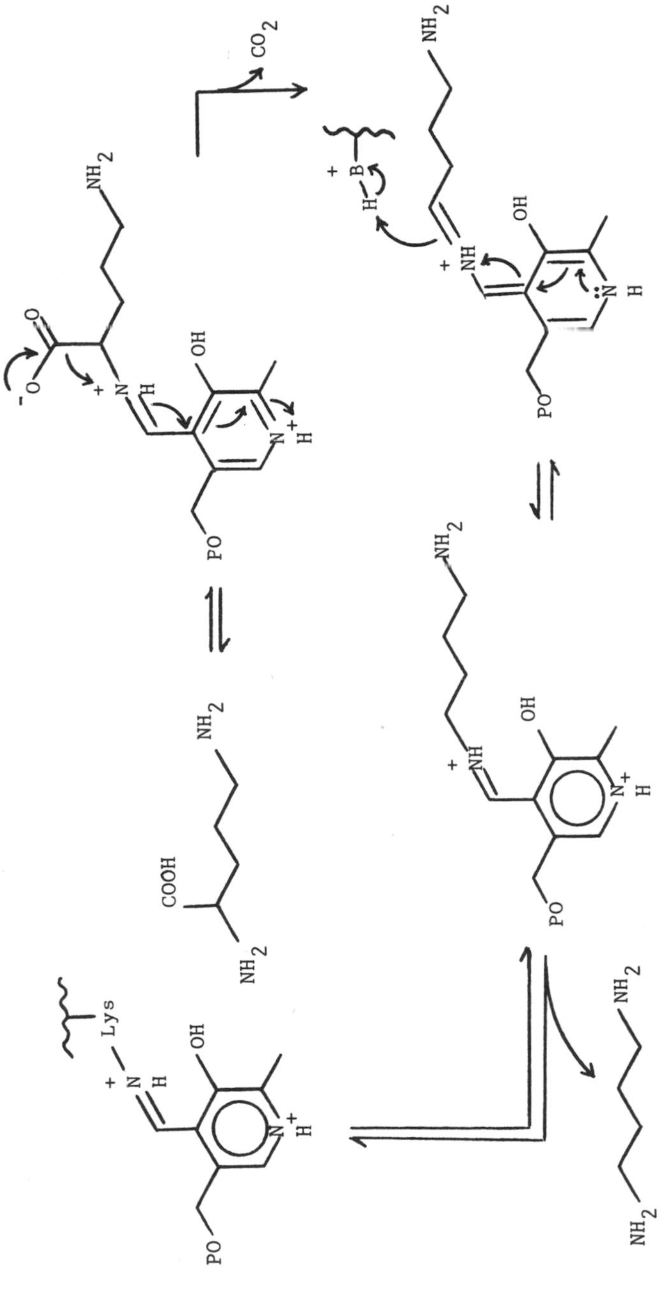

FIG. 11 *Mechanism for ornithine decarboxylase.*

biosynthesis, is associated with the onset of growth in all eukaryotic and prokaryotic organisms. Increases in this enzyme are related to cell proliferation and occur just before the maximum peak of DNA synthesis. Because of the importance of ornithine decarboxylase to rapid cell growth, α-difluoromethylornithine (**5**) (27) was designed as a mechanism-based inactivator of

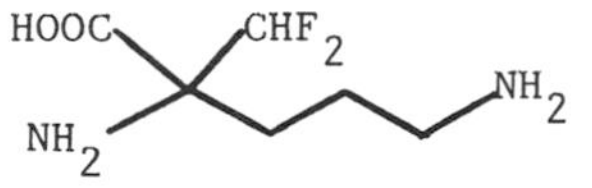

HOOC, CHF_2, NH_2, NH_2

5 *α-difluoromethylornithine*

the enzyme. The mechanism of this pyridoxal 5'-phosphate-dependent enzyme is shown in Fig. 11. The mechanism of inactivation of ornithine decarboxylase by α-DFMO has been suggested to be that shown in Fig. 12 (28). No experimental evidence has been offered to corroborate this proposal nor has the group X been elucidated. Another mechanistic possibility is based on the work of Metzler's group (29,30) and is shown in Fig. 13, starting with the intermediate in Fig. 12 formed by initial fluoride ion release. α-DFMO has remarkably low toxicity and has been shown to reduce the incidence of experimental colon cancers and inhibit the growth of human lung cancer cells and human leukemia cells in tissue culture (31). It also is effective against the growth of Lewis lung carcinoma and the development of secondary lung metastases (32). However, early clinical trials with cancer patients have not been very successful (33). The most promising results are found for the treatment of protozoal infections, including *Trypanosoma brucei gambiense* (5) that causes African sleeping sickness and *Pneumocystis carinii* (5) a protozoa that causes pneumonia in patients with acquired immune deficiency syndrome (AIDS).

E. *CLAVULANIC ACID*

There are several mechanisms by which bacteria can become resistant to β-lactam antibiotics (e.g., penicillins); the most

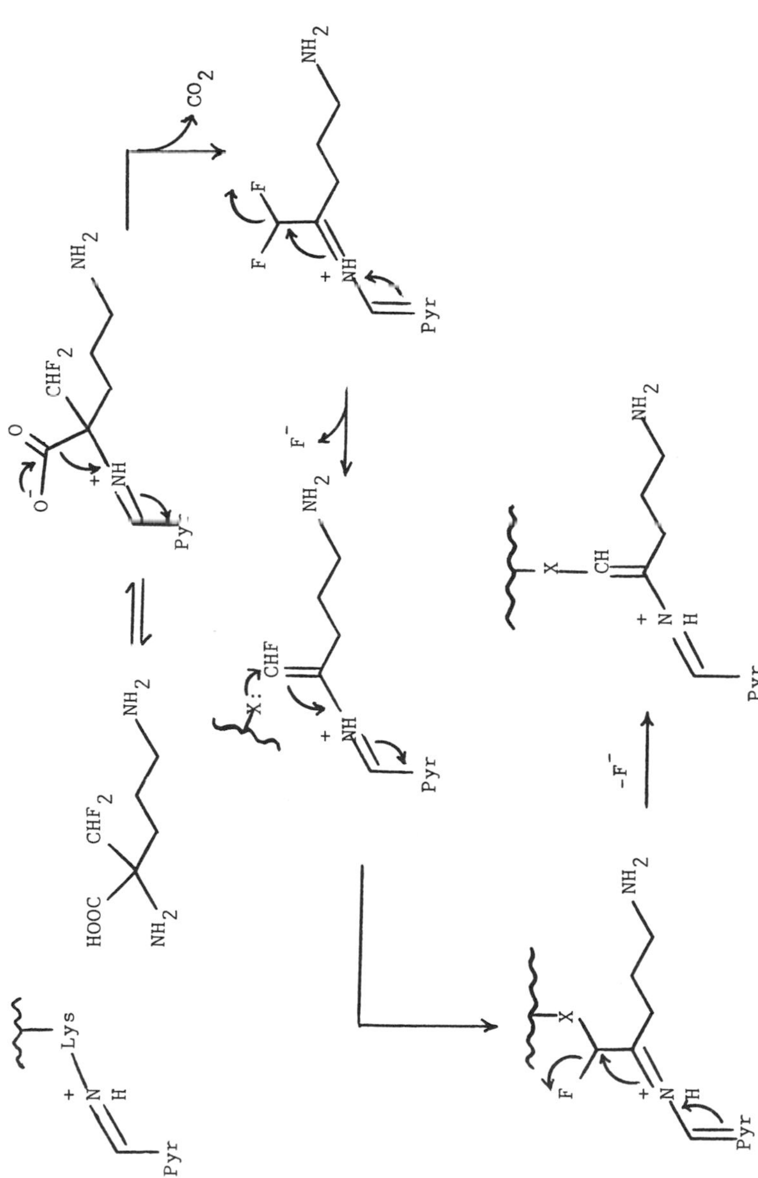

FIG. 12 *Proposed mechanism for inactivation of ornithine decarboxylcse by α-DFMO. Pyr is the pyridoxal 5'-phosphate.*

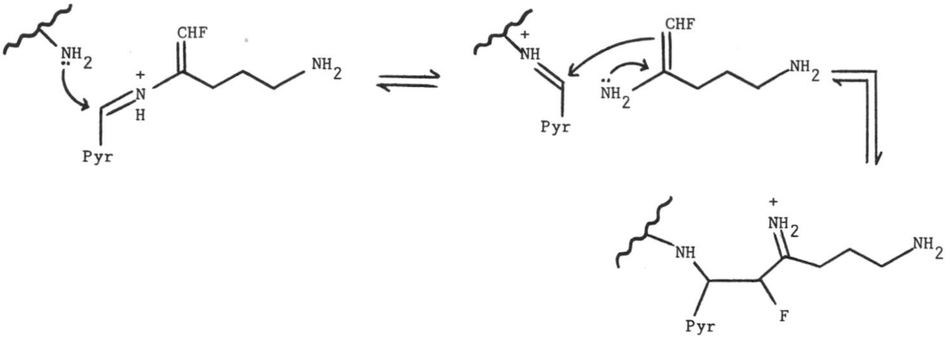

FIG. 13 *An alternative mechanism for inactivation of ornithine decarboxylase by α-DFMO. Pyr is the pyridoxal-5'-phosphate.*

important way is for the bacterium to produce a β-lactamase, an enzyme that can hydrolyze the β-lactam ring and, thereby, deactivate the drug. Clavulanic acid (**6**) is a natural product, produced by *Streptomyces clavuligerus*, that is a mechanism-based inactivator of β-lactamases (34,35). Clavulanic acid, then, is not itself an antibiotic, but, rather, is used in conjunction with β-lactam antibiotics to protect them from β-lactamase hydrolysis. Augmentin is the trade name for a synergistic drug

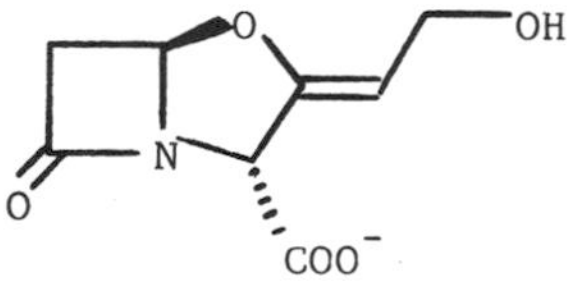

6 *clavulanic acid*

combination of the penicillin derivative, amoxicillin, and the β-lactamase inhibitor clavulanic acid.

The mechanism for β-lactamase-catalyzed hydrolysis of penams to the penicilloic acid derviative, is believe to proceed by covalent catalysis via a serine ester (Fig. 14). Clavulanic acid was shown to inactivate β-lactamase by a related mechanism (Fig. 15) (36). Since the discovery of clavulanic acid, several other β-lactam structures have been shown to be β-lactamase inac-

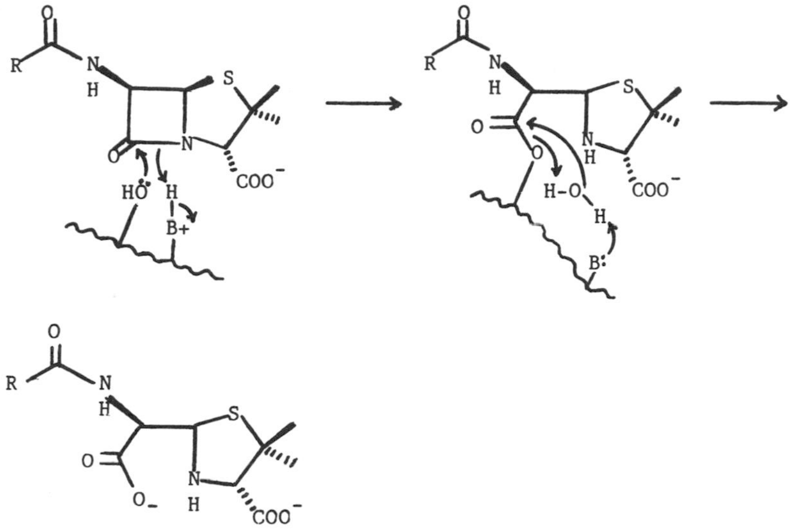

FIG. 14 *Mechanism for β-lactamase-catalyzed hydrolysis of penicillin analogues.*

tivators (36). The two key features of clavulanic acid which allow it to be an inactivator are, first, that hydrolysis of the acyl enzyme is slow so that tautmerization to give the transiently-inhibited enzyme or transimination which leads to inactivation have time to occur; second, the five-membered ring contains a good leaving group so that β-elimination can occur (37). In the case of penams, hydrolysis is fast and the leaving group is not good enough. Oxidation of the sulfide to a sulfone, however, transforms the penam into a β-lactamase inactivator (38).

F. *CHLORAMPHENICOL*

Chloramphenicol (**7**) is an antibiotic, produced by *Streptomyces venezuelae*, that possesses a wide spectrum of antimicrobial activity. It has been shown to potentiate the action of various other drugs (39) and to protect against toxicity and tumorigenicity of several different xenobiotics such as polycylic aromatic hydrocarbons, polychlorinated hydrocarbons, and nitroso com-

hydrolysis

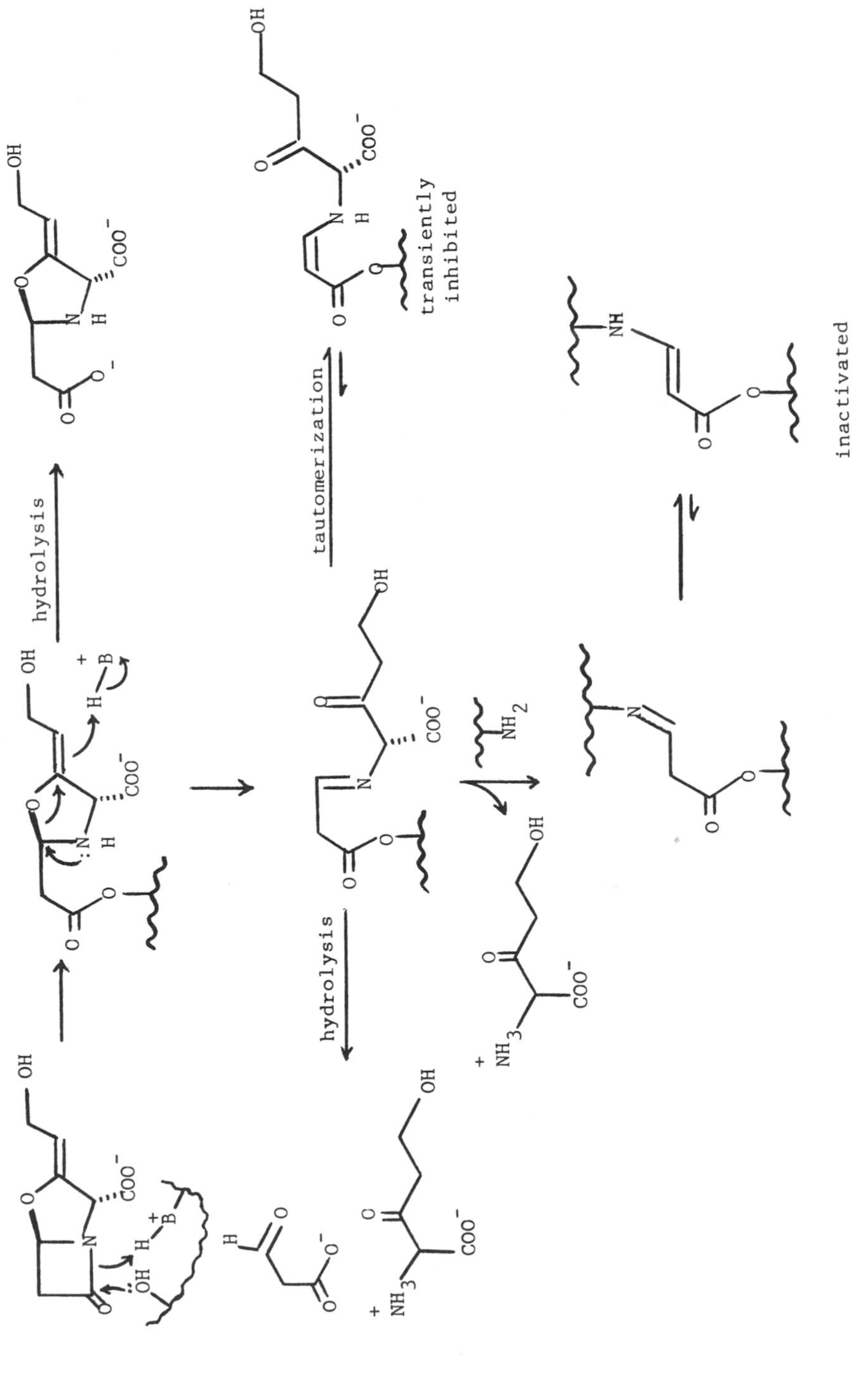

hydrolysis

tautomerization

transiently inhibited

inactivated

FIG. 15 *Proposed mechanism for inactivation of β-lactamases by clavulanic acid (36).*

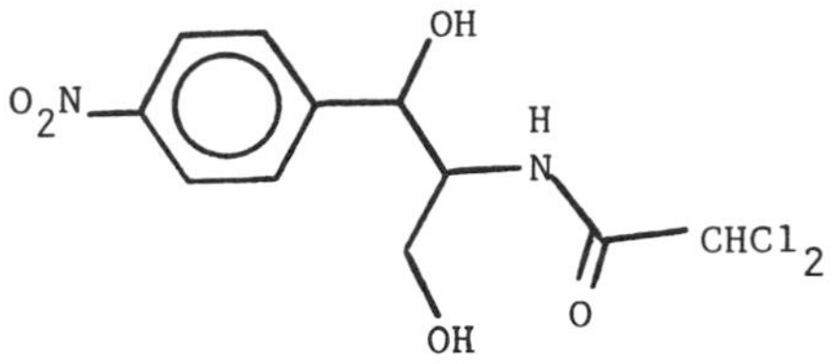

7 *chloramphenicol*

pounds. It also is effective in preventing heptatoxicity by carbon tetrachloride and the development of adrenal necrosis caused by 7,12-dimethylbenz[a]anthracene (40). It is believed that all of these actions (excluding its antimicrobial mechanism, which is unrelated) result from mechanism-based inactivation of cytochrome P-450 by chloramphenicol. Cytochrome P-450 is one of the heme-containing, non-specific mixed function oxidases responsible for the metabolism of xenobiotics in the liver. The mechanism for the enzyme still is unclear, but the active oxygenating species is believed to be an iron (V) oxene (Fig. 16) (41). Chloramphenicol has been shown to become oxygenated by cytochrome P-450, which leads to inactivation (Fig. 17) (42,43). An active site lysine residue was found to be acylated during enzyme inactivation.

G. 4-*AMINOHEX*-5-*ENOIC ACID* (γ-*VINYL GABA; VIGABATRIN*)

It is estimated that 0.5-1.0% of the world population has epilepsy (44). Many seizures have been shown to result from an

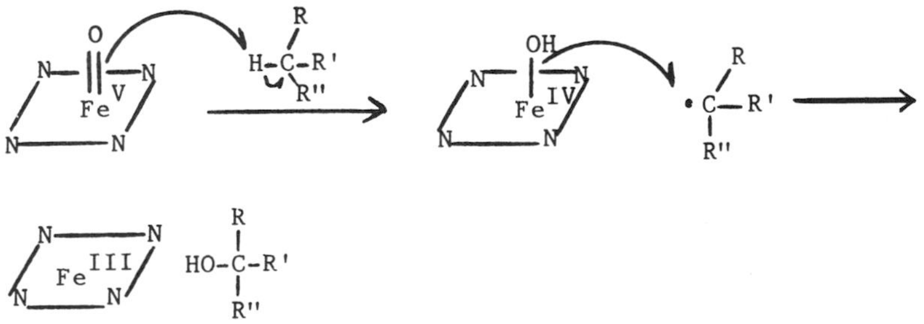

FIG. 16 *Proposed mechanism for cytochrome P-450* (*41*). *The box with four N's is the heme cofactor.*

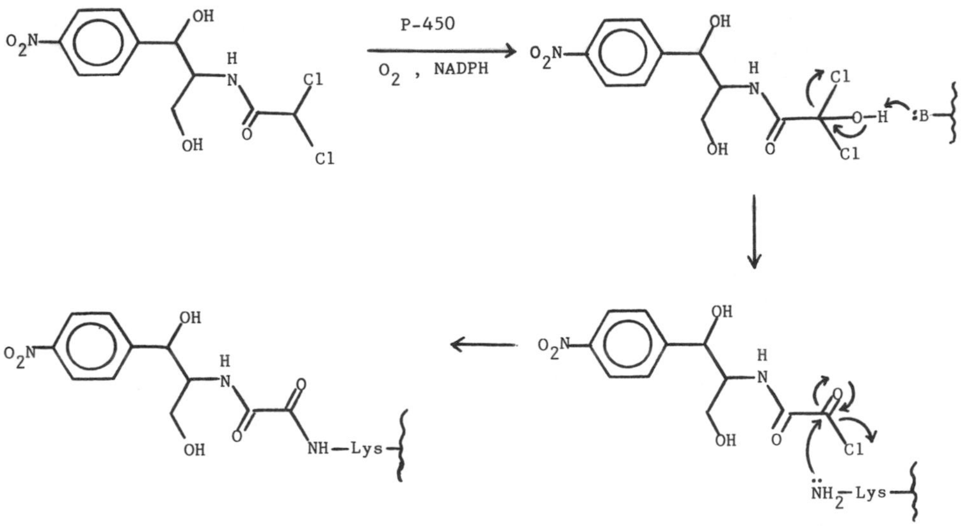

FIG. 17 *Proposed mechanism for inactivation of cytochrome P-450 by chloramphenicol* (*42*,*43*).

imbalance in the brain concentrations of the inhibitory neurotransmitter, γ-aminobutyric acid (GABA) and the excitatory neurotransmitter, glutamic acid. When GABA levels in the brain diminish, convulsions result; return of normal GABA levels terminates the seizure (45,46). It has been shown that whole brain GABA concentration is not important; rather, the GABA levels at the nerve terminals of the substantia nigra is what influences convulsions (47). GABA is biosynthesized from L-glutamic acid

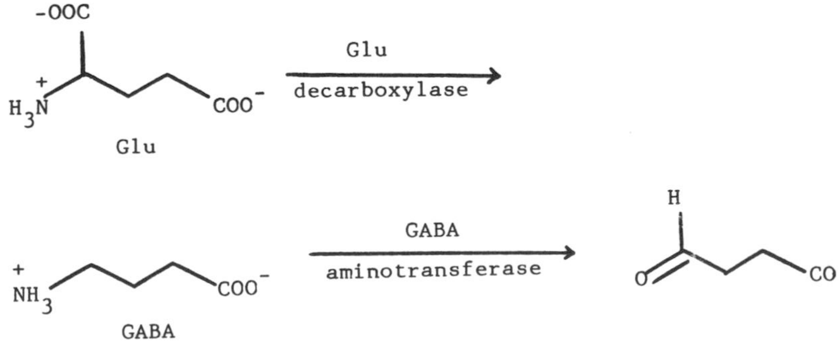

FIG. 18 *Brain biosynthesis and degradation of GABA.*

and is degraded to succinic semialdehyde. L-Glutamic acid decarboxylase catalyzes the biosynthetic reaction and γ-aminobutyric acid aminotransferase catalyzes the catabolism of GABA (Fig. 18). Diminished levels of GABA can arise from either an underactive L-glutamate decarboxylase or an overactive GABA aminotransferase. Although GABA has anticonvulsant properties when injected directly into the brain, it is inactive when administered peripherally. The reason for this is that GABA does not cross the blood-brain barrier (48). However, if a compound could be prepared that crossed the blood-brain barrier and specifically inhibited GABA aminotransferase, this would block the breakdown of GABA. Provided that L-glutamate decarboxylase is not inhibited, the concentration of GABA should increase. This is the approach that was taken, and the first rationally-designed and commercially viable mechanism-based inactivator, 4-aminohex-5-enoic acid (**8**), was conceived (49). Presumably, the vinyl group

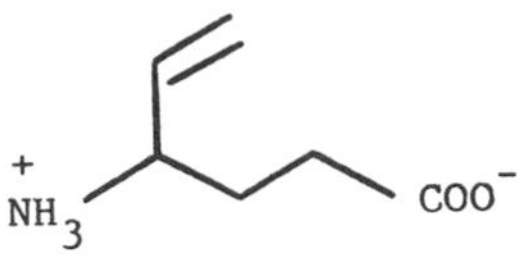

$\overset{+}{N}H_3$ COO^-

8 *γ-vinyl GABA*

increases the lipophilicity, and, because of its electron-withdrawing properties, it lowers the pK_a of the amino group sufficiently to affect the zwitterionic population. These two effects, although minor, are sufficient to permit γ-vinyl GABA to cross the blood-brain barrier, to irreversibly inactivate GABA aminotransferase (50), and to have a potent anticonvulsant effect (51). Tardive dyskinesia also is reduced by this drug (52).

GABA aminotransferase is a pyridoxal 5'-phosphate-dependent enzyme; its mechanism is shown in Fig. 19. γ-Vinyl GABA has been proposed to inactivate the enzyme by the mechanism shown in Fig. 20 (49). This mechanism is only hypothetical; no in-depth studies have been reported.

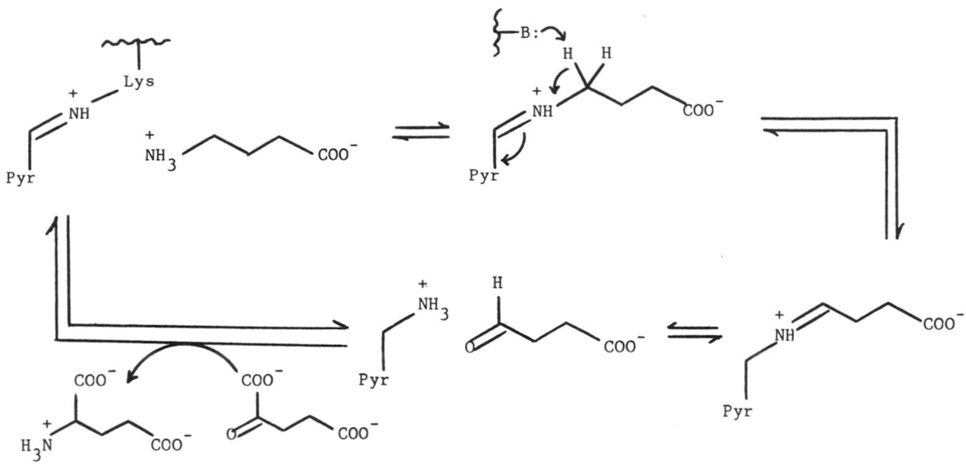

FIG. 19 Mechanism for GABA aminotransferase.

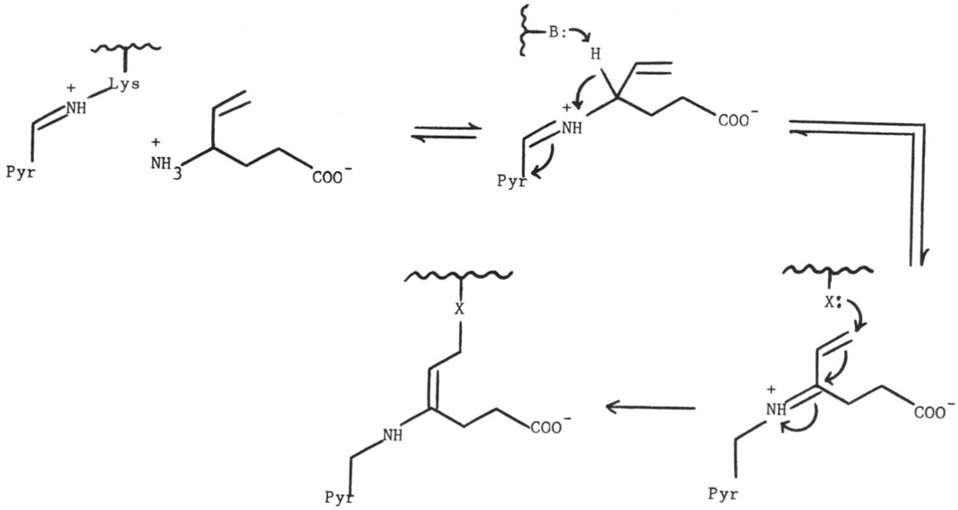

FIG. 20 *Proposed mechanism for inactivation of GABA aminotransferase by γ-vinyl GABA.*

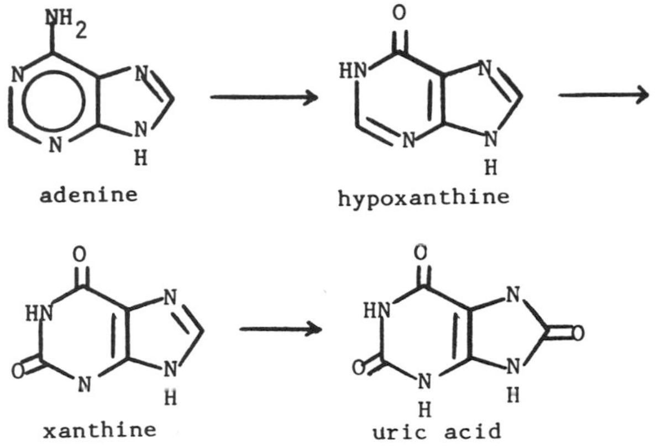

adenine → hypoxanthine → xanthine → uric acid

FIG. 21 *Biosynthesis of uric acid*

H. *ALLOPURINOL*

Adenine is hydrolyzed to hypoxanthine which is converted by xanthine oxidase first to xanthine then to uric acid (Fig. 21), which is excreted through the kidneys. Xanthine oxidase is a flavoenzyme that contains a tightly-bound Mo(VI) atom and four atoms of iron; its mechanism of action is unclear. At physiological pH uric acid is poorly soluble, but it tends to form supersaturated solutions. Crystals of uric acid then begin to form in the joints and connective tissues resulting in the disease, gout. Inactivation of xanthine oxidase prevents uric acid production and allows the crystallized uric acid to redissolve. Allopurinol (**9**) was originally prepared as an antineopolastic agent, but was found, instead, to be a potent inhibitor of xanthine oxidase. Its mechanism of inactivation was determined almost ten years later (53). Allpurinol is an unusual mechanism-based inactivator

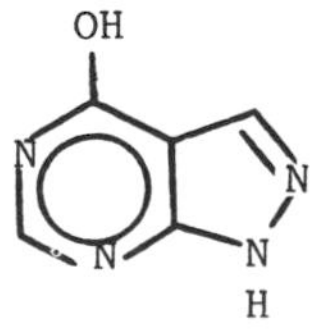

9 *allopurinol*

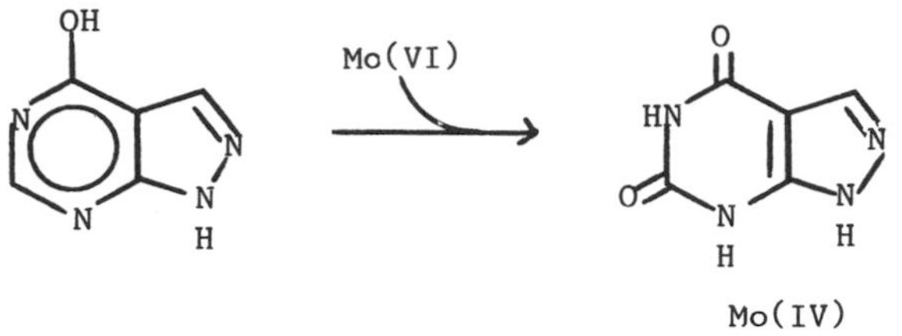

FIG. 22 *Proposed mechanism for inactivation of xanthine oxidase by allopurinol* (*53*).

because it does not form a covalent bond to xanthine oxidase. The oxidation product of allopurinol, alloxanthine, binds exceedingly tightly to the reduced form of the enzyme (Fig. 22). It is not essential for a mechanism-based inactivator to form a covalent bond to an enzyme; only that it be converted by the target enzyme into another compound which, without prior release from the active site, inactivates that enzyme.

III. CONCLUSION

I have shown here several mechanism-based enzyme inactivators that already are important pharmaceuticals. The future potential for these types of inactivators as low toxicity, high potency drugs is even greater now that the first generation of rationally-designed mechanism-based enzyme inactivators is currently in the final stages of clinical trials (γ-vinyl GABA and α-DFMO). When these compounds become commercially available, it should be a strong impetus for many pharmaceutical companies to implement or expand programs with an emphasis on the rational design of mechanism-based enzyme inactivators for medical uses.

REFERENCES

1. K. Endo, G.M. Helmkamp, Jr., and K. Bloch, *J. Biol. Chem.* *245*: 4293 (1970).
2. N. Seiler, M.J. Jung, and J. Koch-Weser, *Enzyme-Activated Irreversible Inhibitors*, Elsevier/North-Holland, Amsterdam (1978).
3. R.H. Abeles and A.L. Maycock, *Acc. Chem. Res.* *9*: 313 (1976).

4. J.M. Buchanan, *Adv. Enzymol.* *39*: 91 (1973).

5. A. Sjoerdsma, J.A. Golden, P.J. Schechter, J.L.R. Barlow, and D.V. Santi, *Trans. Assoc. Am. Phys.* *97*: 70 (1984).

6. E.A. Zeller, J. Barsky, J.P. Fouts, W.F. Kirshheimer, and L.S. Van Orden, *Experientia* *8*: 349 (1952).

7. R.J. Baldessarini in *Goodman and Gilman's The Pharmacological Basis of Therapeutics*, 7th Edit. (A.G. Gilman, L.S. Goodman, T.W. Rall, and F. Murad, eds.), Macmillan Publishing Co., New York, p. 387 (1985).

8. C.L. Zirkle, C. Kaiser, D.H. Tedeschi, R.E. Tedeschi, and A. Burger, *J. Med. Pharm. Chem.* *5*: 1265 (1962).

9. R.B. Silverman, S.J. Hoffman, and W.B. Catus III, *J. Am. Chem. Soc.* *102*: 7126 (1980).

10. R.B. Silverman and S.J. Hoffman, *Biochem. Biophys. Res. Commun.* *101*: 1396 (1981).

11. R.B. Silverman, *J. Biol. Chem.* *258*: 14766 (1983).

12. R.B. Silverman and R.B. Yamasaki, *Biochemistry* *23*: 1322 (1984).

13. R.B. Silverman, *Biochemistry* *23*: 5206 (1984).

14. R.B. Silverman and P.A. Zieske, *Biochemistry* *24*: 2128 (1985).

15. M.L. Vazquez and R.B. Silverman, *Biochemistry* in press.

16. C. Paech, J.I. Salach, and T.P. Singer, *J. Biol. Chem.* *255*: 2700 (1980).

17. J.P. Johnston, *Biochem. Pharmacol.* *17*: 1285 (1968).

18. M.B.H. Youdim and J.P.M. Finberg, *Mod. Probl. Pharmacopsychiatry* *19*: 63 (1983).

19. P. Bey, J. Fozard, J.M. Lacoste, I.A. McDonald, M. Zreika, and M.G. Palfreyman, *J. Med. Chem.* *27*: 9 (1984).

20. R.G. Alken, M.G. Palfreyman, M.J. Brown, D.S. Davies, P.J. Lewis, and P.J. Schechter, *Brit. J. Clin. Pharmacol.* *17*: 615P (1984).

21. M. Zreika, I.A. McDonald, P. Bey, and M.G. Palfreyman, *J. Neurochem.* *43*: 448 (1984).

22. D.V. Santi and P.V. Danenberg in *Folates and Pteridines* (R.L. Blakely and S.J. Benkovic, eds.) Wiley, New York, Vol. 1, p. 343 (1984).

23. C. Heidelberger in *Antineoplastic and Immunosuppressive Agents* (A.C. Sarorelli and D.G. Johns, eds.) Springer-Verlag, Heidelberg, Part 2, p. 193 (1975).

24. C.A. Lewis, Jr. and R.B. Dunlap in *Topics in Molecular Pharmacology* (A.S.V. Burgen and G.C.K. Roberts, eds.), Elsevier/North-Holland, New York, p. 169 (1981).

25. G.F. Maley and F. Maley in *Biochem. Metab. Processes, Proc. Steenbock-Lilly Int. Symp.* (D.L. Lennon, F.W. Stratman and R.N. Zahlten, eds.), Elsevier/North-Holland, New York, p. 243 (1983).

26. P. Reyes and C. Heidelberger, *Mol. Pharmacol 1*: 14 (1965).

27. P. Bey, J.P. Vevert, V. Van Dorsselaer, and M. Kolb, *J. Org. Chem. 44*: 2732 (1979).

28. B.W. Metcalf, P. Bey, C. Danzin, M.J. Jung, P. Casara, and J.P. Vevert, *J. Am. Chem. Soc. 100*: 2551 (1978).

29. J.J. Likos, H. Ueno, R.W. Feldhaus, and D.E. Metzler, *Biochemistry 21*: 4377 (1982).

30. H. Ueno, J.J. Likos, and D.E. Metzler, *Biochemistry 21*: 4387 (1982).

31. A.N. Kingsnorth, W.E. Russell, P.P. McCann, K.A. Diekema, and R.A. Malt, *Cancer Res. 43*: 4035 (1983).

32. J. Bartholeyns, *Eur. J. Cancer Clin. Oncol. 19*: 567 (1983).

33. A. Sjoerdsma and P.J. Schechter, *Clin. Pharmacol. Ther. 35*: 287 (1984).

34. J. Fisher, R.L. Charnas, and J.R. Knowles, *Biochemistry 17*: 2180 (1978).

35. R.L. Charnas, J. Fisher, and J.R. Knowles, *Biochemistry 17*: 2185 (1978).

36. J.R. Knowles, *Acc. Chem. Res.* 18: 97 (1985).

37. J. Fisher, J.G. Belasco, R.L. Charnas, S. Khosla, and J.R. Knowles, *Phil. Trans. Roy. Soc. (London) B 289*: 309 (1980).

38. J. Fisher, R.L. Charnas, S.M. Bradley, and J.R. Knowles, *Biochemistry 20*: 2726 (1981).

39. L.K. Christensen and L. Skovsted, *Lancet 2*: 1397 (1969).

40. B. Näslund, J. Rydström, M. Bengstsson, and J. Halpert, *Biochem. Pharmacol. 32*: 707 (1983).

41. F.P. Guengerich and T.L. MacDonald, *Acc. Chem. Res. 17*: 9 (1984).

42. J. Halpert, *Mol. Pharmacol. 21*: 166 (1982).

43. L.R. Pohl and G. Krishna, *Biochem. Pharmacol. 27*: 335 (1978).

44. H.H. Jasper, A.A. Ward, and A. Poper in *Basic Mechanism of the Epilepsies) Little, Brown and Co., Boston (1969).*

45. E. Roberts, T.M. Chase, and D.B. Tower, *GABA in Nervous System Function*, Raven Press, New York (1976).

46. P. Mandel and F.V. DeFeudis, *Advances in Experimental Biology and Medicine: GABA*, Plenum Press, New York (1979).

47. K. Gale and M.J. Iadarola, *Science 208*: 288 (1980).

48. B.S. Meldrum and R.W. Horton in *Epilepsy* (P. Harris and C. Mawdsley, eds.), Churchill Livingston, Edinburgh, p. 55 (1974).

49. B. Lippert, B.W. Metcalf, M.J. Jung, and P. Casara, *Eur. J. Biochem. 74*: 441 (1977).

50. M.J. Jung, B. Lippert, B.W. Metcalf, P. Bohlen and P.J. Schechter, *J. Neurochem. 29*: 797 (1977).

51. P.J. Schechter, N.F.J. Hanke, J. Grove, N. Huebert, and A. Sjoerdsma, *Neurology 34*: 182 (1984).

52. P.A. Lambert, P. Cantiniaux, J.-P. Chabannes, G.P. Tell, P.J. Schechter, and J. Koch Weser, *L'Encéphale 8*: 371 (1982).

53. V. Massey, H. Komai, G. Palmer, and G.B Elion, *J. Biol. Chem. 245*: 2837 (1970).

9

Probing the Active Site of a Steroid Isomerase with a Solid Phase Reagent

William F. Benisek

Department of Biological Chemistry
School of Medicine
University of California
Davis, California

Maureen Hearne

Cancer Research Institute
University of California
San Francisco, California

I. INTRODUCTION

The technique of photoaffinity labeling has been widely employed for a variety of purposes in biochemical and biological research (1). The areas of application have been at one of three levels of biological, and specifically macromolecular structural organization. Some studies have used affinity labeling as a way of identifying specific macromolecules as the bearers of binding sites for particular ligands whose structures are mimicked by corresponding affinity reagents. In this application, the target macromolecule is present in a complex, crude mixture such as that of an intact cell, cell organelle, or cell extract. The mixture may also be an extracellular fluid. The second and closely related area of application is at the level of protein quaternary structure; here affinity labeling may be used to identify which subunit in a multisubunit protein carries the targeted ligand binding site. Studies conducted at these levels are technically feasible as a result of two circumstances. One of these is the availability of some reagents as radiolabeled forms of high specific activity. This usually requires an organic synthesis specifically adapted for radiochemical synthetic purposes. The other circumstance is the availability of a sufficiently high resolution method of analyzing the photolabeling reaction mixture in order to observe the incorporation of the reagent's radiolabel into specific macromolecular species. The most widely employed method of analysis is SDS-gel electrophoresis, which separates denatured polypeptides according to molecular weight. Radioautography of the gels or scintillation counting of gel slices obtained from aliquots of the labeling reaction mixture can serve to identify ligand binding site bearing polypeptides, whether free or complexed to other subcellular structures or to soluble macromolecular species at the moment of labeling.

The third area of application of photoaffinity labeling has been at a much finer level of biological structure. Here the goal has been to use the technique to specifically tag and thereby identify specific amino acid residues in the ligand binding site of a purified protein. This kind of application generally requires the use of a radiolabeled reagent and a labeling efficiency of more than 50% in order to make it feasible to identify the labeled residue. Unfortunately, the efficiency of covalent attachment in photoaffinity labeling studies is usually much lower than this because the reactive species formed by photolysis undergo rearrangement reactions and reactions with the solvent which proceed more rapidly than do the attachment reactions. In addition, the reactive species can initiate reactions which do not couple the reagent to the protein. The potential paths that an affinity labeling experiment may follow are shown in Fig. 1.

The identification of the residue at which covalent attachment occurs requires the digestion of the inefficiently labeled protein with chemical or enzymatic agents followed by the separation of the resulting fragment peptides by a suitable high resolution technique

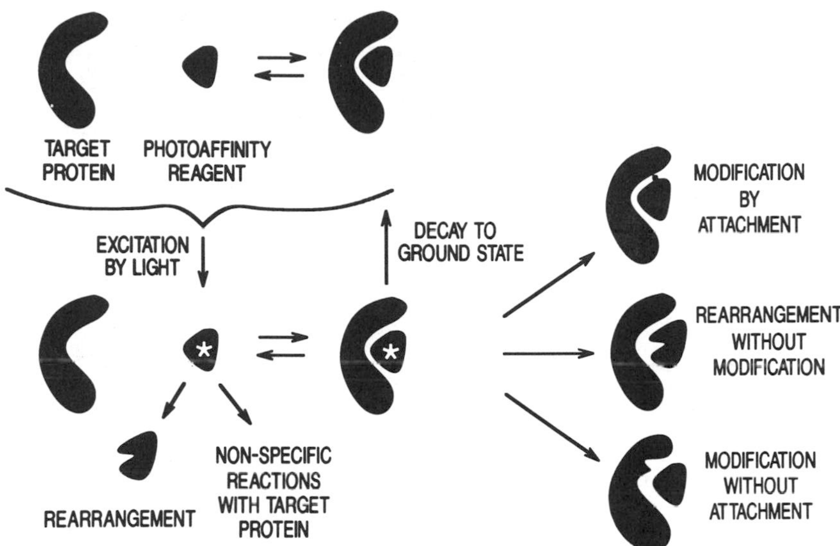

FIG. 1 *COMPETING PROCESSES IN PHOTOAFFINITY LABELING. The photoaffinity reagent binds to the target binding site in a reversible binding reaction. Irradiation of the equilibrium binding mixture creates a homologous system in which the photoaffinity reagent excited/reactive state may participate in a similar binding equilibrium. If the reactive state of the photoaffinity reagent is created while the reagent is in the binding pocket, then several competing processes can occur. These are dissociation from the target site, rearrangement, reaction with solvent molecules, relaxation to the ground state (in the case of an electronically excited state as the reactive species), coupling to a nearby residue, and modification of the protein without coupling of the reagent to a residue. Reagent molecules off of the target site may be excited by the UV light and react with the binding protein in various ways. These include noncovalent binding to the target protein, nonspecific reaction with the protein, rearrangement, reaction with solvent molecules, and relaxation to the ground state.*

such as hplc on reverse phase media. Since the labeled peptide is only a minor component in a complex mixture of unlabeled peptides, its purification is difficult. This has discouraged the routine application of photoaffinity labeling to this level of structural analysis.

The utility of photoaffinity labeling in the identification of ligand binding site residues might be significantly enhanced if a method were available to facilitate the specific retrieval of peptide fragments which contain the labeled residue. Having been efficiently retrieved, the peptide could be subjected to classical Edman sequence analysis in order to pinpoint the site of attachment of the reagent in terms of the primary structure. Such a technique was proposed by us (2) in connection with our effort to identify a minor reaction which occurs during the photoaffinity modification of the delta-5-3-ketosteroid isomerase (EC 5.3.3.1) from *Pseudomonas testosteroni*. The reagents which have been employed have been delta-6-testosterone, 19-nortestosterone acetate, and testosterone. In the latter two cases it was found that the major photoreaction did not involve covalent attachment of the steroid reagent to the enzyme. Yet for the inactivations supported by these reagents, a minor reaction was detected which involved a covalent union between reagent and enzyme. These minor processes appeared to be active site specific since they could be suppressed by non-UV absorbing competitive inhibitors. The coupling efficiency was low since only 10-15% of the inactivating events involved covalent attachment. However, the low efficiency precluded the ready identification of the site or sites of attachment.

It occurred to one of us that the features of solid phase synthetic chemistry which have so benefitted peptide and oligonucleotide synthesis could also be advantageously applied to the problem of isolating the small population of peptides which had become covalently attached to a reagent in a photoaffinity labeling experiment. Quite independently, Singh *et al.* (3) recognized these same attributes of a solid phase methodology in affinity labeling experiments.

The present studies were undertaken in order to pinpoint the site of the covalent attachment reaction detected in photoaffinity inactivation of the steroid isomerase by steroid ketones.

A. *PRINCIPLE OF SOLID PHASE PHOTOAFFINITY LABELING*

The principle of the solid phase form of photoaffinity labeling is presented in Fig. 2. In the solid phase method the photoaffinity

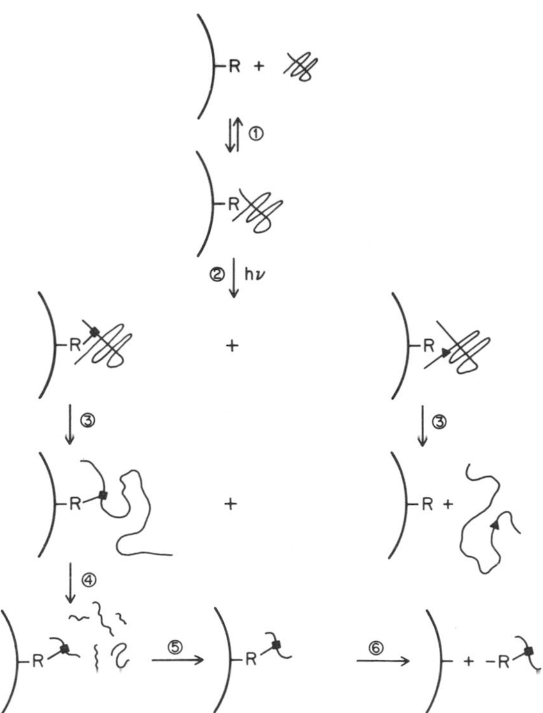

FIG. 2 *PRINCIPLE OF SOLID PHASE PHOTOAFFINITY LABELING. The photoreactive affinity reagent, R, is linked to an insoluble matrix,), by an arm, -, which contains a cleavable bond. Step 1: A solution containing the target protein is applied to the solid phase reagent and the target protein binds via noncovalent interaction with R. Step 2: Photolysis results in excitation of R to a reactive species (the electronically excited state or a reactive species derived, therefrom). The reactive species from R will react with the protein to produce two classes of modified protein: one class is modified (▲ = modified residue) without covalent attachment to R, the other class becomes covalently attached (■ = attachment site) to R. Some protein molecules escape modification due to rearrangement of R or reaction of R with the solvent (not shown). Step 3: The covalently attached subclass of protein molecules is separated from the other protein species by elution of the solid phase with a denaturing solvent. Only the covalently attached species remains with the solid phase. Step 4: A fragment of the attached polypeptide which contains the attachment site can be isolated after digestion of the solid phase-bound protein with a protease or with a chemical specific cleavage agent. Step 5: Unattached fragments are washed away with a strongly denaturing solvent. Step 6: The fragment containing the site of attachment is released from the solid phase by cleavage of the labile bond in the arm. The site of attachment of R is established by sequence analysis of the released fragment using Edman degradation or mass spectrometry. (From Ref. 29).*

reagent is presented to the target protein as a covalent conjugate with an insoluble, but solvated, macromolecular matrix. The target protein binds to this solid phase reagent to form a non-covalent complex exactly as would be the case in affinity chromatography. The solid phase complex is then irradiated with light - normally containing components having a wavelength below 400 nm - in order to induce photochemical reactions between the reagent moieties and the protein. This is shown in step 2 of Fig. 2. The system will undergo all of the reactions that would occur in the solution phase experiment. The excited state of the reagent chromophore may relax to the ground state or may undergo rearrangement. These processes will not modify the target protein. Alternatively, the excited state of the reagent may partition between two paths which modify the protein. These reactions are indicated in step 2 of Fig. 2. The various protein species which do not become covalently attached to the solid phase reagent can be removed by washing the solid phase with buffer, buffer plus competitive inhibitor, and denaturing solvents such as SDS and 6 M guanidinium chloride. The results of these washing procedures is indicated in step 3 of Fig. 2. Only the protein which has undergone a reaction leading to covalent attachment remains bound to the solid phase reagent.

In order to facilitate the identification of the site of covalent attachment, the solid phase reagent is constructed to possess a labile bond in the arm which links the photoaffinity ligand, R, to the solid phase matrix. The covalently attached polypeptide(s) could be released at this point by cleavage of this labile bond. This option would be selected if the goal was to identify or isolate a ligand-binding polypeptide from a complex mixture. However, if the goal was to identify the residue of attachment this would not necessarily be the optimal option. As indicated in step 4 of Fig. 2, a more efficient way to achieve this goal is to fragment the attached protein by chemical or enzymatic means, wash away the unattached fragments (step 5), and release the peptide fragment(s) which remain due to covalent bonding by cleavage of the labile bond in the arm (step 6). Structural analysis of the released fragment(s) by amino acid analysis and sequence determination by Edman degradation or mass spectroscopy may be used to pinpoint the site of covalent attachment in terms of primary structure.

B. *DESIGN OF SOLID PHASE REAGENT*

In designing a solid phase reagent consideration should be given to the structure and properties of the three parts of the reagent, the photoaffinity group, the arm, and the insoluble matrix. The photoaffinity group will normally have a structure closely similar to that of the natural ligand which binds in the target site to be labeled. It may, in fact, be the natural ligand, itself, if the ligand contains a UV light-absorbing group. This is the case for some steroids, for mono- and polynucleotides, and for several antibiotics. It is more common that the natural ligand does not contain such a group and so a suitable photoreactive group needs to be incorporated into the structure. Nitrene precursors such as aryl azides have been widely employed in soluble photoaffinity reagents (4). Carbene precursors such as diazo ketones (5) and diazirines (6) have been used by several investigators. It is important to check that appended photoreactive groups do not interfere with the specific binding of the ligand to the target site. The photoaffinity group also needs a second functional group which can serve as the point of attachment to the arm.

The arm is a chain of atoms which links the photoaffinity group to the matrix. One of its functions is the same as the arms used in affinity chromatography media; to space the photoaffinity group far enough from the matrix to avoid steric inhibition of binding by the macromolecular matrix. Thus, the arm should be of sufficient length to achieve this objective. The length required will vary from case to case as indicated by experience with affinity chromatography media (7) and will usually be in excess of 4 atoms. The arm should not, by itself, interfere with the specific binding of the photoaffinity group and this can be checked by binding or inhibition studies using arm-photoaffinity group conjugates in the solution phase. Another function of the arm is to provide a chemically cleavable bond which can be ruptured when it is desired to release covalently bound material from the photolyzed solid phase reagent. This bond as well as the rest of the arm must be stable to the conditions of photolysis, especially to the ultraviolet light employed. This eliminates disulfide bonds and diazo bonds as cleavage points since these structures are photolabile (8,9). Cis-diols are a possible cleavable link which is UV-stable, but should be used with caution since the periodate which is used to oxidatively cleave the carbon-carbon bond of the cis-diol has been shown to oxidize methionine residues to methionine sulfoxide (10); tryptophan, also, is subject to oxidation by periodate (10). These modifications, if not recognized, might cause ambiguity in the interpretation of amino acid analyses or sequence data. A cleavable bond which appears to meet most of the criteria is the ester linkage. This is readily cleaved by mild base hydrolysis or by super-nucleophiles such as hydroxylamine at neutral pH. Mild base hydrolysis at pH 10.5 will quantitatively cleave ester linkages without significant hydrolysis of peptide bonds, but there is a danger that disulfide bonded half-cystine residues might undergo beta elimination. This could be avoided by reduction and alkylation of disulfides prior to esterolysis with base. With hydroxylamine there exists the possibility of the cleavage of asn-gly peptide bonds, but this may not be a significant problem at pH values below 9 (11). In our work on the steroid isomerase we have used the ester bond and base hydrolysis, since this protein contains no disulfide bonds (12).

The matrix should possess all of the properties which are required of an affinity chromatography medium. Thus, it should have a porous and open structure in an aqueous environment, possess adequate mechanical stability to survive elution protocols and contain readily derivatized functional groups which will allow attachment of the arm by a UV light-stable linkage. In addition, the matrix must be UV-light transparent at the excitation wavelengths

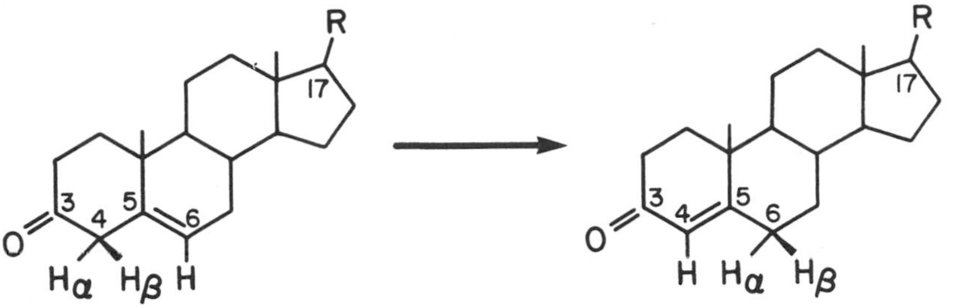

FIG. 3 *STEROID ISOMERASE REACTION. The steroid isomerase of P. testosteroni catalyzes the migration of the 5-6 double bond of the substrate to the 4-5 position of the product. The 4 beta proton of the substrate is transferred, with partial exchange with solvent protons, to the 6 beta position of the product. A wide variety of substituents, R, on C-17 are tolerated.*

of the photoaffinity group. Fortunately, these qualities are provided by the popular affinity chromatography matrix, agarose. Crosslinked polyacrylamide appears to offer similar properties, but we have not investigated it in this application. In our studies of steroid isomerase we have used O-carboxymethyl agarose rather than agarose as the starting material for the synthesis of solid phase photoaffinity reagents because of the convenience of its commercial availability.

C. *PROPERTIES OF STEROID ISOMERASE FROM PSEUDOMONAS TESTOSTERONI*

The steroid isomerase of *Pseudomonas testosteroni* catalyzes the allylic isomerization of delta-5-3-ketosteroids to the conjugated delta-4-isomers. The reaction is shown in Fig. 3. Other properties of the enzyme which are relevant to our photoaffinity labeling work are summarized in Table 1. The enzyme is composed of two, apparently identical, 13,394 dalton, polypeptide chains of known primary structure (12,13). It uses no cofactors or metal ions in the performance of the catalytic act. The protein noncovalently binds a wide variety of steroids, including androgens, progestins, and estrogens forming complexes having dissociation constants ranging from 1 to several hundred micromolar (14). The facts that

Table 1 *Properties of steroid isomerase*

Structure	Function
Mr = 26788 (2 x 13394)	*Binds wide variety of steroids*
Known amino acid sequence	*Ki = 1 to ca. 300 micromolar*
No known cofactors	*kcat/Km = 2.28 x* 10^8 M^{-1} sec^{-1}
6-Angstrom structure available	*pH dependence of kcat/Km indicates groups of pKa of 4.7 and 9.2 involved in the catalytic act*

the isomerase has a small size of known covalent structure, is a relatively stable protein in its native conformation, and is available in multimilligram quantities via a published purification protocol (15), have made it an attractive model for steroid binding proteins in efforts to develop methodology generally applicable to the study of this functional class of protein (2,16). In addition, the enzyme has been intensively studied in efforts to elucidate its catalytic mechanism and to identify amino acid residues at its active site (2,13,17-21). Affinity labeling studies by Pollack and his coworkers (20,21) have suggested that the enzyme can bind steroids in at least two orientations, a catalytically productive one in which the A-ring is juxtaposed to the catalytic groups of the substrate, and a nonproductive one in which the D-ring is in proximity to the catalytic groups. The pseudo-two-fold rotational symmetry of the androgen skeleton could allow such alternative binding modes. A recent crystallographic structure determination of the isomerase and its complex with 4-mercuri-17-beta-estradiol has located the probable steroid binding site on each protomer (22).

D. *A SOLID PHASE REAGENT FOR STEROID ISOMERASE*

Consideration of the design criteria (see above) for solid phase reagents and the steroid-binding properties of steroid isomerase, led us to delta-6-testosterone agarose as a potential reagent. The selection of delta-6-testosterone as the photoreactive group is based on several factors. This steroid was found to be a competitive inhibitor of the isomerase (Ki=60 micromolar), and its 17-hemisuccinate also competitively inhibited the enzyme (Ki=85 micromolar), indicating that the enzyme tolerated substitution on the 17-hydroxyl oxygen (2). In addition, solution phase photoinactivation studies showed that delta-6-testosterone supported a rapid photoinactivation of the enzyme and that the rate of photoinactivation was reduced in the presence of the UV-transparent competitive inhibitor, sodium cholate (Hearne and Benisek, unpublished experiments). Thus, delta-6-testosterone, like 19-nortestosterone acetate (16), behaved as an active site photoaffinity reagent. The acetyl, hemisuccinyl ethylenediamine arm was selected for its ease of synthesis and for the presence of hydrophilic amide groups which would tend to suppress nonspecific hydrophobic binding which is sometimes observed in affinity chromatography media bearing purely hydrocarbon type arms (7). Agarose was the matrix of choice due to its mechanical stability, transparency to UV light of wavelengths capable of exciting delta-6-testosterone and its availability as the O-carboxymethyl derivative, which facilitated the stepwise construction of the arm. The structure of delta-6-testosterone agarose is shown in Fig. 4. Its synthesis has been described by us earlier (2).

Hearne and Benisek (2) found that delta-6-testosterone agarose supported a rapid photoinactivation of the enzyme and that the inactivation rate was greatly reduced by the presence of sodium cholate in the fluid phase. These investigators also observed that 15-20% of the isomerase protein became covalently coupled to the agarose phase during photolysis in the presence of delta-6-testosterone agarose. Moreover, in subsequent experiments, Hearne and

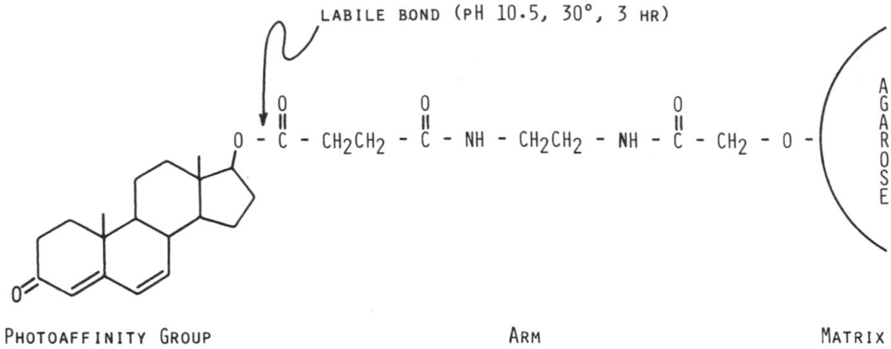

LABILE BOND (pH 10.5, 30°, 3 HR)

$O-\overset{O}{\overset{\|}{C}}-CH_2CH_2-\overset{O}{\overset{\|}{C}}-NH-CH_2CH_2-NH-\overset{O}{\overset{\|}{C}}-CH_2-O-$ AGAROSE

PHOTOAFFINITY GROUP ARM MATRIX

FIG. 4 *STRUCTURE OF DELTA-6-TESTOSTERONE AGAROSE. The solid phase photoaffinity reagent, delta-6-testosterone agarose, is composed of delta-6-testosterone as the photoaffinity group, N-acetyl-N′-hemisuccinyl-ethyenediamine as the arm and agarose as the matrix. The cleavable bond is the ester linkage to the 17-hydroxyl of the steroid.*

Benisek (2a) found that sodium cholate in the fluid phase greatly reduced the amount of the protein which suffered covalent attachment to the agarose phase. No attachment, or inactivation, occurred if photolysis was conducted in the presence of acetyl agarose as a control. Acetyl agarose has an acetyl group in place of the delta-6-testosterone hemisuccinate group. Moreover, deoxycholyl agarose, in which a deoxycholyl moiety replaces the delta-6-testosterone hemisuccinyl moiety, did not support photoinactivation of the enzyme (2) even though the isomerase non-covalently binds to this agarose derivative. These results indicated that the covalent attachment reaction involved one or more residues at the active site. Further experiments described in this paper were designed to identify the site(s) of covalent attachment.

II. MATERIALS AND METHODS

A. *MATERIALS*

Delta-5-3-ketosteroid isomerase was purifed from progesterone-induced cells of *P. testosteroni* by the procedure of Jarabak *et al.*

(15), as modified by Benson *et al.* (23), and Ogez *et al.* (13). The present studies were conducted using the major isozyme species having an isoelectric point of 4.75 and a specific activity of 40-50 kilounits per milligram (23). Androst-5-ene-3-beta-ol-17-one and delta-6-testosterone-17 beta-hemisuccinate were obtained from Steraloids. Androst-5-ene-3,17-dione was synthesized from the 3 beta alcohol by the method of Djerrassi *et al.* (25). The solid phase photoaffinity reagent, delta-6-testosterone-succinyl-ethylenediamine-O-carboxymethyl-agarose (delta-6-testosterone-agarose), was synthesized by coupling delta-6-testosterone-17 beta-hemisuccinate to the product of coupling ethylenediamine to O-carboxymethyl agarose, as described by Hearne and Benisek (20). O-carboxymethyl agarose was obtained as CM-Biogel A from BioRad. Sigma was our supplier of guanidinium chloride (grade 1) and TLCK-treated alpha-chymotrypsin. Triethylamine was the Sequenal grade of Pierce. This was redistilled from dansyl chloride before use. Phenol used in acid hydrolysis was the crystalline product of Mallinckrodt. Ethylenediamine, also, was obtained from this supplier. Constant boiling HCl was prepared from 6N HCl by double distillation. Sephadex G-15 was obtained from Pharmacia.

B. *PHOTOLYSIS PROCEDURE*

The method of photolysis was that described in detail by Hearne and Benisek (2). All photolyses were conducted under anaerobic conditions in order to minimize oxygen-dependent, nonspecific inactivation of the enzyme (2). Following the irradiation period, the minicolumns were eluted with 0.3 M potassium phosphate, pH 7.0, the same buffer containing 0.5 mM sodium cholate, 6 M guanidinium chloride, and then water. Control irradiations without applied isomerase were conducted in parallel to the experimental irradiations and were eluted by the same protocol.

C. *ISOLATION OF COVALENTLY BOUND PEPTIDE FRAGMENTS*

Following the elutions described above, the solid phases from forty "experimental" minicolumns and twelve "control" minicolumns were combined into pools of 5-6 minicolumns. These were subjected to chymotryptic digestion by application of a solution composed of 3.3 ml of 0.1 M sodium phosphate, 0.1 mM CaCl2, pH 7.6, containing 2 microliters of toluene and 25 microliters of 0.25 mg/ml chymotrypsin in 1 mM HCl. The digestions were allowed to proceed for 24 hr at 30 degrees with continuous stirring.

The digestion mixtures from the experimental and control minicolumns were combined into separate pools and these were washed with an additional 40 ml of 6 M guanidinium chloride followed by 300 ml of water. Then, attached peptide material was cleaved from the solid phase by hydrolysis with 25 ml of 1% TEA-HOAc, pH 10.5, for 3 hr at 30 degrees. The soluble hydrolysate was recovered by centrifugation and the solid phase was washed with 2 additional 15 ml portions of the TEA buffer used for the hydrolysis. The combined hydrolysate and washings were lyophilized. This hydrolytic procedure releases large quantities of amines from the solid phase reagent. These need to be separated from the peptide material since they interfere with sequence analysis by the Edman degradation.

The separation of peptide material from the low molecular weight amino compounds was achieved by chromatography on a column of Sephadex G-15 which was equilibrated and eluted with 0.1 M ammonium bicarbonate. The column effluent was monitored at 210 nm. Six peaks (A-F) were obtained. Amino acid analysis of acid hydrolysates of aliquots of the pooled peaks showed that only the first eluted peak, Peak A, contained a significant quantity of peptide. This was true, also, for the irradiated enzymeless control.

D. *IDENTIFICATION OF THE COVALENTLY BOUND PEPTIDE AND THE SITE OF PHOTOAFFINITY LABELING*

The identification of the nature of the peptide material released by the pH 10.5 treatment was made by amino acid analysis and amino acid sequencing. Acid hydrolysis prior to amino acid analysis was conducted *in vacuo* in sealed tubes containing constant boiling HCl and a small crystal of phenol at 110 degrees for 24 hr. The hydrolysate residues were subjected to amino acid analysis using either a Durrum D-500 or a Beckman 6300 amino acid analyzer. More detailed evidence on the nature of the site of covalent attachment was obtained by Edman degradation of the peptide peak, Peak A, of the Sephadex chromatography step, described above. Attempts to sequence the released peptide material without the gel filtration step were unsuccessful due to the large amount of low molecular weight amines which contaminated the peptide material. These amines reacted efficiently with phenylisothiocyanate and diverted this reagent from its intended reaction with the N-terminal residues of any released polypeptides. Sequence analysis of Peak A of the chromatographic eluate of the Sephadex G-15 chromatography was performed by a Beckman Model 890M sequencer using the 0.1 M Quadrol program, 050783. The identification and quantitation of PTH-amino acids was accomplished by reverse phase chromatography on C-18 loaded silica (26) or on the IBM Cyano reverse phase column (27). For Edman degradation cycles 7, 8 and 9 the PTH-amino acids were additionally analyzed by the chromatographic system of Mahoney and Hermodsen (28). Details of the sequence analysis of Peak A have recently been described by Hearne and Benisek (29).

III. RESULTS

The isomerase was photolyzed as its complex with delta-6-testosterone agarose until 20% of the applied isomerase activity was recovered in a sodium cholate eluate; the photolyzed isomerase protein was found to be distributed into 3 elution fractions. These are fraction 1: eluteable by 0.3 M potassium phosphate buffer, pH 7.0 containing 0.5 mM sodium cholate; fraction 2: eluteable by 6 M guanidinium chloride or sodium dodecyl sulfate but not by buffer plus cholate; fraction 3: eluteable by 1% triethylamineacetate buffer, pH 10.5, but not by any of the preceding treatments. A typical distribution of protein among these fractions is shown in Table 2. Native, unphotolyzed isomerase is found in fraction 1, as indicated in the dark control experiment, while isomerase which had

Table 2 *Partitioning of isomerase protein among elution fractions: Effect of sodium cholate*[a]

Experiment	Irradiation[b]	[cholate] mM	Recovery of activity in cholate fraction %	Recovery of protein in fraction eluted by: 1 Cholate %	2 SDS[c] %	3 TEA[d] %
1	+	0	20.6	41.1	22.5	20.0
2	-	0	99.3	94.2	2.6	2.7
3	+	3.4	102.2	87.4	1.1	5.2
4	-	3.4	105.4	94.4	1.0	0.1

[a] *Data are the mean of two experiments. Source: Ref. 2a.*

[b] *Samples were irradiated for 4 hr.*

[c] *SDS: sodium dodecyl sulfate. In these experiments, SDS was used to elute fraction 2.*

[d] *TEA: triethylamine acetate buffer, pH 10.5.*

undergone covalent attachment to the reagent is found in fraction 3. The nature of the protein found in fraction 2 is at present unknown; it might contain the decarboxylation product of aspartate-38 which was described by Ogez *et al.* (13) as the major protein photoproduct in the photoinactivation of the isomerase by 19-nortestosterone acetate. It might also contain unmodified isomerase subunits from isomerase dimers which had become covalently attached to the reagent via a single steroid-binding site. Sodium cholate is a competitive inhibitor of the isomerase, and it does not absorb light of the wavelengths which impinge on the photolysis reaction mixture. If the covalent attachment reaction is active site specific, then the inclusion of cholate in the photolysis reaction mixture should reduce the amount of isomerase protein which is found in fraction 3. The results of Table 2 show that this is the case and give us assurance that the coupling reaction involves an isomerase residue located in the active site.

In the effort to identify the site of covalent attachment, 40 minicolumns of reagent-isomerase complex were irradiated under anaerobic conditions as described in "Materials and Methods" and by Hearne and Benisek (2). Twelve similar minicolumns to which isomerase was not applied served as an irradiated, enzymless control. Fractions 1 and 2 were removed by the successive elutions described above. The covalently attached peptide material was not hydrolytically detached at this point, but was subjected to chymotryptic digestion as described in "Materials and Methods". Released chymotryptic peptides were removed by washing as described in "Materials and Methods" and the still attached peptide material was detached from the solid phase by the mild base hydrolytic procedure. Initial characterization of the released peptide material was made by amino acid analysis of the material released from the experimental and enzymeless control solid phases. The results of these analyses are given in Table 3. After taking into account the amino acids found to be present in the acid hydrolysate of the control sample, it was found that the amino acid composition of the released peptide material approximated that expected for peptide C-7 of the

Table 3 *Amino acid composition of Peaks A*[a]

Amino Acid	amount (nanomoles)[b] Peak A (experimental)	Peak A (control)	Peak A (net)	mole ratio (pro=2) Peak A (net)	C-7	C-7+C-8	C-7+C-9
Asp	6.3	0.7	5.6	2.0	3	3	3
Thr	5.5	0.2	5.3	1.9	2	2	2
Ser	5.2	0.7	4.5	1.6	2	2	2
Glu	7.7	1.0	6.7	2.4	2	3	3
Pro	5.8	0.3	5.5	2.0	2	2	2
Gly	7.3	2.5	4.8	1.7	2	2	2
Ala	9.1	0.5	8.6	3.1	2	4	4
Val	5.5	0.0	5.5	2.0	2	2	2
Met	0.0	0.0	0.0	0.0	0	0	0
Ile	1.5	0.1	1.4	0.5	0	1	1
Leu	0.6	0.3	0.3	0.1	0	0	0
Tyr	0.9	0.0	0.9	0.3	0	0	1
Phe	1.6	0.0	1.6	0.6	0	1	1
His	0.0	0.0	0.0	0.0	0	0	0
Lys	0.3	0.3	0.0	0.0	0	0	0
Arg	4.6	0.0	4.6	1.7	1	2	2

[a] *Source: Ref. 29*

[b] *The quantities of amino acids from experimental and control samples are normalized to equal amounts of delta-6-testosterone agarose for direct comparison and calculation of the net quantities*

chymotryptic digest obtained by Benson *et al.* (12) in their determination of the sequence of the enzyme. Significant discrepancies from the theoretical composition for C-7 were noted, however; the observed composition was 1 residue low in aspartic acid and was approximately one-half residue high in glu, ile, tyr, phe, and arg. Moreover, 3.1 instead of 2 residues of ala were found in the hydrolysate. Chymotryptic peptides other than C-7 had theoretical amino acid compositions which did not resemble that observed. The sequence (12,13) of the isomerase polypeptide in the vicinity of C-7 is shown in Fig. 5. The sequence includes that of the neighboring chymotryptic peptides, C-8 and C-9. We noted that the residues which appeared in excess amount were those contained in C-9, suggesting to us that the peptide material released from the solid phase might be an approximately equimolar mixture of C-7 and the double chymotryptic peptide, C-7 + C-9. Another possibility was that the released material was a mixture of C-7 and C-9. The presence of tyrosine in the analysis of the released material suggested that C-8 or the double chymotryptic peptide, C-7 + C-8, could be only a minor component of the released material.

As mentioned in "Materials and Methods", attempts to sequence the material released from the solid phase by the pH 10.5 buffer without further purification were unsuccessful, the problem being traced to the presence of large quantities of low molecular weight amines in the sample. In order to obtain sequence data on the peptide material, the pH 10.5 eluted material was freed of low molecular weight solutes by gel filtration chromatography on Sephadex G-15, as described in "Materials and Methods". The results of the Edman degradation are summarized in Fig. 6. A single sequence was obtained which agreed, with the exception of one residue, with that expected for the double peptide, C-7 + C-9. The fact that a single sequence was obtained eliminated the possibility that the released material was a mixture of C-7 and C-9. We obtained a sequence which was that of C-7 + C-9 with the exception of a blank residue at cycle 8, corresponding to aspartate 38 of the intact isomerase polypeptide. We had experienced no difficulty in the Edman degra-

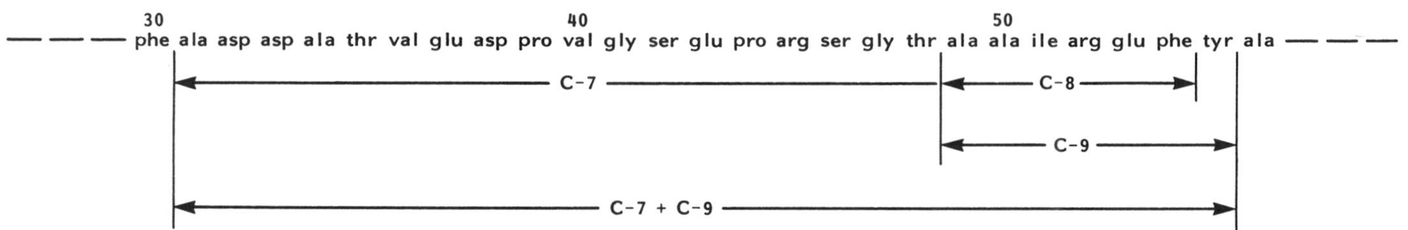

FIG. 5 *SEQUENCE OF ISOMERASE IN THE VICINITY OF PEPTIDE C-7. The sequence shown is that of Benson, et al. (12) as modified by Ogez, et al. (13). The nomenclature is that of Benson, et al. (12). (From Ref. 29).*

ala asp asp ala thr val glu x pro val gly ser glu pro arg ser gly thr ala ala ile arg glu phe tyr — — —

FIG.6 *SEQUENCE OF PEAK A. The sequence was obtained by automated Edman degradation as described in Materials and Methods. 4600 pmoles of peptide from Peak A were subjected to 28 cycles of sequence analysis. PTH-amino acids were identified by two independent hplc methods as described in Materials and Methods. No PTH-amino acid was detected at cycle 8 or at cycles 26-28. (From Ref. 29).*

dation of C-7 in our earlier investigation of the solution phase photoaffinity inactivation of isomerase (13). PTH-asp was readily identified at cycle 8 of the degradation of C-7 of the native isomerase in that study. Since all other residues of C-7 + C-9 were identified as those of the native enzyme, we conclude from our results that the site of photomodification, presumably by covalent attachment of the steroid, is at aspartic acid-38. These results have been obtained twice using two completely independent photoaffinity modification reactions.

IV. DISCUSSION

The present studies have provided evidence that the site of photoaffinity covalent attachment of steroid isomerase to the solid phase reagent, delta-6-testosterone agarose, is at residue 38 of the polypeptide chain. This conclusion is based on the amino acid analysis and amino acid sequencing of the chymotryptic peptide which was retained on the agarose phase even after extensive washing with the strong protein denaturant and solublizing agent, 6 M guanidinium chloride. The identity of the covalently attached peptide as a mixture of C-7 and the double chymotryptic peptide, C-7 + C-9, which are adjacent in the isomerase primary structure, is based on the observed amino acid composition which was well accounted for by this mixture as shown in Table 3, columns 4, 5 and 7. The fact that the observed composition was one residue low in the amount of aspartic acid expected for the unmodified peptide (columns 5-7) indicated that the site of covalent attachment was at one of the three aspartic acid residues. These are located at positions 32, 33, and 38. In order to pinpoint the particular aspartate, the peptide was sequenced by automated Edman degradation. The sequence obtained confirmed the hypothesis that the peptide was C-7 + C-9 or a mixture of C-7 and C-7 + C-9. Cycles 2 and 3 of the degradation gave good yields of PTH-asp, indicating that residues 32 and 33 were not modified. On the other hand, no PTH-amino acid could be detected at cycle 8, which corresponds to asp-

38 of the intact isomerase polypeptide. In previous studies we have sequenced C-7 obtained from native isomerase (13) and clearly demonstrated the presence of asp at position 38. Moreover, in the present study, no difficulty was encountered in the identification of residues 32 and 33 as asp. We conclude that our failure to detect a PTH-amino acid at cycle 8 of the degradation is due to its having been modified by the photoaffinity group of the solid phase reagent. Attachment to a steroid moiety may inhibit extraction of the anilinothiazolinone from the spinning cup of the sequencer; it is possible that the steroidal PTH-amino acid binds irreversibly to the C-18 columns used for identification of the PTH-amino acids by hplc.

It should be noted that asp-38 is the same residue which was identified by us (13) as the major site of photoaffinity modification by the photoreagent, 19-nortestosterone acetate. In this instance, the modification was found to be the decarboxylation of asp-38 to yield an alanine residue, which was identified by sequence analysis of C-7. No trace of PTH-ala was detected in the present study.

The affinity labeling studies of Pollack and his colleagues (20,21) have also demonstrated the presence of asp-38 at the steroid binding site. These workers found that a 17-beta-oxiranyl steroid alkylated asp-38 and that a 3 beta oxiranyl steroid alkylated an acidic amino acid in the vicinity of asp-38, possibly asp-38, itself. They raised the interesting possibility that steroids might bind in two different orientations in the steroid binding site, a normal mode and a retro mode differing by a 180 degree rotation about the pseudo two-fold rotation axis which is perpendicular to the B/C ring juncture of androgens.

Westbrook *et al.* (22) have determined a 6 Angstrom resolution crystal structure of the isomerase and its complex with the competitive inhibitor, 4-mercuri-17 beta-estradiol. They have described in general terms the low resolution geometry of the steroid binding site and the orientation of this steroid in the site. The site appears to be a deep pit, closed at the bottom; the steroid is oriented such that the A-ring with its appended mercury atom lies at the open end of the pit. It is not yet known whether this binding mode corresponds to the productive binding mode of substrates or to their retro binding mode. The location of asp-38 in the steroid binding pit is of interest since it has been suggested to be an important catalytic group (13,20,21). Is it near the base of the pit or near its mouth?

The results of the presently reported work provide evidence that asp-38 is located somewhere in the lower half of the pit. If one considers the structure of delta-6-testosterone agarose, one notes that the D-ring is attached through the succinyl ethylenediamine arm to a macromolecular matrix. Such a bulky, hydrophilic substituent on the D-ring should prevent the steroid from binding in a mode which places the D-ring at the base of the pit and the A-ring at the surface. A much more plausible binding mode would place the A-ring at the base of the pit and the D-ring at the surface. Since the photoreactive group of delta-6-testosterone is located in the A and B rings, we would expect that asp-38 lies close to these rings in the complex. Thus, we propose that asp-38 lies deep within the steroid binding pit and that the binding mode observed for 4-mercuri-17 beta-estradiol is a retro mode. Further experiments are planned to test this hypothesis.

ACKNOWLEDGMENTS

The authors are indebted to Alan Smith and the personnel of the UCD Protein Structure Laboratory for the amino acid analyses and amino acid sequencing reported in this work. We also wish to thank the National Institutes of Health for a research grant, AM-14729, which supports this research. We also wish to thank Professor Robert E. Feeney for valuable discussions relating to this research.

REFERENCES

1. V. Chowdry and F. H. Westheimer, Photoaffinity labeling of biological systems, *Ann. Rev. Biochem. 48:* 293 (1979).

2. M. Hearne and W. F. Benisek, Photoaffinity modification of delta-5-3-ketosteroid isomerase by light-activatable steroid ketones covalently coupled to agarose beads, *Biochemistry 22:* 2537 (1983).

2a. M. Hearne and W. F. Benisek, Modifications of delta-5-3-ketosteroid isomerase induced by ultraviolet irradiation in the presence of the solid phase photoaffinity reagent delta-6-testosterone agarose, *J. Protein Chem. 3*: 87 (1984).

3. P. Singh, S. D. Lewis, and J. A. Shafer, A support for affinity chromatography that covalently binds amino groups via a cleavable connector arm, *Arch. Biochem. Biophys. 193*: 284 (1979).

4. H. Bayley, Photogenerated Reagents in Biochemistry and Molecular Biology, *Elsevier Science Publishers, Amsterdam,* p. 29 (1983).

5. H. Bayley, Photogenerated Reagents in Biochemistry and Molecular Biology, *Elsevier Science Publishers, Amsterdam,* p. 36 (1983).

6. M. Nassal, 4′-(1-Azi-2-2-2-trifluoroethyl)-phenylalanine, a photolabile carbene-generating analogue of phenylalanine, *J. Am. Chem. Soc. 106*: 7540 (1984).

7. C. R. Lowe, An introduction to affinity chromatography, *Laboratory Techniques in Biochemistry and Molecular Biology, Vol. 7,* (T. S. Work and E. Work, eds), North-Holland, Amsterdam, p. 319 (1979).

8. A. D. McLaren and D. Shugar, Photochemistry of Proteins and Nucleic Acids, *Pergamon Press, Oxford,* p. 121 (1964).

9. R. J. Brewer, The photochemistry of the hydrazo, azo, and azoxy groups, The Chemistry of Hydrazo, Azo, and Azoxy Groups, Vol. 2, (S. Patai, ed), Wiley, New York, p. 936 (1975).

10. R. B. Yamasaki, D. T. Osuga and R. E. Feeney, Periodate oxidation of methionine in proteins, *Analyt. Biochem. 126*: 183 (1982).

11. P. Bornstein, Structure of alpha 1-CB8, a large cyanogen bromide produced fragment from the alpha 1 chain of rat collagen. The nature of the hydroxylamine-sensitive bond and composition of tryptic peptides. *Biochemistry 9:* 2408 (1970).

12. A. M. Benson, R. Jarabak, and P. Talalay, The amino acid sequence of delta-5-3-ketosteroid isomerase of *Pseudomonas testosteroni, J. Biol. Chem. 246*: 7514 (1971).

13. J. R. Ogez, W. F. Tivol, and W. F. Benisek, A novel chemical modification of delta-5-3-ketosteroid isomerase occurring during its 3-oxo-4-estren-17-beta-yl acetate dependent photoinactivation, *J. Biol. Chem. 252*: 6151 (1977).

14. H. Weintraub, F. Vincent, E. E. Baulieu, and A. Alfsen,

Interaction of steroids with *Pseudomonas testosteroni* 3-oxosteroid delta-4-delta-5-isomerase, *Biochemistry 16:* 5045 (1977).

15. R. Jarabak, M. Colvin, S. H. Moolgavkar, and P. Talalay, Delta-5-3-ketosteroid isomerase of *Pseudomonas testosteroni*, *Methods in Enzymol. 16:* 642 (1969).

16. R. J. Martyr and W. F. Benisek, Affinity labeling of the active sites of delta-5-3-ketosteroid isomerase using photo-excited natural ligands, *Biochemistry 12: 2172* (1973).

17. A. Viger, S. Coustal, and A. Marquet, A reinvestigation of the mechanism of *Pseudomonas testosteroni* delta-5-3-ketosteroid isomerase, *J. Am. Chem. Soc. 103:* 451 (1981).

18. T. M. Penning, D. F. Covey, and P. Talalay, Irreversible inactivation of delta-5-3-ketosteroid isomerase of *Pseudomonas testosteroni* by acetylenic suicide substrates, *J. Biol. Chem. 256:* 6842 (1981).

19. T. M. Penning and P. Talalay, Linkage of an acetylenic secosteroid suicide substrate to the active site of delta-5-3-ketosteroid isomerase: isolation and characterization of a tetrapeptide, *J. Biol. Chem. 256*: 6851 (1981).

20. R. H. Kayser, P. L. Bounds, C. L. Bevins, and R. M. Pollack, Affinity alkylation of bacterial delta-5-3-ketosteroid isomerase: identification of the amino acid modified by 17-beta-oxiranes, *J. Biol. Chem. 258*: 909 (1983).

21. C. L. Bevins, S. Bantia, R. M. Pollack, P. L. Bounds, and R. H. Kayser, Modification of an enzyme carboxylate residue in the inhibition of 3-oxo-delta-5-steroid isomerase by (3S)-spiro[5-alpha-androstane-3,2′-oxirane]. implications for the mechanism of action, *J. Am. Chem. Soc. 106*: 4957 (1984).

22. E. M. Westbrook, O. E. Piro, and P. B. Sigler, The 6-Angstrom crystal structure of delta-5-3-ketosteroid isomerase; architecture and location of the active center, *J. Biol. Chem. 259*: 9096 (1984).

23. A. M. Benson, A. J. Suruda, R. Shaw, and P. Talalay, Affinity chromatography of 3-oxosteroid-delta-4-delta-5-isomerase of *Pseudomonas testosteroni*, *Biochim. Biophys. Acta 348:* 317 (1974).

24. L. Schriefer and W. F. Benisek, An inexpensive computerized enzyme kinetics system based on a Gilford spectrophotometer and an Apple IIe microcomputer, *Analyt. Biochem. 141*: 437 (1984).

25. C. Djerassi, R. R. Engle, and A. Bowers, The direct conversion of steroidal delta-5-3-beta-alcohols to delta-5- and delta-4-ketones, *J. Org. Chem. 21:* 1547 (1956).

26. A. S. Bhown, J. E. Mole, A. Weissinger, and J. C. Bennett,

Methanol solvent system for rapid analysis of phenylthiohydantoin amino acids by high-pressure liquid chromatography, *J. Chromatog. 148*: 532 (1978).

27. M. W. Hunkapiller and L. E. Hood, Analysis of phenylthiohydantoins by ultrasensitive gradient high-performance liquid chromatography, *Methods in Enzymol. 91*: 486 (1983).

28. W. C. Mahoney and M. A. Hermodsen, Separation of large denatured peptides by reverse phase high performance liquid chromatography; trifluoroacetic acid as a peptide solvent, *J. Biol. Chem. 255*: 11199 (1980).

29. M. Hearne and W. F. Benisek, Use of a solid phase photoaffinity reagent to label a steroid binding site: application to the delta-5-3-ketosteroid isomerase of *Pseudomonas testosteroni*, *Biochemistry 24*: in press (1985).

10

Entry of Protein Toxins into Cells

Kirsten Sandvig and Sjur Olsnes

Department of Biochemistry
Norsk Hydro's Institute for Cancer Research
Montebello, Oslo, Norway

I. INTRODUCTION

A number of protein toxins including the bacterial toxins diphtheria toxin, *Pseudomonas aeruginosa* toxin, *Shigella* toxin, cholera toxin, pertussis toxin, anthrax toxin and the heat labile toxin from *E. coli*, as well as the plant toxins abrin, ricin, modeccin and viscumin (for review, see 1) exert their effect on cells after binding to the cell surface and penetration of an enzymatically active part of the molecule through the cell membrane and into the cell cytosol. Most of these toxins have two different functional domains. One polypeptide chain, designated the A-moiety, has enzymatic activity and is linked by a disulfide bond to the B-moiety which binds the toxin to receptors at the cell surface. The binding moiety consists of one or several polypeptides held together by noncovalent bonds. There is now good evidence that many of these protein toxins enter the cytosol from intracellular compartments which they reach after receptor mediated endocytosis. In spite of the structural similarities between the toxins, their entry mechanisms are not identical, and the mechanism of protein transfer across biological membranes is still not known in detail. The toxins have different requirements for entry, and they seem to enter the cytosol from different intracellular compartments.

Recently there has been an increasing interest in protein toxins with intracellular sites of action. A main reason for this interest is the cancerostatic properties of the plant toxins abrin and ricin (2,3,4) as well as the use of these toxins in the construction of target-specific cytotoxic conjugates (for review, see 5). In order to construct conjugates which are highly toxic more knowledge about the mechanism of action of the naturally occurring protein toxins is important.

In the present article recent information about the entry of the bacterial toxin diphtheria toxin as well as of the plant toxins abrin, ricin and modeccin is presented.

II. STRUCTURE-FUNCTION RELATIONSHIP IN DIPHTHERIA TOXIN

Diphtheria toxin is synthesized as a single polypeptide chain (Fig. 1). This form of the toxin is easily cleaved by proteases into two fragments linked by a disulfide bridge. Fragment A is an enzyme that inhibits protein synthesis by inactivation of elongation factor 2 after entry into the cytosol. The inactivation of elongation factor 2 is achieved by ADP-ribosylation of the protein. Fragment B binds the toxin to receptors at the cell surface and in addition it plays a role in the entry of the A fragment (for review, see 1).

There is now good evidence that diphtheria toxin enters the cytosol from the early acidic vesicles that are formed by endocytosis and that the low pH is required for the entry. Thus, it is possible to inhibit entry of diphtheria toxin by adding ammonium chloride and the ionophore monensin, compounds that both

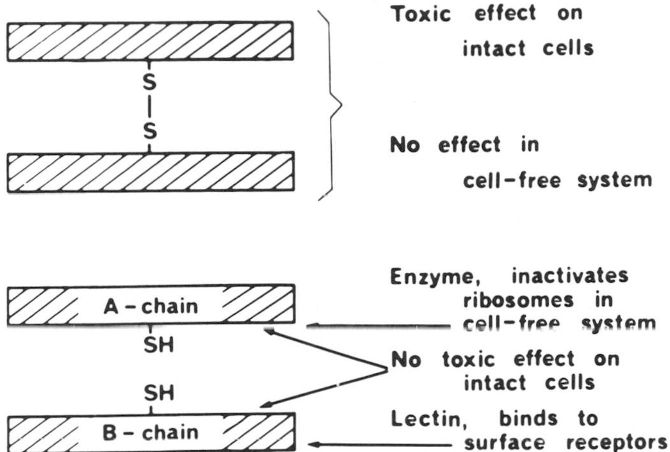

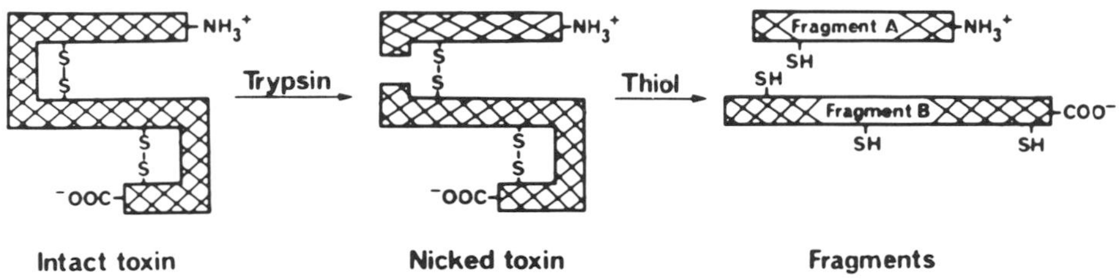

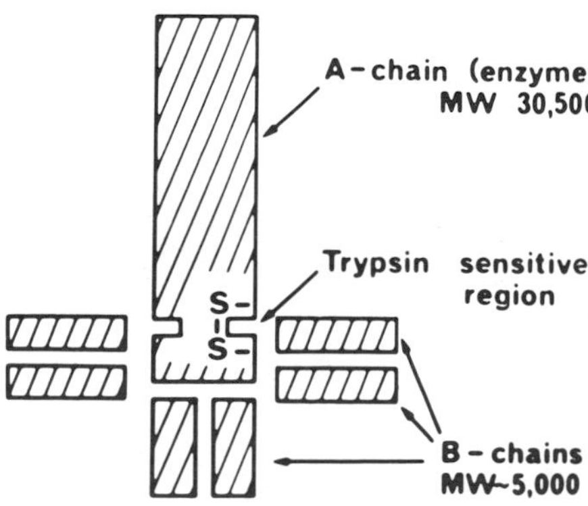

FIG. 1 *Schematic models of toxin structure.*

increase the pH in intracellular compartments. As shown in Fig. 2, it is even possible to induce a rapid entry of diphtheria toxin into cells by exposing cells with surface bound diphtheria toxin to low pH (6-8).

An important feature of the B-fragment is that it contains a hydrophobic domain. This domain is normally not exposed (9). However, at acidic pH a conformational change takes place in the toxin and the hydrophobic domain is then exposed. As a consequence, the toxin becomes able to bind nonionic detergents (8). Experiments with artificial lipid bilayers have shown that the hydrophobic fragment is inserted into the lipid bilayer as a transmembrane protein inducing the formation of voltage-dependent ion-permeable channels in the bilayer at low pH (4,5,10,11).

A sufficient low pH to expose the hydrophobic domain of the B-fragment i.e. pH < 5.4 is, under normal conditions, only found in intracellular vesicles. However, when toxin entry is induced by low pH in the medium the toxin apparently inserts itself directly into the surface membrane and the enzymatically active A-fragment is then transferred to the cytosol. In the following experiments we have to a large extent taken advantage of this possibility, since it is much easier to control ion gradients and the electrical gradient across the surface membrane than across membranes of intracellular vesicles.

A. *BINDING OF DIPHTHERIA TOXIN TO CELLS*

A number of conditions affect the binding of diphtheria toxin to cells. Thus, when cells were treated with low concentrations of tumor promoting phorbol esters like TPA, the ability to bind diphtheria toxin was strongly reduced (not shown). Also treatment with vanadate and fluoride strongly decreased the binding ability of the cells. To induce the reduction in binding ability, it was necessary to incubate the cells at physiological temperature, and there was no reduction in cells depleted for ATP. Since TPA activates protein kinase C (12) and since vanadate and fluoride

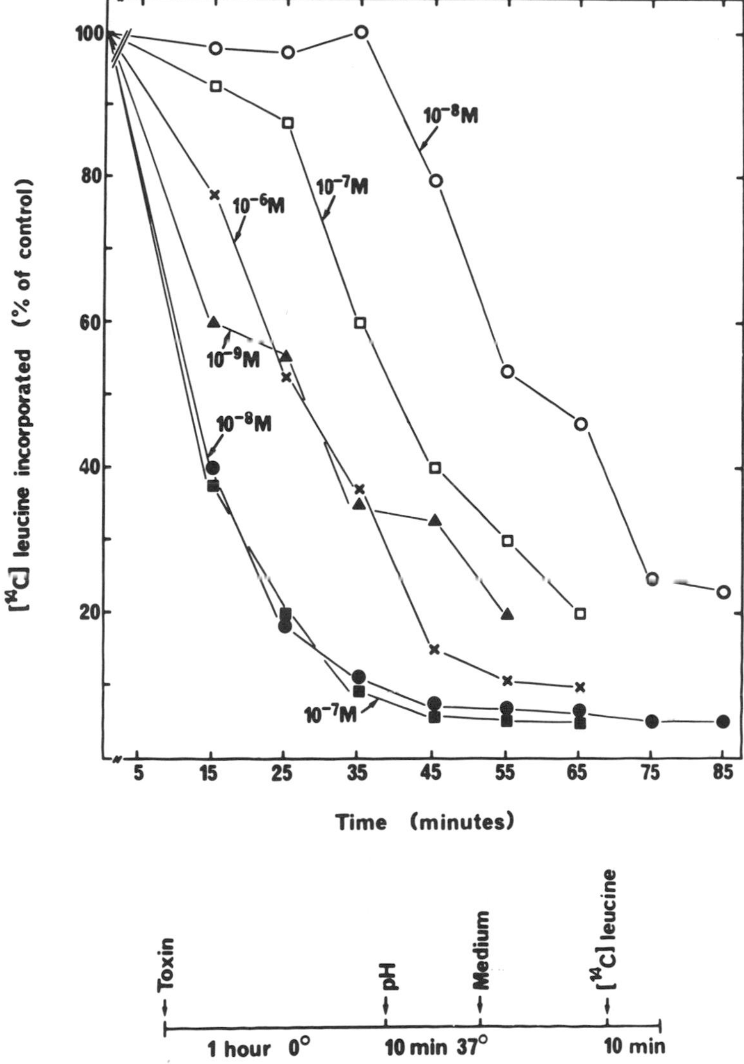

FIG. 2 *Effect of pH on the rate of protein synthesis inhibition after exposure of cells to diphtheria toxin. The indicated amounts of toxin were added to Vero cells in 24-well disposable trays, and after 1 hour incubation at 0°C the cells were incubated for 10 min at 37°C in medium with pH 7.2 (open symbols) or pH 4.5 (closed symbols). Prewarmed medium (pH 7.2) was then added and the ability of the cells to incorporate [^{3}H] leucine was measured after different periods of time.*

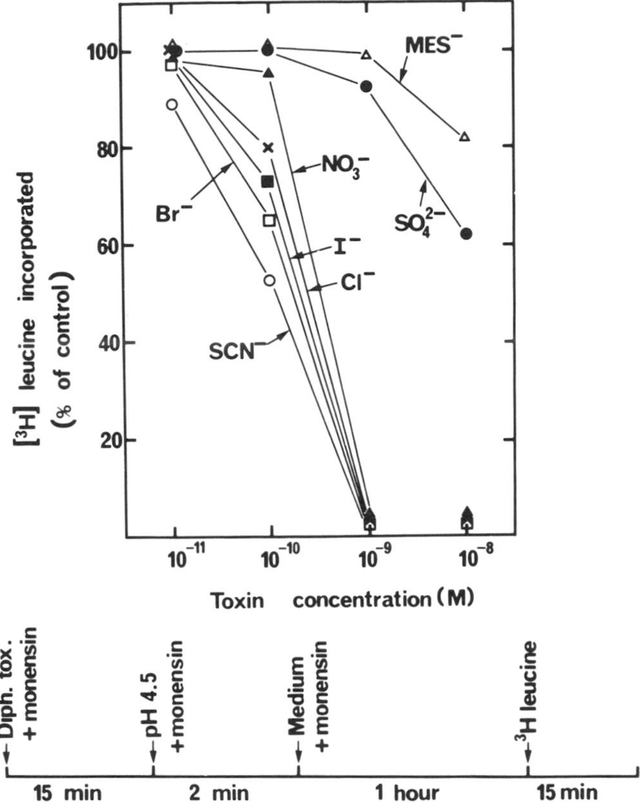

FIG. 3 *Ability of different anions to support diphtheria toxin entry at low pH. Vero cells were incubated for 15 min at 37°C with increasing concentrations of diphtheria toxin in Hepes medium containing 10 μM monensin. Then the medium was removed and a buffer, containing 20 mM MES, adjusted to pH 4.5 with Tris, 10 μM monensin, 0.14 M of the sodium salt of the anion indicated or 260 mM mannitol was added to the cells. After 2 min at 37°C this buffer was removed and Hepes medium containing 10 μM monensin was added. One hour later the rate of protein synthesis was measured.*

inhibit dephosphorylation reactions (13,14), it is therefore tempting to speculate that phosphorylation of the binding site for diphtheria toxin eliminates the binding capability of the cells.

For binding to occur it is also necessary that permeable anions are present in the buffer. Thus, when cells were incubated with ^{125}I-labelled diphtheria toxin in a buffer osmotically balanced with mannitol or with Na_2SO_4 or NaSCN, very little binding took place. Also treatment of cells with phospholipase C or trypsin decreased the binding of diphtheria toxin (1).

III. REQUIREMENTS FOR DIPHTHERIA TOXIN ENTRY AT LOW pH

There is now evidence for a role of permeant anions not only for binding of diphtheria toxin to the cell surface receptors, but also for the actual transport of fragment A across the membrane. When cells with prebound diphtheria toxin were exposed to medium without permeant anions, the major part of the bound toxin remained bound to the cells. In spite of this, we were unable to induce entry of the A-fragment into the cytosol at low pH when there was not permeant anions in the medium (Fig. 3). This supports the view that anions are required for the translocation across the membrane. Furthermore, when cells with prebound toxin were exposed to inhibitors of anion transport like DITC and piretanide, the cells were protected against the intoxication even in the presence of permeant anions, indicating that it is not sufficient for toxin entry that permeant anions are present, but that anion entry must be allowed to take place (Fig. 4).

When cells with prebound diphtheria toxin were exposed to low pH, the toxin was efficiently inserted into the cell membrane. Thus, after such treatment the toxin became inaccessible to labelling with ^{125}I with the lactoperoxidase method (Fig. 5) and both Cl^- transport and SCN^- transport into the cells were reduced (Fig. 6). Furthermore, the ability of the cells to take up $^{35}SO_4{}^{2-}$ and $^{36}Cl^-$ by antiport was strongly reduced (not shown).

Thus, the J_{max} for Cl^- was reduced from 1.6×10^8 ions x $cell^{-1} \times sec^{-1}$ to 6.2×10^7 ions x $cell^{-1} \times sec^{-1}$, while the J_{max} for $SO_4{}^{2-}$ was reduced from 3.1×10^6 ions x $cell^{-1} \times sec^{-1}$ to 3.7×10^5 ions x $cell^{-1} \times sec^{-1}$. On the other hand the K_m for the ions was the same before and after toxin treatment. This indicates that the inserted toxin somehow interferes with the anion antiporter in the cells. It is an interesting possibility

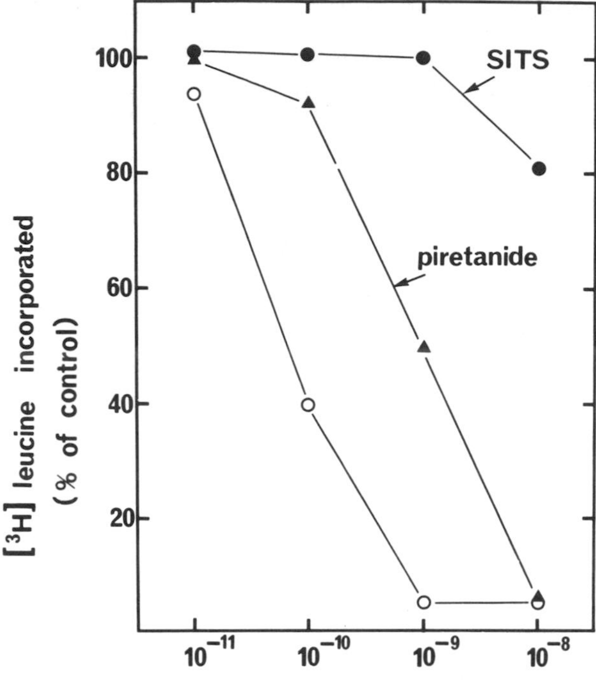

FIG. 4 *Ability of piretanide and SITS to inhibit entry of diphtheria toxin at low pH. Vero cells were incubated with 10 μM monensin and increasing concentrations of diphtheria toxin for 15 min. Then the medium was removed and the cells were incubated for 2 min with and without 2 mM piretanide or 0.1 mM SITS in a buffer containing 20 mM MES, adjusted to pH 4.5 with Tris, 10 μM monensin and 0.14 M NaCl. The buffer was then removed, Hepes medium with 10 μM monensin was added, and 1 hour later the rate of protein synthesis was measured.*

that the anion antiporter is identical with the diphtheria toxin receptor in Vero cells.

Diphtheria toxin appears to be inserted into the membrane at low pH even in the absence of permeant anions, although there was no evidence for transfer of the A-fragment to the cytosol. However, when the cells were subsequently treated with low pH in the presence of permeant anions, the cells were intoxicated (not

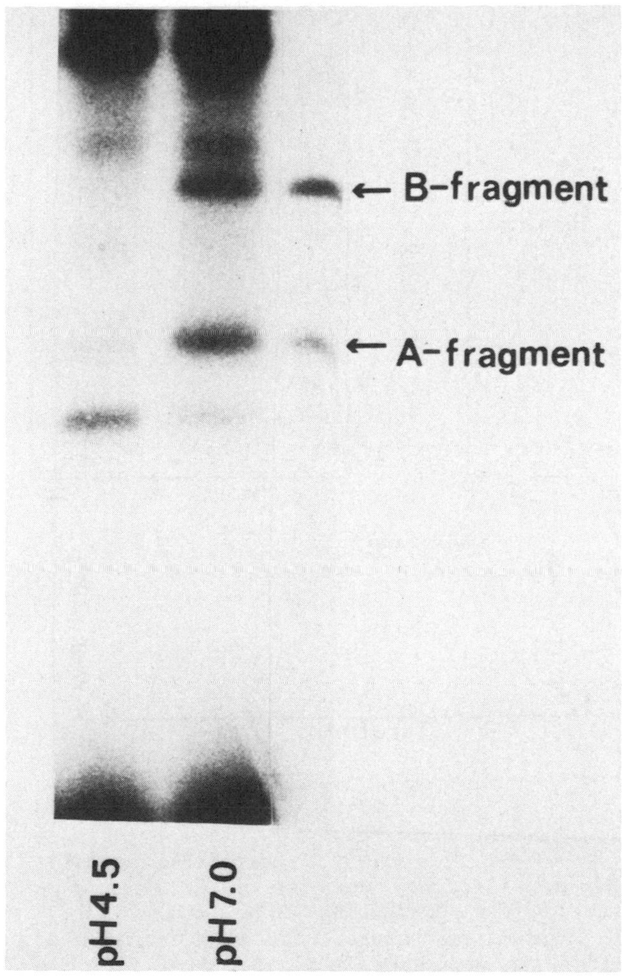

FIG. 5 *Insertion of diphtheria toxin into cells at low pH. Vero cells were incubated with diphtheria toxin (10 μg/ml) for 1 hour at 0°C before the medium was removed and the cells were exposed to pH 4.5 or pH 7.0 for 5 min at 37°C. The cells were then iodinated by using lactoperoxidase/H_2O_2 at pH 7.0, dissolved with Triton X-100, nuclei were removed by centrifugation, and antidiphtheria toxin and staphylococci were added. The immunoprecipitated proteins were then resolved by polyacrylamide electrophoresis under reducing conditions, the gel was dried and an autoradiogram was made. The migration of diphtheria toxin alone is shown in the right lane.*

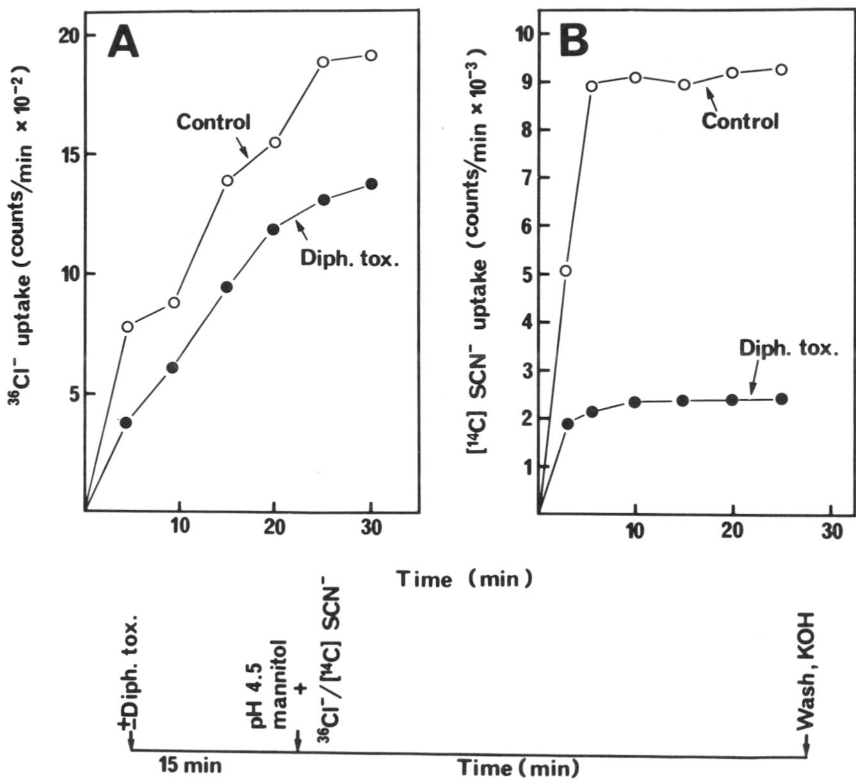

FIG. 6 *Ability of diphtheria toxin to inhibit chloride and thiocyanate uptake at low pH. Vero cells were incubated for 15 min with and without diphtheria toxin (5 μg/ml). The cells were washed in a buffer (20 mM MES/Tris pH 4.5) containing mannitol (260 mM) instead of NaCl (140 mM) and the cells were then incubated in the same buffer containing 0.17 μCi/ml $^{36}Cl^-$ or 1 μCi/ml [^{14}C] SCN^- at room temperature. The cells were washed in ice cold buffer after the indicated incubation times and the radioactivity was measured.*

shown). Apparently therefore, it is possible to divide the entry process into two parts. It should be noted that the entry during the second exposure to low pH did not occur unless the cells had been exposed to medium with neutral pH for a short interval between the two treatments with low pH. We think the reason for this requirement is that the cytosol had been acidified during the first exposure to low pH as the experiment was done in the presence of monensin, and that entry requires neutral cytosolic pH. The experiments therefore suggest that a pH gradient across the membrane is necessary for toxin entry. Possibly, the pH gradient acts as a driving force for the translocation.

If anti-diphtheria toxin was added before the second exposure to low pH, intoxication was inhibited. This indicates that the toxin is still exposed to the surrounding medium after insertion of the B-fragment into the membrane at low pH in the absence of permeant anions.

In order to further elucidate the requirement for transfer of the A-fragment across the membrane, we studied the role of the normal membrane potential. Since at the low pH fragment A has a positive charge, it was conceivable that the negative interior of the cell might pull the A-fragment across the membrane. To test this we incubated cells with prebound diphtheria toxin in a buffer containing isotonic KCl, which was shown to depolarize the cells, and then we exposed the cells to low pH to induce toxin entry. The results (Fig. 7) showed that there was no difference in the extent of intoxication whether the cells were incubated in the presence of NaCl or KCl. It therefore appears that a normal membrane potential is not required for the transfer of the toxin A-fragment across the membrane.

The cell interior has normally close to neutral pH. When the cells are exposed to low pH there is therefore a large pH gradient across the membrane. This gradient could act as a driving force for entry of the A-fragment.

To test if the pH gradient is indeed required for entry, we dissipated the pH gradient in Vero cells and then tried to induce toxin entry by exposing the cells to medium with low pH. For this purpose we incubated the cells with weak acids, such as acetic acid, which, in their protonated form, are able to penetrate the membrane. When the acids enter the cytosol they dissociate and thus decrease the internal pH.

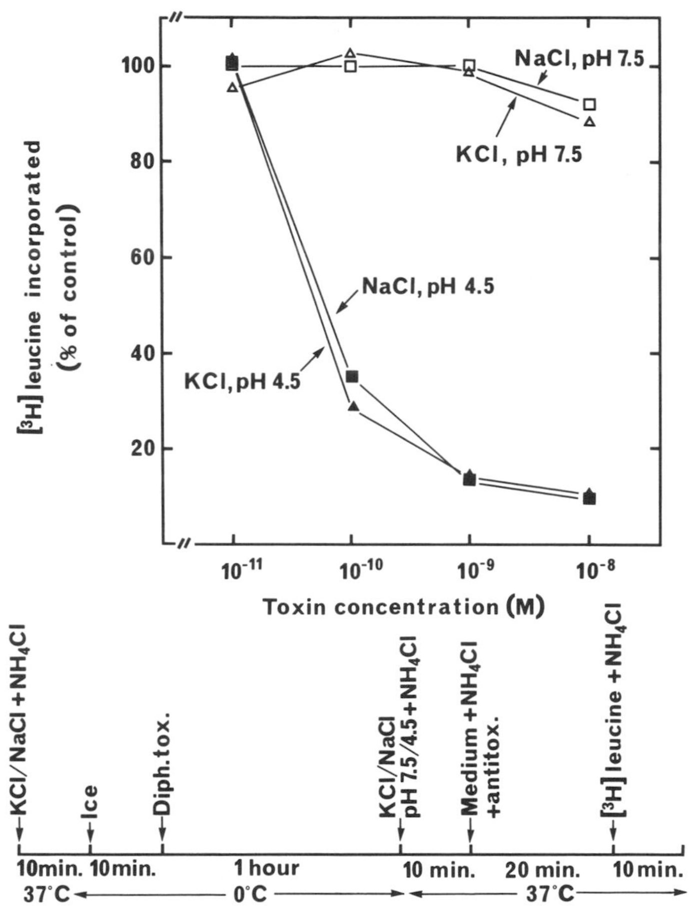

FIG. 7 *Entry of diphtheria toxin into depolarized cells. Diptheria toxin was added to Vero cells at 0°C in Hepes (20 mM) containing buffers pH 7.5 with 0.14 M KCl or 0.14 M NaCl. The cells were after 1 hour incubation at 0°C exposed to pH 7.5 or pH 4.5 as indicated, and after 10 min at 37°C normal medium containing antitoxin and ammonium chloride was added. Protein synthesis was then measured after a 20 min incubation at 37°C.*

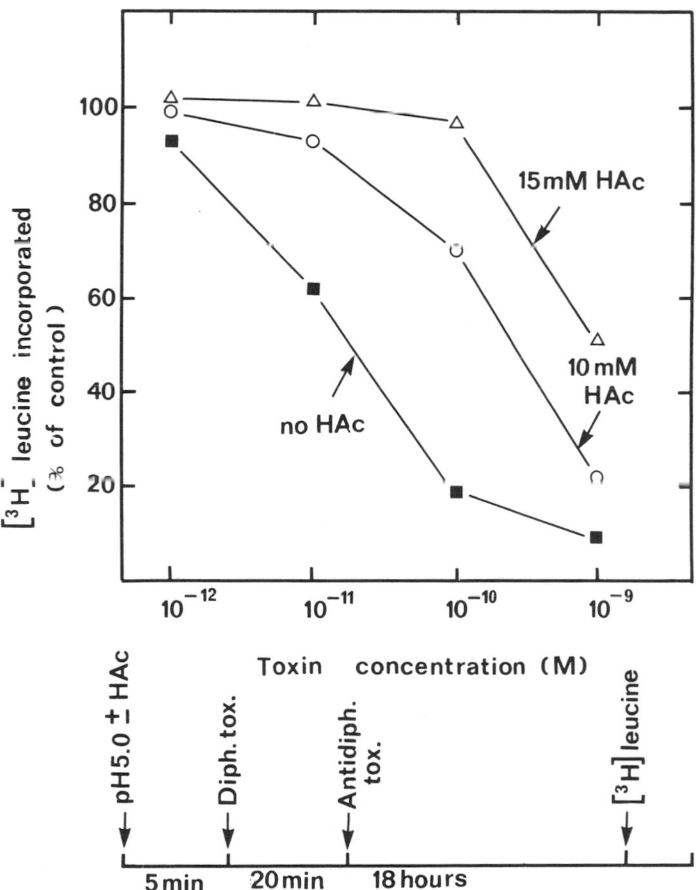

FIG. 8 *Ability of acetic acid to protect against diphtheria toxin. Vero cells were incubated in medium with pH 5.0 with and without the indicated concentrations of acetic acid for 5 min before increasing concentrations of diphtheria toxin were added. After 20 min further incubation growth medium with antitoxin and the cells were incubated overnight before protein synthesis was measured.*

The data in Fig. 8 show that when we had acidified the cytosol in this way, entry of diphtheria toxin A-fragment induced by exposure of the cells to medium with low pH, was strongly reduced (Fig. 8). Clearly, it is not sufficient for toxin entry that the exterior of the cell is exposed to low pH, and it appears that a certain pH gradient across the membrane is required for entry to occur.

IV. ENTRY OF MODECCIN INTO CELLS

As with diphtheria toxin, also the plant toxin modeccin seems to require low pH for entry. Thus, cells are protected against modeccin by NH_4Cl, chloroquine, and the protonophores FCCP and CCCP (15), which all increase the pH of acidic vesicles in the cells. This further indicates that also modeccin must be endocytosed before it is able to enter the cytosol.

The entry mechanism for modeccin seems to be more complex than the one described for diphtheria toxin. There is no rapid entry of modeccin from the cell surface into the cytosol when cells with bound modeccin are exposed to medium with low pH. Furthermore, when modeccin is added to cells, the protein synthesis in the cells declines after a much longer period of time than in the case of diphtheria toxin. Since both binding and endocytosis of modeccin occur rapidly, it appears that it is the transport of modeccin from the endocytic vesicle into the cytosol which is the most time consuming process. This is supported by the finding that compounds known to raise the pH of endocytic vesicles are able to protect well against modeccin even if they are added to the cells as much as 1 hour after the addition of modeccin (15).

When Vero cells were incubated with modeccin and diphtheria toxin at 20°C rather than at 37°C, the toxic effect of modeccin was much more strongly reduced than that of diphtheria toxin. It has been demonstrated that fusion of endocytic vesicles with lysosomes as well as transport within the Golgi apparatus are

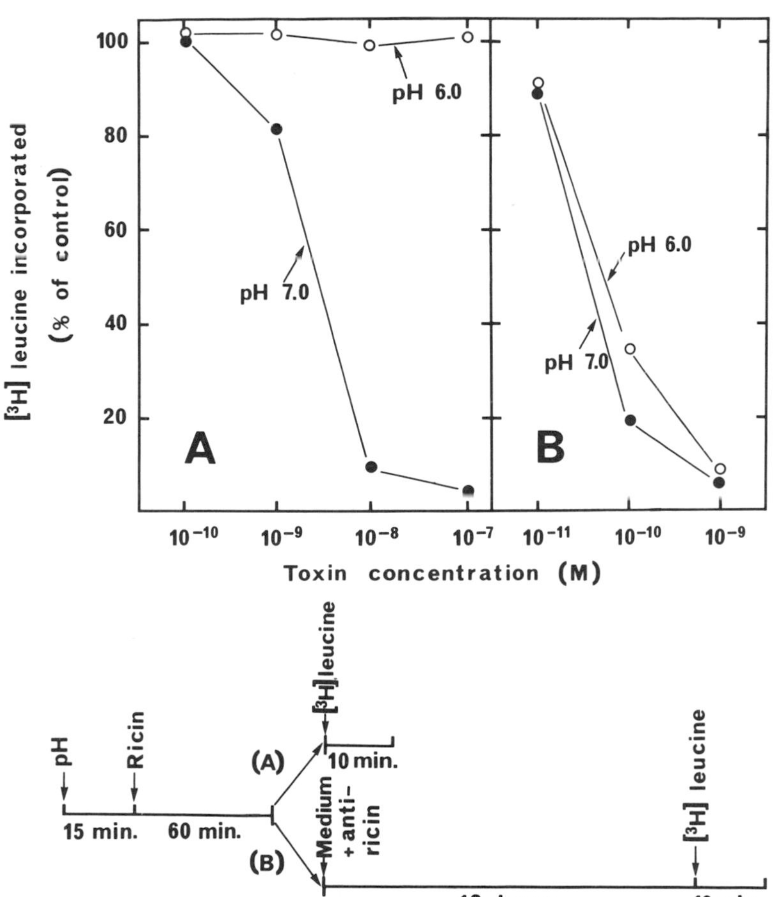

FIG. 9 *Ability of low pH in the medium to protect against ricin (A), and the ability of endocytosed ricin to enter the cytosol (B). Vero cells were incubated in a Hepes (20 mM) buffered medium at pH 6 or pH 7 for 15 min before ricin was added. After 1 hour incubation at 37°C protein synthesis was measured (A) or growth medium containing antiricin was added and the cells were incubated for 18 hours before protein synthesis was measured (B).*

inhibited at low temperature (16-18). The data therefore suggest that fusion of modeccin-containing vesicles with another cellular compartment is required for entry to occur.

As described earlier, the low concentrations of monensin that interfere with normal glycosylation and transport within the Golgi apparatus (19), strongly protect against modeccin (20). Possibly, modeccin must be transported to another cellular compartment, such as the Golgi apparatus, before it can enter the cytosol. We have earlier found that modeccin is much less effective in inhibiting cell-free protein synthesis than one would expect from its extreme toxicity to cells. This suggests that the molecule is altered after entry into the cells, a process which could take place in the Golgi apparatus.

V. REQUIREMENTS FOR ENTRY OF ABRIN AND RICIN

In the case of abrin and ricin there is no evidence for a requirement for intracellular low pH. Amines and ionophores known to increase the pH of acidic compartments do not protect cells against these plant toxins. On the contrary, NH_4Cl and low concentrations of monensin sensitize cells to abrin and ricin (20,21).

In fact, neutral or slightly alkaline pH in the medium was found to be required for entry of the plant toxins. The cells gradually became more protected when the pH was reduced from 7 to 6 (Fig. 9), although there was no corresponding reduction in the binding and endocytosis of the toxin at the low pH (20). The reason for the protection is not clear. However, ricin that was endocytosed during the incubation at low pH is able to enter the cytosol when the cells are incubated overnight in normal medium, suggesting that also this toxin is able to enter the cytosol from an intracellular compartment (Fig. 9).

The low concentrations of monensin required for the sensitization are unlikely to affect endosome acidification. However, such low concentrations of monensin have been found to

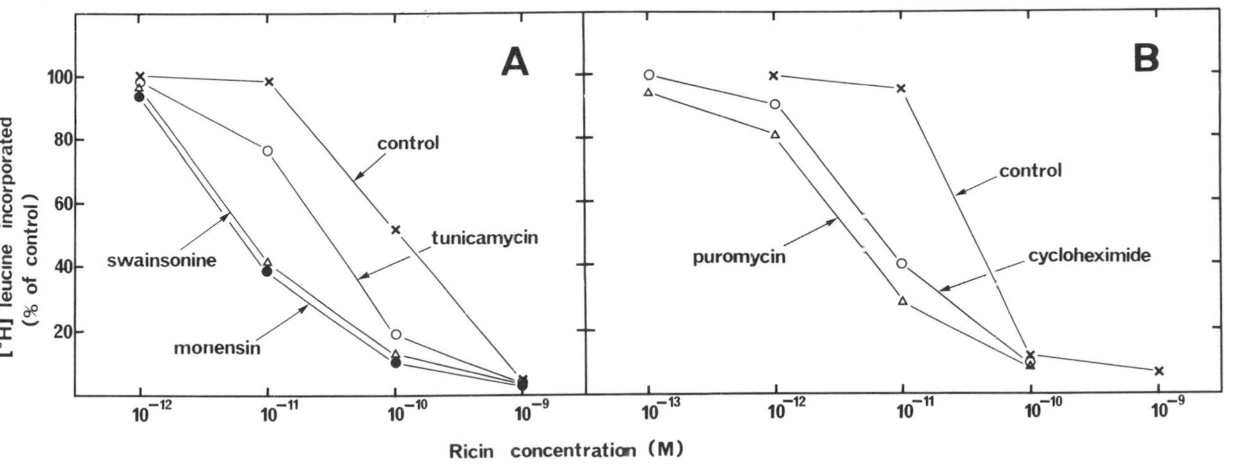

FIG. 10 *Ability of tunicamycin, swainsonine, monensine, puromycin, and cycloheximide to sensitize Vero cells (A) and HeLa cells (B) to ricin. Cells were preincubated for 30 min in the absence and presence of tunicamycin and for 15 min with a μg/ml swainsonine, 0.1 μM monensin, 1 mM puromycin, and 10 μg/ml cycloheximide. Then increasing concentrations of ricin were added and the incubation was continued at 37°C for 3 hours. Finally, protein synthesis was measured during a 15 min interval.*

interfere with protein glycosylation in the Golgi apparatus. We therefore tested if other compounds known to interfere with glycoprotein synthesis and processing sensitized cells to abrin and ricin. As shown in Fig. 10, also tunicamycin, that inhibits additions of carbohydrate to asparagine (22), and swainsonine that inhibits processing of carbohydrates on glycoproteins (23), also sensitized the cells to ricin.

We considered the possibility that ricin competes with newly formed glycoproteins for modification by enzymes in the Golgi apparatus or for transport between the different Golgi elements. To test this, we studied the effect of ricin on cells where protein synthesis had been blocked with cycloheximide or puromycin. As shown in Fig. 10, these compounds sensitized the cells to ricin. Similar results were obtained with abrin (data not shown). When the localization of ricin in the cells was studied by electron microscopy using ricin coupled to horseradish peroxidase, we found ricin in <u>trans</u> Golgi elements. However, at the present time we can not conclude that transport of abrin and ricin to the Golgi apparatus is an obligatory step in the entry of these toxins into the cytosol.

In further attempts to characterize the entry process, we have studied the ion requirements for entry. There seems to be a requirement for Ca^{2+} transport. Thus, when Ca^{2+} was removed from the medium, substituted with Co^{2+}, or when the Ca^{2+} entry was blocked with verapamil, the cells were protected against abrin and ricin as well as against modeccin. Control experiments showed that this protection was not due to reduced binding or endocytosis of the toxins. The calcium requirement appeared to be associated with the A-fragment. However, it is not known in which cellular compartment the required calcium transport takes place (24).

VI. CONCLUSION

Diphtheria toxin entry involves binding of the toxin to cell surface receptors and transfer of the enzymatically active fragment A across the membrane. The binding of toxin to the cells is reduced by several compounds known to increase the level of phosphorylation in cells. Furthermore, permeable anions are required for optimal binding. Also the transport of fragment A across the membrane requires permeable anions. Although insertion of fragment B into the membrane occurred in buffers osmotically balanced with mannitol, it was necessary to expose the cells to low pH and permeable anions simultaneously for entry of fragment A to occur. Also a proton gradient across the membrane appears to be required for entry.

The plant toxin modeccin also seems to require low pH for entry. However, the entry mechanism differs from that of diphtheria toxin. Exposure of cells with surface-bound toxin to low pH does not induce entry of modeccin.

Also abrin and ricin appear to enter the cytosol from intracellular vesicles. However, these toxins do not appear to require low pH for entry and they enter most efficiently when the pH of the medium is slightly alkaline. The entry of abrin and ricin appears to depend upon a Ca^{2+} flux. This is also the case with modeccin. Whereas both binding to cell surface receptors and endocytosis are involved in the normal entry of all toxins here studied, the toxins apparently utilize different mechanisms to gain access to their target in the cytosol.

REFERENCES

1. S. Olsnes and K. Sandvig, Entry of polypeptide toxins into animal cells *Endocytosis* (I. Pastan and M. C. Willingham, eds.), Plennum Press, p. 195 (1985).
2. J.-Y. Lin, K.-Y. Tserng, C.-C. Chen, L.-T. Lin and T.-C. Tung, *Nature (Lond.)* 227: 292 (1970).
3. Ø. Fodstad, G. Kvalheim, A. Godal, J. Lotsberg, S. Aamdal, H. Høst and A. Pihl, *Cancer Res. 44*: 862 (1984).
4. Ø. Fodstad, S. Olsnes and A. Pihl, *Cancer Res. 37*: 4559 (1977).
5. E. S. Vitetta, K. A. Krolick, M. Miyama-Inaba, W. Cushley and J. W. Uhr, *Science 219*: 644 (1983).

6. R. K. Draper and M. I. Simon, *J. Cell Biol. 87*: 849 (1980).

7. K. Sandvig and S. Olsnes, *J. Cell Biol. 87*: 828 (1980).

8. K. Sandvig and S. Olsnes, *J. Biol. Chem. 256*:9068 (1981).

9. P Boquet, M. S. Silverman, A. M. Pappenheimer Jr. and W. B. Vernon, *Proc. Natl. Acad. Sci. USA 73*: 4449 (1976).

10. B. L. Kagan, A. Finkelstein and M. Colombini, *Proc. Natl. Acad. Sci. USA 78*: 4950 (1981).

11. J. J. Donovan, M. I. Simon, R. K. Draper and M. Montal. *Proc. Natl. Acad. Sci USA 78*: 172 (1981).

12. Y. Nishizuka, *Nature 308*: 693 (1984).

13. G. Carpenter, *Biochem. Biophys. Res. Commun. 102*: 1115 (1981).

14. M. Mumby and J. A. Traugh, *Biochemistry 18*: 4548 (1979).

15. K. Sandvig, A. Sundan and S. Olsnes, *J. Cell Biol. 98*: 963 (1984).

16. W. A. Dunn, A. L. Hubbard and N. N. Aronson Jr., *J. Biol. Chem. 255*: 5971 (1980).

17. J. Saraste and E. Kuismanen, *Cell 38: 535 (1984)*.

18. K. S. Matlin and K. Simons, Cell 34: 233 (1983).

19. A. M. Tartakoff, *Cell 32*: 1026 (1983).

20. K. Sandvig and S. Olsnes, *J. Biol. Chem. 257*: 7504 (1984).

21. K. Sandvig, S. Olsnes and A. Pihl, *Biochem. Biophys. Res. Commun. 90*: 648 (1979).

22. A. D. Elbein, *Trends Biochem. Sci. 6*: 219 (1981).

23. D. R. P. Tulsiani, T. M. Harris and O. Touster, *J. Biol. Chem. 257*: 7936 (1982).

24. K. Sandvig and S. Olsnes, *J. Biol. Chem. 257*: 7495 (1982).

11

Immunotoxins: Chemical Variables Affecting Cell Killing Efficiences

Jon W. Marsh and David M. Neville, Jr.

Laboratory of Molecular Biology
National Institute of Mental Health
Bethesda, Maryland

I. INTRODUCTION

The concept of an antibody-toxin conjugate was developed to give a naturally occurring cytotoxic protein specificity, through the coupled antibody, for a pathological or undesirable cell line. Native protein toxins lack cellular specificity, and in general, are toxic to all cells within a sensitive species. Even with this complication of non-specificity, the choice of natural protein toxins over other toxic compounds (radionuclides and antimetabolities for example) for antibody conjugates is due to two unique features: 1) the protein toxins are highly potent (LD50: 10 ng - 1000 ng/kg); 2) protein toxins contain their own mechanisms for gaining entrance to cells. Extensive use of the plant toxin ricin and the bacterial diphtheria toxin has occurred since the imparted toxicity is due to an enzymatic activity, and thus a very limited number of toxin molecules are required to reach their site within a cell to destroy it. These toxins have remarkable structural-functional similarities: they are synthesized as a single polypeptide (1-3), and are proteolyzed to yield two subunits, A and B. These subunits are joined only by a native disulfide bond, and once internalized into the cell, the host's cleavage of the disulfide appears to be essential for cytotoxicity (4). The A subunit possesses the enzymatic capability to inhibit protein synthesis. Ricin A chain is capable of specifically inactivating the 60s subunit of eukaryotic ribosomes (5) whereas the A chain of diphtheria toxin catalyzes the transfer of the ADP-ribosyl moiety of NAD to elongation factor 2, thus inactivating it (6). These enzymes are so potent that a single molecule within the cytosol will totally inactivate the cell's protein synthesis capability (7, 8), thus their high potential for use in the construction of clinically relevant antibody conjugates.

Following the binding of toxins to their respective cellular receptors, the intoxication process involves a minimum of four steps. The first is endocytosis, which begins immediately once the ligand is bound (9). Endocytosis is followed by a dose and pH dependent processing step which is obligatory for the translocation step, where the enzymatic A chain or whole toxin crosses the membrane barrier that separates the extracellular and intravesicular space from the cytosol. The entry of the enzymatic element of the toxin into the cytosol leads to the forth step: enzymatic inactivation of the protein synthesizing machinery.

One can measure cellular protein synthesis by adding a radio-labeled amino acid to the cellular milieu, and by adding the label at timed intervals to aliquots of cells, the development of cellular intoxication can be observed kinetically. Typically, there is a lag period during which no toxic effects are evident, followed by an exponential decline in protein synthesis. This (pseudo)first order inactivation can be made linear by placement on a semi-log plot (see Fig. 1). This figure, to be discussed in more detail later, is typical of the toxic process. Since endocytosis of toxins begins immediately on incubation, the lag period, and potentially, the first order rate-limiting step, appear to occur as an intracellular process. Recent work in this laboratory has demonstrated for diphtheria toxin that the third step, or translocation of the toxic A chain into the cytosol, is the rate-limiting process (10). Thus improvements in the efficacy and potency of toxin-conjugates would be easily detected by alteration of this rate-limiting process, as demonstrated on kinetic curves.

The initial interaction of a toxin with a cell is through the B, or binding, subunit. The B subunits of these toxins bind to membrane components, and in the case of ricin, specificity for galactosyl-terminating complex carbohydrates has been established (11). In order to gain cell specificity, that is, limit conjugate uptake to the specificity of the antibody, the B subunit must be blocked or removed. By including lactose in the medium, a conjugate of ricin and an anti-mouse leukemia antibody has been demonstrated to possess high specificity and toxicity toward a

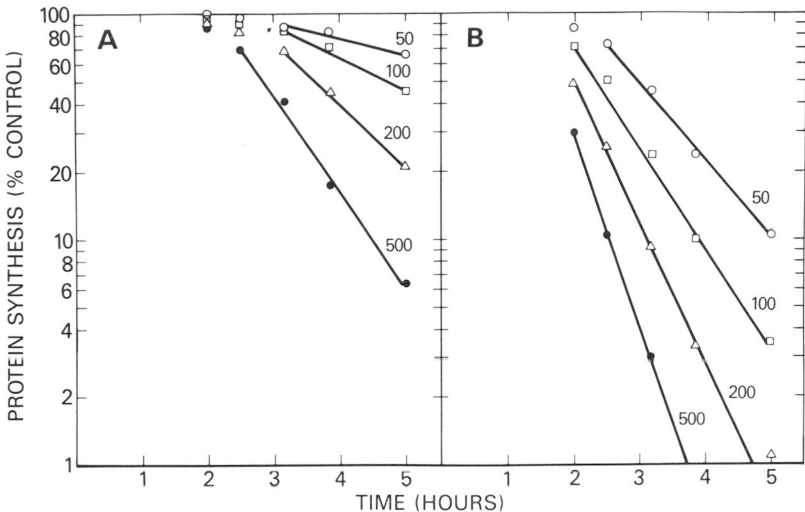

FIG 1. *Kinetics of protein synthesis inhibition by two different immunotoxins. Varying levels of conjugate (expresssed in ng/ml) were exposed to target cells for the specified times followed by a short pulse of* ^{14}*C-leucine. The data points, placed at the center of the 60 min pulse, represent the percent incorporated label as compared to a control lacking toxins. A. Conventional conjugate. B. Peptide-linker conjugate. From reference 30.*

cultured mouse leukemia cell line, 10^2 to 10^3 times more toxic than a non-target cell line (12).

Alternatively, one can remove the B subunit and achieve high specificity in an antibody-ricin A chain conjugate (13, 14); however, the importance of the B chain for optimal cytotoxicity has been established by comparing antibody-ricin and antibody-A chain conjugates and from the demonstration that antibody-ricin A chain conjugates are highly potentiated, 5 fold, by the addition of free B chain (12).

Although the binding of diphtheria toxin to a cell can be partially inhibited by various polyphosphate compounds such as various nucleotides and inositol hexaphosphate (15, 16), this approach has not been taken with this toxin. Most of the conjugate work has involved the A chain only (see Table 1). Additionally, since most of the human population has been immunized against diphtheria toxoid, the eventual clinical application of a conjugate utilizing diphtheria toxin is unlikely.

This review will limit itself to the protein tailoring involved in the development of efficient immunotoxins. Readers wishing to learn more about the nature of toxins,

Table 1 *Summary of Constructed Toxin-antibody Conjugates*

Antibody-Toxin Linkage		Activated Species	Thiolated Species	Reference
I. Thioether	A.	Toxin	Antibody	
		Ricin + MBS	IgG + DTT	12, 47
		Ricin + MBS	IgG + 2-imino-thiolane	24
		Ricin + N-hydroxysuccinimide ester of iodoacetic acid	SPDP thiolated Ig	64
	B.	Antibody	Toxin	
		$F_{(AB')2}$ + MBS	Ricin A	69
		IgG + MBS	Ricin A	67
		IgG + MBS	SPDP thiolated PAP	67
		F_{AB}+ N,N'-o-phenyldimaleimide	Ricin A	69
II. Disulfide	A.	Toxin	Antibody	
		S-sulfonated DTA	$F_{AB'}$	63
		Pseudomonas +2-iminothiolane + DTNB	IgG + 2-iminothiolane	76
	B.	Antibody	Toxin	
		IgG + SPDP	Ricin A	13, 77-80
		"	DTA	14, 78, 80
		"	Abrin A	68
		"	SPDP thiolated Ricin	81
		"	SPDP thiolated Gelonin	72
		"	SPDP thiolated PAP	67
		$F_{(AB')2}$ + SPDP	Ricin A	67, 69, 83
		"	SPDP thiolated PAP	67
		$F_{AB'}$ + DTNB	Ricin A	54, 69, 82
		F_{AB} + SPDP	Ricin A	82
		IgG + methyl 3-merceptopropionimidate + DTNB	Ricin A	61
		IgG/IgM + 3-(2-pyridyldithio) propionic acid coupled via carbodiimide	Ricin A	58, 59
		IgG + cystamine dihydrochloride coupled via carbodiimide	DTA	57

as well as biological and clinical features of immunotoxins, should examine some recent reviews on this matter (17, 18).

II. ANALYTICAL TECHNIQUES

As it is with most scientific advances, much of the progress in the field of toxin conjugates over the past decade was due to progress outside of the field. The development of monoclonal antibody methodologies (19) as well as synthesis of the various heterobifunctional cross-linking reagents discussed in this review have brought enough definition to the work so that questions could be both asked and answered. By combining these new developments with the established knowledge of toxins and cell-culturing techniques, experimentalists have been able to adequately improve the specificity and toxicity of immunotoxins to the point that the conjugates are now viable adjuncts to certain *in vitro* clinical procedures (17).

A measure of the immunotoxin's ability to kill cells can be accomplished by two different methods. One method, the clonogenic assay (20), involves the dilution of cells into a semi-solid medium, such as agar or methylcellulose, after exposure to the immunotoxin. After an incubation of several days, growth of visible colonies from single "surviving" cells then becomes a measure of the immunotoxin's ability to kill cells. This assay is particularly sensitive. By using serially diluted cells, one can measure the killing of 99.9999% of a target cell population (21). This extent of kill is also described as a "six log" kill. The second method commonly used is to follow treated cells in their ability to synthesize protein. Operationally, cells exposed to immunotoxin are pulsed with a radio-labeled amino acid, then the cellular protein is collected and the level of label incorporated is determined. This pulse can occur at the end of an 18 - 48 hr incubation after exposure to the immunotoxin at various concentrations (see Fig. 2) and is called a "titration" or dose response study. This figure is a comparison of two different conjugates possessing different potencies. Alternatively, one can pulse aliquots of cells at timed intervals, resulting in an exponential decline curve of protein synthesis (22). An example of a kinetic experiment, comparing two different conjugates, can be seen in Fig. 1A and B. The slopes of the first order inactivation curves (made linear by plotting on a semi-log scale) represent the rate of protein synthesis inactivation. The slope, measured in logs/hr, is generally given a positive sign. The rate of this process relates to the level of bound toxin or immunotoxin through a power function (23, 24), $y = a x^b$,

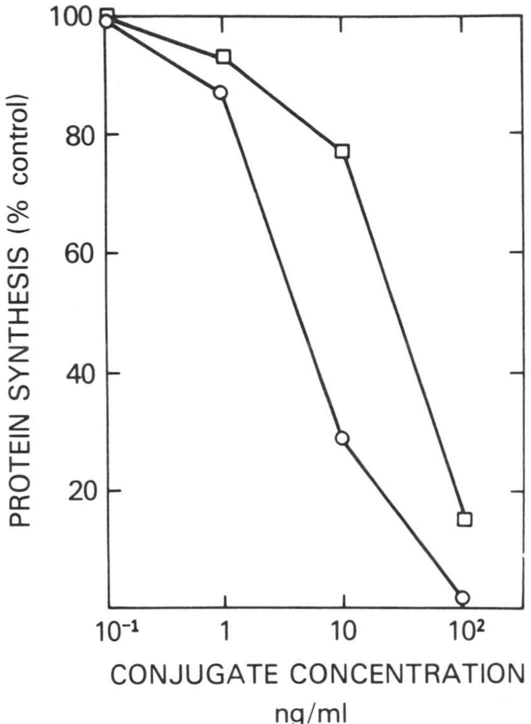

FIG. 2 *Immunotoxin titration assay. Cells processing the Thy-1.1 antigen, towards which the antibody is directed,were exposed to two different ricin-antibody conjugates at the specified concentrations, for 18 hrs, followed by a short pulse of ^{14}C-leucine. The extent of label incorpora- tion into cellular proteins was determined and compared to a control which lacked toxin. Circles: peptide-linker conjugate. Squares: conventional conjugate.*

where y is the slope of inactivation, and x is the bound toxin or immunotoxin level. This type of function can be made linear on a log-log plot. The power function plot of the data from Fig. 1 is shown in Fig. 3A. This is a comparison of the two conjugates previously compared in Fig. 2 by the titration experiment. Figure 3B is a comparison of the same conjugate, but in the presence of 10 mM NH_4Cl. The potentiation of the conjugate toxicity by ammonia will be referred to later on. Both methods, the titration and kinetic study, demonstrate that the potency of the two different conjugates varies by a factor of ~ 10.

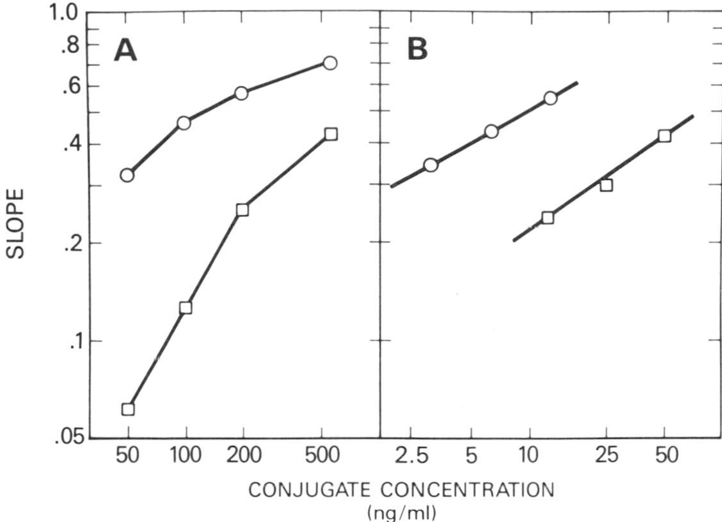

FIG. 3. *Power function plot of two different immunotoxins. A. The slopes (given a positive sign) of the inactivation curves of the two conjugates from Fig. 1 are plotted against the conjugate concentrations on a log-log scale. Circles: peptide-linker conjugate. Squares: conventional conjugate. B. Assay of same conjugates at lower concentrations, but in the presence of 10 mM NH_4 Cl. From reference 30.*

While the titration experiments have been useful in studying gross differences between toxins or conjugates, the meaningful development of immunotoxins as therapeutic agents dictates that the assay systems be capable of measuring or predicting the killing of cells over several orders of magnitude or "logs". As an example consider the experimental system of injecting tumor-causing BCL_1 cells into mice (25). BCL_1 cells consistently caused the development of tumors, if the number of cells injected was 10 or more per mouse. If the pretreatment of the cells with an immunotoxin were to have any therapeutic effect on the injection of 10^4 BCL_1 cells, then a 99.9% (3 log) kill of cells must occur. For another example, consider the clinical application of reducing T cells from donor allogenic bone marrow transplants for the prevention of graft-versus-host disease (26, 27). Donor marrow is given at a dose of 10^8 nucleated cells per kg of recipient body weight and 25% of these cells are unwanted T cells. A 50 kg recipient will receive ~10^{10} T cells. If one donor T cell caused the disease a 10 log kill would be required. Fortunately good therapeutic results are obtained using immunotoxins with modest efficacy (3-4 log kill).

The above discussion demonstrates the problem with the commonly used titration experiment. The data below the 1 to 5% protein synthesis (99 to 95% inhibition) are less reliable, and yet this is the area of most interest. As mentioned above, the clonogenic can go 6 logs (0.0001% of control values) of killing. The utility of kinetic studies hinges on the finding that the first order decline in protein synthesis after exposure to toxins (8), or A chains, or whole ricin conjugates (28) continues for 15 to 20 hrs; that is, the kinetic rate (or slope) that one measures early on in the intoxication process continues until a maximal inactivation is seen. Extrapolation of the rate of inactivation through the 15 to 20 hr period yields the total kill. For example, an immunotoxin with an inactivation rate of 0.3 log/hr would after 20 hrs kill 6 logs of cells.

Under some circumstances it is essential to demonstrate that the enzymatic activity of the coupled toxin is intact. This is done by introducing the conjugate to a cell-free protein synthesizing system (29). This type of assay does not address the binding or entry functions of a conjugate; only the toxin A chain enzymatic function and its accessibility to the enzymatic substrate (i.e., ribosome or EF2) are measured.

The utilization of these bioassays to determine the optimal structural features in toxin-antibody conjugates requires that attention be given to several variables. Indeed, the unknown or inherently uncontrollable variables in such a complex biological system dictate that the experimentalist examine only one introduced (or controlled) variable at a time. By supplying greater definition to the nature of the conjugate, better data can be acquired.

The use of monoclonal (vs. polyclonal or serum derived) antibodies permits: 1) specificity for a single cellular antigen of choice, 2) selection of a single (and measurable) high binding affinity, 3) elimination of variant antibody classes found in polyclonal preparations, 4) ability to generate large amounts of the identical antibody from time to time, as well as from laboratory to laboratory, 5) control over (or elimination of) interaction of the antibody with complement, and 6) as a consequence of uniformity, meaningful biochemical studies can be accomplished; features such as binding affinity, antigen number per cell, and biophysical characterizations of the conjugates can be accomplished.

In comparing different classes of antibodies, it is important to compare equally bound material, since variation in binding affinity is likely, and resultant differential toxicity may not be due to the structural differences that are under question. This feature also holds true for comparing toxin-antibody conjugates to the native toxins. Frequently, these comparisons are done at equivalent concentrations in the medium. The equally frequent conclusion that the conjugate is more toxic than the unconjugated toxin is misleading, since the binding affinity constant of the antibody is commonly magnitudes of order greater than the toxin. To speak in general terms, the efficacy of immunotoxins is less than that of the native toxins: the maximal inhibition rate of native toxins has not yet been even approximated by conjugates. When comparing conditions where the immunotoxin may have saturated its cellular receptors, and when an equivalent concentration of toxin results in low receptor occupancy, the conjugate may (erroneously) appear to be more efficient in killing the cells. Indeed, the affinity constant of some monoclonal antibodies is in the order of $10^{10}M^{-1}$, such that the target cells essentially deplete the medium of antibody up to the point of antigen saturation. Furthermore, comparisons of the cytotoxicity of toxins and conjugates often involves incubation in the presence of serum. Free toxins, such as ricin, bind to the glycoproteins found in serum, limiting their interaction with the cells.

With the recent finding that whole ricin conjugates may utilize the toxin moiety's normal route of entry (24, 38), the choice of a non-target (lacking the antigen towards which the antibody is directed) cell line should involve a comparison of its sensitivity of the native toxin with that of the target cell. Non-target cells which have low sensitivity to the toxin would potentially display lower "non-specific" sensitivity to the conjugate.

Cells will also display a day to day variation (up to ~50%) in their sensitivity to toxins and immunotoxins; when comparing conjugates, it is essential to carry out assays at the same time using a single batch of cells.

III. CHEMISTRY OF ANTIBODY-TOXIN CONJUGATION

The nature of the chemical linkage can greatly alter the cytotoxicity of an antibody-toxin conjugate. The evolution of heterobifunctional reagents brought about optimal cross-linking methodologies. The earliest toxin-polyclonal antibody conjugates utilized reagents such as toluene diisocyanate (31), glutaraldehyde (32, 33), and diethyl malonimidate (34). These homobifunctional reagents resulted in not only the desired toxin-antibody conjugates, but also in large aggregates, homopolymers, as well as intramolecular cross-linking which would diminish the toxicity of most natural toxins. A slight advancement was accomplished with chlorambucil derivatives (35, 36), but the yields were generally low, and undesirable cross-linking still persisted. Highly specific conjugation, that is cross-linking between only two different molecular species, was accomplished with the development of the reagents such as m-maleimidobenzoyl-N-hydroxysuccinide ester (MBS; ref. 37, 38), N-succinimidyl 3-(2-pyridyldithiolpropionate) (SPDP; ref. 39), methyl 3-mercaptopropionimidate (40), 2-iminothiolane (40-42), and methyl-5-bromovalerimidate (43).

In general, the efficient and specific cross-linking methodologies have followed one direction: first generate a free sulfhydryl on one protein species, and then establish on the second protein a chemical moiety that will selectively react with the free thiol group. This review will define the second species as the "activated" protein. The success of this methodology is largely due to the fact that free reactive sulfhydryl groups do not normally exist on native proteins, and therefore, one can control the direction and extent of cross-linking. The earlier use of homobifunctional reagents involved the cross-linking of amino groups, which are found in abundance on all protein species, thus resulting in non-specific cross-linking.

A. GENERATION OF THIOL GROUPS ON FIRST PROTEIN

The generation of free sulfhydryl groups has been accomplished by two means. The first is reductive cleavage with agents such as dithiotheitol (DTT; ref. 44) of the native cystine residues in proteins. The second is the chemical addition of thiol groups (thiolation).

1. Reduction of native disulfides An example of cleavage of native disulfides is the cleavage of the bond linking the A and B chains of toxins. After reduction, the A and B toxin chains are separated, in the case of ricin, by a combination of affinity chromatography on Sepharose 4B, in which only the B chain and whole ricin become bound, and ion exchange chromatography (45, 46). The purified, reduced A chain, possessing a single reactive sulfhydryl group, is then coupled to the activated antibody as discussed later in this chapter (see Fig. 4). In that the B subunit binds to practically all cells, this approach has the added benefit of increasing the immunotoxin specificity: binding is limited to the antibody.

Another example of utilizing the protein's native cystine residues is to reduce the disulfides of the antibody (47). The non-covalent interactions between the separate peptide chains of an antibody are sufficient enough that mild reduction does not significantly alter the functional features of the immunoglobulin (48, 50). In order to get a reasonable yield of cross-linking of the reduced immunoglobulin to an activated toxin, approximately five cystines (generation of 10 thiol groups) must be reduced (47, 51). Furthermore, with different monoclonal antibodies the concentration of DTT has to be varied: levels too high denature the protein, lower levels can result in inadequate cross-linking. While the monoclonal T101 requires 10 mM DTT, TA1 and UCHT1 are incubated with 100 mM DTT (26, 52).

One can also generate a single free sulfhydryl on the antigen binding portion of an antibody, the F_{AB}' fragment. Pepsin proteolysis of rabbit IgG molecule results in a single, disulfide-linked $F_{(AB')2}$ fragment ($F_{(AB')}$-s-s-$F_{(AB')}$; ref. 53). Selective reduction cleaves the intersubunit disulfide, generating an $F_{AB'}$ moiety (54). The heavy and light chain remain associated, and the free sulfhydryl group is accessible for coupling.

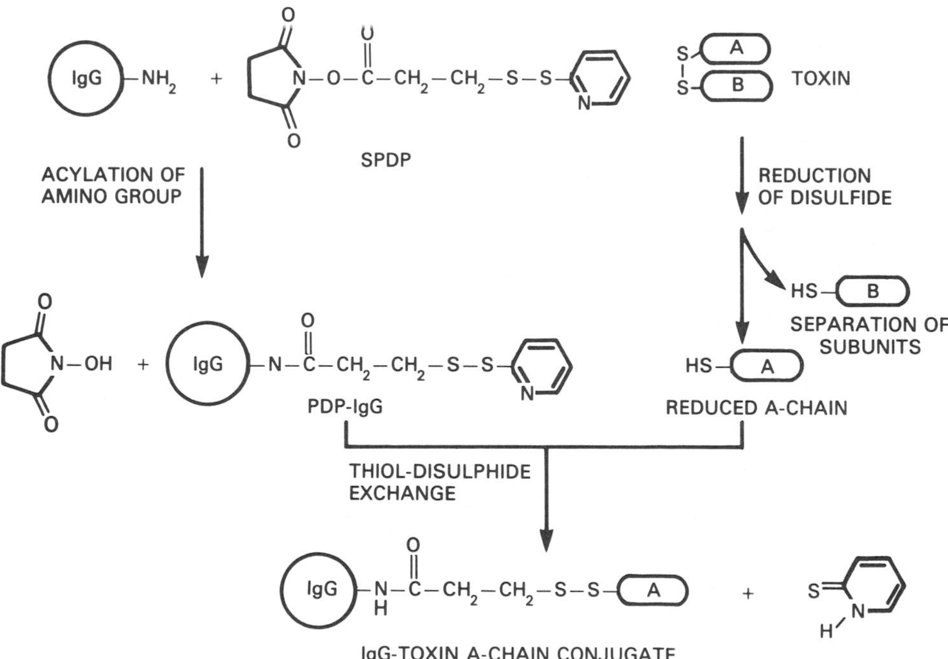

FIG. 4. *Scheme for synthesis of toxin A chain-immunoglobulin disulfide-linked conjugate. Thiolation of toxin A chain is accomplished by reduction of interchain disulfide; activation of antibody is accomplished with the reagent SPDP, succinimidyl 3-(2-pyridyldithiolpropionate).*

I. ADDITION OF THIOL TO ANTIBODY

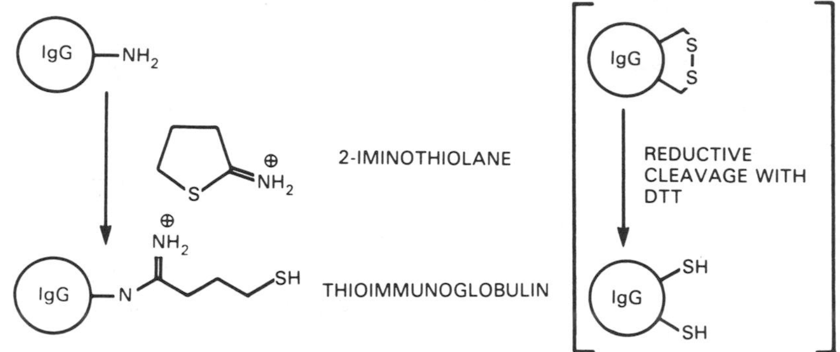

II. ACTIVATION OF RICIN WITH MBS

RICIN NH_2 +

ACYLATION OF
AMINO GROUP

RICIN N—C
H

MB-RICIN

III. COUPLING OF MB-RICIN TO THIOIMMUNOGLOBULIN

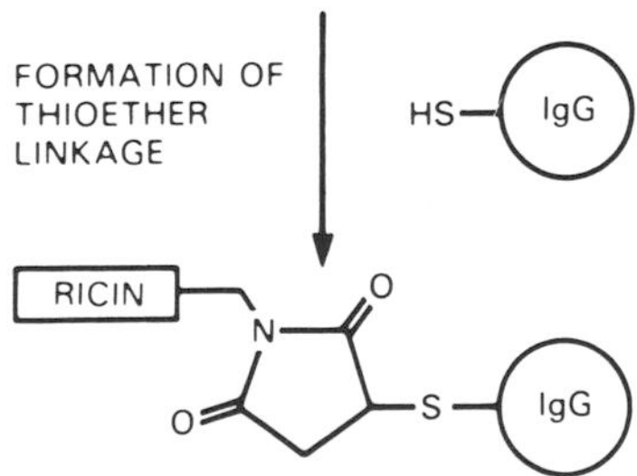

FIG. 5. *Scheme for synthesis of whole toxin-immunoglobulin conjugate with a thioether linkage. Thiolation of IgG can be accomplished by either the chemical addition with 2-iminothiolane or by limited reduction with dithiothreitol (DTT). Activation of the toxin ricin is accomplished with m-maleimidobenzoyl-N-hydroxysuccinimide ester (MBS).*

2. Chemical introduction of thiol groups The second method for the introduction of free sulfhydryls is through the use of reagents such as SPDP and 2-iminothiolane. SPDP reacts with the protein's amino groups through the succinimidyl moiety, generating proteins with exposed dithiopyridyl groups (see Fig. 4). The thiopyridyl moiety is a good leaving group, and under controlled reduction conditions the disulfide is cleaved; a new free sulfhydryl is generated, and the protein's native disulfides are largely left intact (39). The reagent 2-iminothiolane also reacts with the protein's amino groups. Decyclization of the reagent results in the generation of a free sulfhydryl group (see Fig. 5). In order to determine the level of 2-iminothiolane necessary to generate 10 thiol groups per immunoglobulin, previously demonstrated to be necessary for optimal antibody-ricin conjugation, the antibody at 80 μM was incubated with varying levels of 2-iminothiolane, then assayed for the presence of new thiol groups (see Fig. 6). An examination of the thiolated antibody preparation's ability to couple to "activated" ricin indicated that only one or two 2-iminothiolane generated thiols were necessary. The reason for the lower levels of

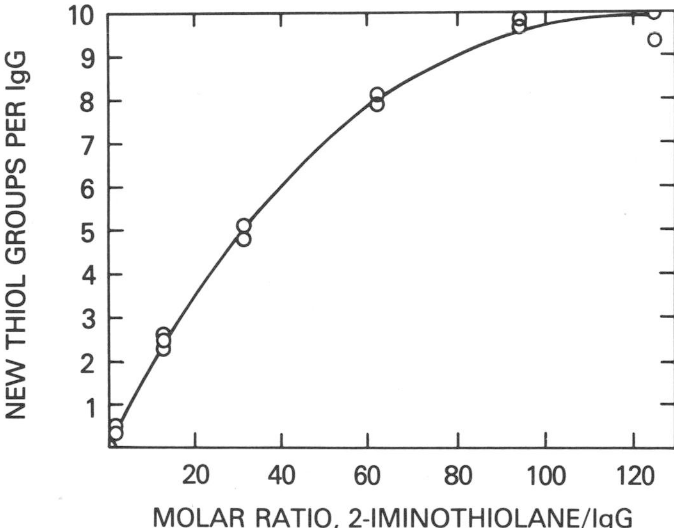

FIG. 6. *Generation of thiol groups on immunoglobulin with 2-iminothiolane. Varied levels of 2-iminothiolane were incubated with a constant concentration of immunoglobulin (80 μM). Thiol groups were determined with 5,5'-dithio-bis-(2-nitrobenzoic acid) (DTNB). From reference 24.*

thiols required is not clear; perhaps because of steric problems, certain cystine residues, although accessible to reducing agents, are not accessible to the activated second protein species. The thiols generated by the reagent, 2-iminothiolane, presumably are all reactive.

B. ACTIVATION OF THE SECOND PROTEIN

Once the first protein has been thiolated, the second protein must be "activated", or made specifically reactive towards the sulfhydryl group of the first protein. The means by which the second protein is modified can result in either a reductively cleavable disulfide, or a stable thioether linkage between the two proteins.

1. Activation that results in disulfide coupling Formation of a disulfide between the thiolated (first) protein and the second protein requires that the second protein possess a dithiol group that contains a good leaving moiety. The introduction of thiol leaving groups has been accomplished in several ways. One is through the use of the heterobifunctional reagent, SPDP. The use of this reagent, mentioned earlier in this review, is outlined in Fig. 4. The thiopyridyl leaving group is displaced by the thiol-containing (first) protein, resulting in the formation of a disulfide bond. Alternative methods of chemically introducing dithiols have been accomplished in the synthesis of immunotoxins, but they involve the use of a reagent, water-soluble carbodiimide, which permits the coupling of most carboxyl and amino groups (55), and thus, for our need, can be viewed as non-specific. One example is the coupling of cystamine dihydrochloride to the protein's carboxyl groups with carbodiimide (56, 57); another is the coupling of 3-(2-pyridyldithiolpropionic acid) to the protein's amino groups with the same reagent (58, 59).

In addition to chemically introducing dithiols, one can also introduce a thiol leaving group to proteins already possessing a reactive sulfhydryl group, such as treatment of a thiolated protein with 5,5'-dithio-bis-(2-nitrobenzoic acid) (DTNB; ref. 60). The resultant thio-nitrobenzoic acid leaving group, like the thio-pyridyl group of SPDP, permits an efficient development of conjugates linked through a reducible disulfide bond. In one report activation of the antibody was achieved by methyl 3-mercaptopropionimidate followed by the addition of DTNB (61). An undesirable side reaction in "activating" thiol-containing proteins with DTNB is the cross-linking of a then activated protein with a remaining thiol-containing protein (62). There is one example in the literature involving S-sulfonation of a toxin A chain ($A\text{-}S\text{-}SO_3^-$), where the sulfite serves as the leaving group (63).

2. *Activation resulting in thioether linkage* A non-reductive thioether is generated when the second protein is activated by coupling a maleimide or iodoacetyl group, both of which have demonstrable specificity for thiol groups. These additions have been achieved by the reagents m-maleimidobenzoyl-N-hydroxysuccinimide ester (MBS; ref. 37, 38) and the N-hydroxysuccinimide ester of iodoacetic acid (64), respectively. Both of these reagents react with the amino groups of one protein through the succinimide ester moiety, resulting in the covalently coupled maleimidyl or iodoacetyl groups which will then react with the free thiols of another protein. An example of how MBS is used can be seen in Fig. 5. We have utilized this reagent extensively in our own laboratory, and to optimize cross-linking, we have examined the efficiency of this reagent by first incubating MBS at various levels with the toxin ricin, then adding thiols in the form of radiolabeled cysteine (see Fig. 7). One must add approximately five moles of MBS to onc mole of ricin, under the conditions

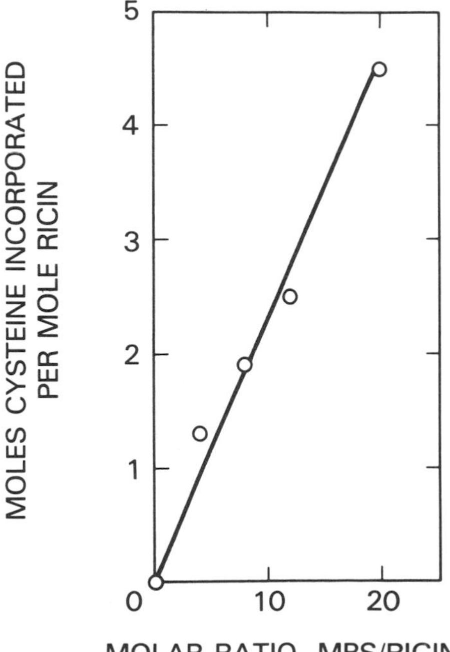

FIG. 7. *Addition of cysteine residues to MBS activated ricin. After incubation with MBS, ricin was exposed to an excess of* ^{14}*C-cysteine, and the incorporation of label was determined. The molar ratio is a measure of MBS input to a constant (10 mg/ml) ricin concentration.*

specified, to effectively couple an average of 1 mole of cysteine. Note that this assay involved two chemical couplings: the succinimidyl moiety's interaction with the ricin's amino groups, then the maleimide reaction with the thiol groups of cysteine. In addition, hydrolysis of both functional groups is taking place (65).

C. PURIFICATION OF CONJUGATES : AN EXAMPLE

Even under the best of circumstances, the cross-linking of two different proteins, perhaps with the exception of coupling reduced A chains to the $F_{AB'}$ fragment, results in heterogeneity. This is due to the fact that the utilized heterobifunctional reagents may react equally with any one of a number of amino groups, or in the case of a a multiple-chained macromolecular, to either chain. Additionally, to achieve a high yield of a conjugate possessing an antibody: toxin ratio of only 1:1, higher polymers will also be produced; therefore, highly resolving techniques, such as ion exchange, result in multiple peaks. Currently, purification is limited to molecular sizing, as accomplished with gel filtration, and affinity chromatography, such as the Sepharose affinity column for ricin containing conjugates (45) or the protein A columns for those immunoglobulins that are capable of binding to protein A (66).

The conjugate preparation is first run on a gel filtration column, and a typical separation is shown in Fig. 8A. The "IgG" peak (45' - 50'), which possesses both free IgG and conjugate, and the shoulder (40' - 45'), representing a conjugate with a larger stokes radius, are then run separately on an affinity Sepharose 4B column. The large peak that goes off-scale is ricin. When the IgG peak is run on the affinity column (Fig. 8B), the free IgG is unretarded. The bound conjugate (bound through the ricin) is competitively displaced by the addition of lactose to the column buffer. This material was shown to be a conjugate species possessing a 1:1, IgG:ricin stoichiometry, where the material eluting from the affinity column, derived from the shoulder (Fig. 8A), was found to contain immunoglobulin possessing two bound ricins (30). Because of the size of the ricin peak on the first gel filtration run, these conjugates are re-run on a gel filtration column to be freed of any residual ricin (Fig. 9). Such preparations can result in single bands on SDS gel electrophoresis.

A compilation of the various modes by which antibody-toxin conjugates have been made is found in Table 1. The list is undoubtedly incomplete. The very important pioneering work briefly described in the Introduction, which involved less than optimal reagents of that day, has been excluded.

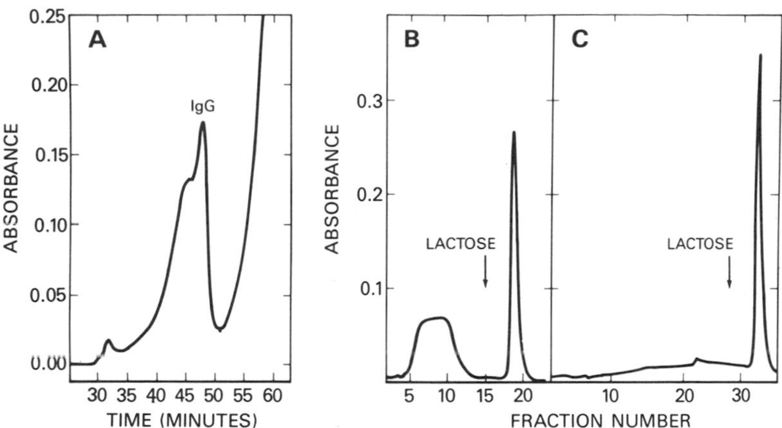

FIG. 8. *Chromatographic purification of ricin-antibody conjugates. A. High performance gel filtration of the conjugate incubation mixture. B. Sepharose 4B affinity column run of IgG peak (45' - 50' from gel filtration). Elution of bound material was achieved by addition of 10 mM lactose to the column buffer. C. Affinity run of shoulder to IgG peak (40' - 50'). From reference 30.*

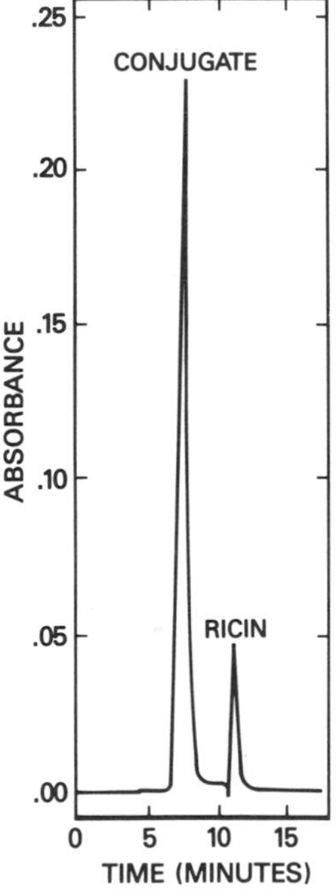

FIG. 9. *Gel filtration chromatography of affinity purified conjugates. Conjugate isolated from lactose eluted peak (Fig. 8c) is re-run over an analytical high performance sizing column to remove residual toxin.*

IV. BIOLOGICAL CONSEQUENCES OF CONJUGATE CONSTRUCTION

The different chemical routes taken to couple natural toxins to antibodies, as described above, actually evolved as experimentalists were confronted with the fact that the conjugation of toxins consistently resulted in a tremendous loss of cytotoxicity; that is, native toxins, although less specific, displayed a proportionally greater ability to kill cells than did the conjugates. The historical development of the chemical aspects of conjugate construction coincided with the development of the biological consequences, but for continuity and ease of appreciation they have been separated in this review.

A. BIOLOGICAL CONSEQUENCES OF THE CHEMICAL COUPLING

As mentioned in the Introduction, in order to gain specificity ricin-antibody conjugates have been constructed by coupling only the enzymatic A subunit to the antibody (the binding subunit is removed), or by coupling the whole toxin, but then including lactose in the incubation of the immunotoxins with the cells. The nature in which the A chain is coupled to the antibody highly affects the resultant cytotoxicity. Coupling of the ricin A chain to the antibody through a non-reducible thioether linkage resulted in the loss of cytotoxicity, although when placed in a cell-free protein synthesizing system, the conjugate displayed toxicity nearly equivalent to free A chain (67). When the antibody and ricin A subunit were linked through a disulfide, a one log kill was seen after 18 hrs at a concentration just under 1 μg/ml. Presumably, target cells must reduce the disulfide linkage between the antibody and ricin A chain in order for the A chain to gain access to the ribosome. The nature of the coupling of whole toxin to antibody is somewhat different. The coupling of whole ricin to antibody through a thioether linkage results in conjugates of high toxicity, reaching an inactivation rate of 0.3 log/hr at receptor saturation (12). When whole abrin (a ricin-like, two-chained toxin) was linked to an antibody, a non-reducible linkage decreased the cytotoxicity, when compared to a reducible disulfide linkage, but only by approximately fifty percent (68). It would appear that coupling via a non-reducible linkage to the B subunit does not as drastically diminish toxicity as does coupling to the A chain; of course the A chain is coupled to the B subunit through a disulfide bond in such a conjugate.

Another feature concerning the nature of the coupling has involved the use of a spacer arm. The introduction of a 29 residue peptide chain between ricin and an anti-mouse leukemia cell antibody has recently been demonstrated to greatly increase cytotoxicity (30). The scheme used to synthesis this peptide-linker conjugate is outlined in Fig. 10. Its purification is like any other ricin-antibody conjugate; in fact, it is this conjugate that was being purified in Fig. 8. When compared to a more conventional conjugate as made by the procedure in Fig. 5, it was found that to reach similar toxicity, only one tenth the concentration of the peptide-linker conjugate was needed. The remarkable increase in potency was attributed to the ricin moiety's increased interactions with the cell's membrane, effecting a more rapid rate of entry into the cytosol.

B. BIOLOGICAL CONSEQUENCES OF MACROMOLECULAR STRUCTURE

In addition to the means by which antibodies and natural toxins are coupled, the nature of the two macromolecules used will greatly affect the cytotoxicity of the conjugate. The fragmentation of the literature in Table 1 involves not only the chemical linkage, but also the two species being coupled. The result that these various states have on cytotoxicity will be discussed here.

1. Antibody structure A comparison of cytotoxicities of ricin A chain linked to either the monovalent $F_{AB'}$ or divalent $F_{(AB')2}$ fragments derived from the same polyclonal serum demonstrated that the $F_{(AB')2}$ conjugates were five times more potent (69). Besides the obvious increase in affinity, it was suggested that the divalent antibodies effected a more efficient internalization into the cell.

Monoclonal antibodies developed against the same antigen do not necessarily result in the same toxicity when used in conjugates. The anti-Thy-1.1 antibodies 31-E6 and 19-E12 demonstrated a ten fold difference in cytotoxicity when coupled to ricin A chain (67). Similarly, the anti-Thy-1.1 OX7, when coupled to pokeweed antiviral proteins (PAP), was not as potent as a similar conjugate using the 31-E6 antibody (67). Although differences in binding affinity for the Thy-1.1 antigen could explain diverse toxicities, this feature was not examined. An examination of an anti-p97 antibody-ricin A chain conjugate's the ability to kill cell lines which had varying levels of the p97 antigen on their surface, resulted in a not unexpected relationship between receptor number and the subsequent toxicity at various conjugate concentrations (70). Cells expressing fewer than 5000 p97 molecules on their surface were largely unaffected by 10^{-9}M conjugate, whereas, under the same conditions, cells possessing 78,000 or greater receptors had gone through one log of kill.

I. SYNTHESIS OF DITHIOPEPTIDE

ϵNH_2

αH_2N

$\overset{+}{NH_2}$ 2-IMINOTHIOLANE

HS $\overset{+}{NH_2}$ N N $\overset{+}{NH_2}$ SH

II. COUPLING PEPTIDE TO RICIN

1) ACIVATION OF RICIN WITH MBS

RICIN $-NH_2$ + N-O-C

ACYLATION OF AMINO GROUP

RICIN N–C MB-RICIN

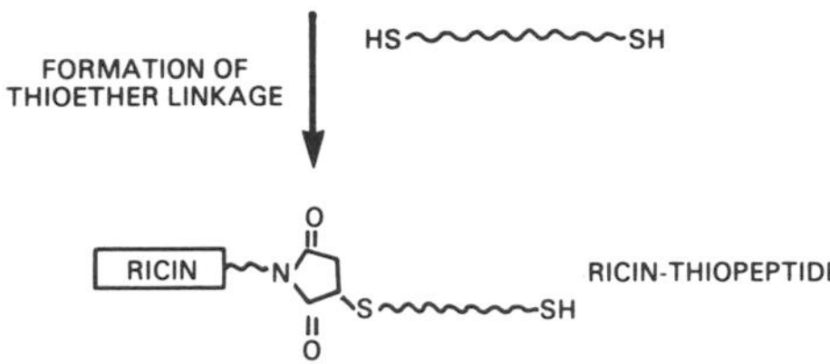

III. RICIN-THIOPEPTIDE COUPLING TO ANTIBODY

1) ACTIVATION OF IgG WITH MBS

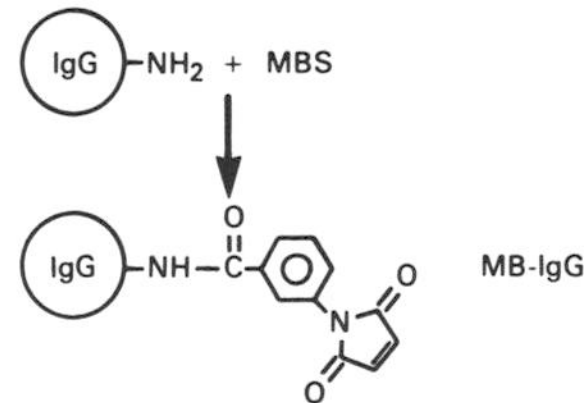

2) COUPLING OF MB-IgG TO RICIN-THIOPEPTIDE

HS~~~~S– RICIN

IgG –S~~~~S– RICIN

FIG. 10. *Scheme for synthesis of peptide-linker conjugate. Thiolation of the two amino groups of oxidized insulin B chain by 2-iminothiolane, is followed by addition of MBS-treated ricin to one end, MBS activated IgG to the other, as fully described in reference 30.*

Differences have been noted for immunotoxins constructed with the IgG or IgM class of immunoglobulin developed against the same cell determinate (71). In comparing the cytotoxicity of the two anti-Thy-1.2 antibody-ricin A chain conjugates, the IgG conjugate was found to be 5 times more toxic than the IgM conjugate, even though conditions were established that placed the ricin A chains bound per cell by both conjugates to be nearly equivalent.

2. *Toxin structure* Although most of the literature that has helped resolve the optimal way of constructing conjugates has involved either ricin or diphtheria toxin, several other natural toxins have been examined. Pokeweed antiviral protein has been coupled to the anti-Thy-1.1 antibody, OX7 (67). At an antigen-saturating level of 16 μg/ml, the inhibition with AKR cells reached a rate of only 0.04 log/hr. A similar rate of inactivation was found for the plant toxin gelonin (72), when coupled to an anti-Thy-1.1 antibody and incubated at 100 μg/ml with a cultured lymphoma cell line. Compare these results with the ricin peptide-linker conjugate (30) which reaches 1 log/hr at 200 ng/ml (~10^{-9}M), or a ricin A chain conjugate (71), which reaches 0.5 log/hr at 2 x 10^{-8}M in the presence of 10 mM NH_4Cl.

As mentioned in the Introduction, the whole ricin conjugates have been demonstrated to be more toxic than A chain conjugates. OX7 antibody coupled to whole ricin was found to be 19 times more toxic than A chain conjugates using the same antibody, when judged by kinetic assays (12). The significance of the ricin B chain has also been demonstrated by the potentiation (5 fold) of A chain conjugate toxicity by the addition of free B chain to the cellular incubation (12).

The number of toxins coupled to the antibody moiety also influence the cytotoxicity. The number of diphtheria toxin molecules on a polyclonal anti-(human lymphocyte) immunoglobulin directly influenced the cytotoxicity of the conjugates on target cells (36). The addition of a second toxin molecule to the antibodies resulted in a 75% loss in toxicity. The addition of a third molecule resulted in a further decrease in toxicity.

Just the opposite effect was found with the coupling of ricin to a monoclonal antibody directed against mouse T cells (see Fig.11). Here, the addition of a second ricin to the immunoglobulin molecule ($IgG \cdot R_2$) greatly enhanced the cytotoxicity (24). In going from one ($IgG \cdot R_1$) to two ricins per antibody molecule, the rate of protein synthesis inhibition went from 0.14 log/hr to 0.4 log/hr at 200 ng/ml. The addition of a third ricin resulted in a decrease to 0.2 log/hr. The near tripling of the

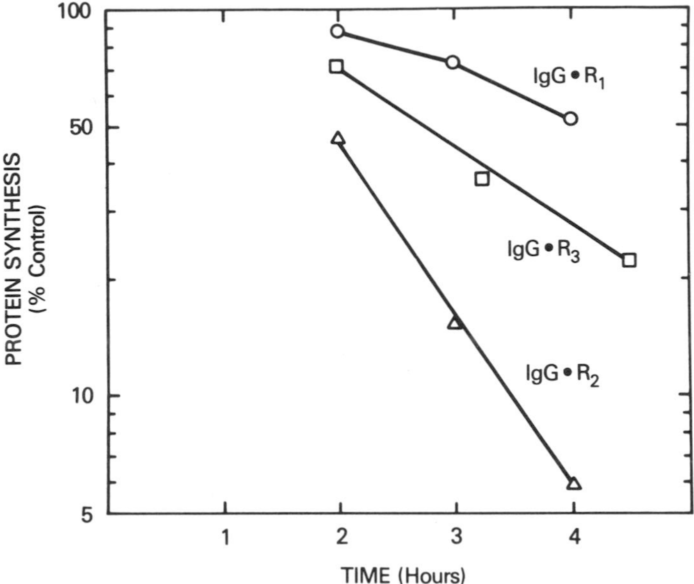

FIG. 11. *Kinetics of protein synthesis inhibition by antibody-ricin conjugates. Three different conjugates possessing the same antibody, but varied levels of ricin (R_1, R_2, R_3 representing one, two and three ricins per IgG, respectively), were assayed for their ability to inhibit protein synthesis. From reference 24.*

inhibition rate could not be attributed just to the increase in toxin molecules per unit mass conjugate. It was suggested that the increased ricin-membrane interaction somehow potentiated the rate-limiting processes of cell toxicity.

V. FUTURE PERSPECTIVE

All discussions here have involved the use of naturally occurring proteins and any tailoring was done with chemical reagents or proteolytic enzymes. Genetically engineered alterations of toxins and antibodies for use in immunotoxins are now in various states of development. Recent work from this laboratory has demonstrated for ricin immunotoxins that the conjugates' potency is related to the ricin moiety's ability to interact with the membrane (24, 30). This is consistent with the finding that the toxicity of ricin internalized by an alternative receptor, the mannose-6-phosphate receptor on fibroblasts, is dependent on the galactose-binding capability (73).

Decreases in the binding capacity were achieved by o-acetylation of the B subunit binding site. Whole ricin conjugates appear to utilize the normal route (anatomical and biochemical) that is taken by ricin, but in a less efficient way. These conjugates can thus be viewed as hindered toxins. As such, the optimal design for immunotoxins should take into account the processing that the native toxin undergoes to gain entry to the target cell's cytosol. These processes are presently unknown, but evidence for a concerted, quantal entry for diphtheria toxin, and potentially similar mechanisms for other toxins, has previously been demonstrated by work in this laboratory (10).

As previously mentioned, A chain conjugates have been found to be less toxic than whole chain conjugates, but a recent finding has demonstrated that certain A chain conjugates are potentiated by the addition of lysosomotropic amines and carboxylic ionophores (71). It has been suggested that these agents prevent delivery of the conjugates to lysosomes (74). Interestingly, this appears to occur with whole ricin conjugates as well (Fig. 3B; ref. 30), but under these circumstances the reason suggested for the potentiation is the enhanced utilization of the highly efficient ricin route of entry.

The necessity to maintain levels of lactose (whole ricin conjugates) or amines (A chain conjugates) limits today's utilization of conjugates to *in vitro* applications, and although *in vitro* clinical situations do exist (75), the need for immunotoxins that work *in vivo* is obvious. The problems in an approach to *in vivo* work have been recently reviewed (17). Undoubtedly, the solutions will come about from an understanding of the cellular processing of natural ligands and the basic mechanism of toxin entry.

REFERENCES

1. F. I. Lamb L. M. Roberts, and J. M. Lord, *Eur. J. Biochem.* 148: 265 (1985).
2. D. M. Gill and L. L. Dinius, *J. Biol. Chem.* 246: 1485 (1971).
3. R. J. Collier and J. Kandel, *J. Biol. Chem.* 246: 1496 (1971).
4. T. Oda and G. Funatsu, *Agric. Biol. Chem.* 43: 547 (1979).
5. S. Olsnes, K. Refsnes, and A. Pihl, *Nature* 249: 627 (1974).
6. T. Honjo, Y. Nishizuka, O. Hayaishi, and I. Kato, *J. Biol. Chem.* 243: 3553 (1968).
7. M. Yamaizumi, E. Mekada, T. Uchida, and Y. Okada, *Cell* 15: 245 (1978).
8. K. Eiklid, S. Olsnes, and A. Pihl, *Exp. Cell Res.* 126: 321 (1980).

9. B. Ray and H. C. Wu, *Mol. Cell. Biol.* 1: 544 (1981).

10. T. H. Hudson and D. M. Neville, Jr. *J. Biol. Chem.* 260: 2675 (1985)

11. J. U. Baenziger and D. Fiete, *J. Biol. Chem.* 254: 9795 (1979).

12. R. J. Youle and D. M. Neville, Jr., *J. Biol. Chem.* 257: 1598 (1980).

13. K. A. Krolick, C. Villemex, P. Isakson, J. W. Uhr, and E. S. Vitetta, *Proc. Natl. Acad. Sci.* 77: 5419 (1980).

14. D. G. Gilliland, Z. Steplewski, R. J. Collier, K. F. Mitchell, T. H. Chang, and H. Koprowski, *Proc. Natl. Acad. Sci.* 77: 4539 (1980).

15. J. L. Middlebrook and R. B. Dorland, *Can. J. Microbiol.* 25: 285 (1979).

16. R. L. Proia, D. A. Hart, and L. Eidels, *Infect. Immun.* 26: 942 (1979).

17. D. M. Neville, Jr., *CRC Critical Review in Therapeutic Drug Carrier Systems,* Vol. 2 (1986; in press).

18. P. Cohen and S. Van Heyningen, eds. *Molecular Action of Toxins and Viruses,* Elsevier Biomedial Press, Amsterdan (1982)

19. G. Kohler and C. Milstein, *Nature,* 256, 495 (1975).

20. A. D. Tepperman, J. E. Curtis, and E. A. McCulloch, *Blood* 44: 659 (1974).

21. F. M. Uckun, R. C. Strong, R. J. Youle, and D. A. Vallera, *J. Immunol.* 134: 3504 (1985).

22. D. M. Neville, Jr. and R. J. Youle, *Immunol. Rev.* 62: 75 (1982).

23. R. S. Esworthy and D. M. Neville, Jr., *J. Biol. Chem.* 259: 11496 (1984).

24. J. W. Marsh and D. M. Neville, Jr., (submitted manuscript).

25. K. A. Krolick, J. W. Uhr, and E. S. Vitetta, *Nature* 295:604 (1982).

26. D. A. Vallera, R. C. Ash, E. D. Zanjani, J. H. Kersey, T. W. LeBien, P. C. L. Beverley, D. M. Neville, Jr., and R. J. Youle, *Science* 222: 512 (1983).

27. A. H. Filipovich, D. A. Vallera, R. J. Youle, D. M. Neville, Jr., and J. H. Kersey, *Transplant. Proc.* 17:442 (1985).

28. R. J. Youle and D. M. Neville, Jr., in *Receptor-Mediated Targeting of Drugs* (G. Gregoriadis, G. Poste, J. Senior, and A. Trouet, eds.) pp. 139-145, Plenum Publ. (1985).

29. H. R. B. Pelham and R. J. Jackson, *Eru. J. Biochem.* 67: 247 (1976).

30. J. W. Marsh and D. M. Neville, Jr. (submitted manuscript).

31. F. L. Moolten and S. R. Cooperband, *Science* 169: 68 (1970).

32. F. L. Moolten, N. J. Capparell, and S. R. Cooperband, *J. Natl. Cancer Inst.* 49:1057 (1972).

33. F. L. Moolten, N. J. Capparell, S. H. Zajdel, and S. R. Cooperband, *J. Natl. Cancer Inst.* 55: 473 (1975).

34. G. W. Philpott, R. J. Bower, and C. W. Parker, *Surgery* 73: 928 (1973).

35. P. E. Thorpe, W. C. J. Ross, A. J. Cumber, C. A. Hinson, D. C. Edwards, and A. J. S. Davis, *Nature* 271: 752 (1978).

36. W. C. J. Ross, P. E. Thorpe, A. J. Cumber, D. C. Edwards, C. A. Hinson, and A. J. S. Davis, *Eur. J. Biochem.* 104: 381 (1980).

37. T. Kitagawa and T. Aikawa, *J. Biochem.* 79: 233 (1976).

38. F.-T. Liu, M. Zinnecker, T. Hamaoka, and D. H. Katz, *Biochemistry* 18: 690 (1979).

39. J. Carlsson, H. Drevin, and R. Axen, *Biochem. J.* 173: 723 (1978).

40. R. R. Traut, A. Bollen, T.-T. Sun, J. W. B. Hershey, J. Sundberg, L. R. Pierce, *Biochemistry* 12: 3266 (1973).

41. H. J. Schramm and T. Dulffer, *Hoppe Seylers Z. Physiol. Chem.* 358: 137 (1977).

42. T. P. King, Y. Li, and L. Kochoumian, *Amer. Chem. Soc.* 17: 1499 (1978).

43. T.-M. Chang and D. M. Neville, Jr., *J. Biol. Chem.* 252: 1505 (1977).

44. W. W. Cleland, *Biochemistry* 3: 480 (1964).

45. S. Olsnes, *Methods Enzymol.* 50: 330 (1978).

46. D. B. Cawley and H. R. Hershman, *Cell* 22: 563 (1980).

47. R. J. Youle and D. M. Neville, Jr., *Proc. Natl. Acad. Sci.* 77: 5483 (1980).

48. D. M. Kranz and E. W. Voss, Jr., *Proc. Natl. Acad. Sci.* 78: 5807 (1981).

49. C. Horne, M. Klein, I. Polidoulis, and K. J. Dorrington, *J. Immunol.* 129: 660 (1982).

50. D. M. Kranz, J. N. Herron, and E. W. Voss, Jr., J. Biol. Chem. 257: 6987 (1982).

51. J. W. Marsh (unpublished result).

52. D. M. Neville, Jr. and R. J. Youle, U.S. Patent Application 456,401.

53. S. Utsumi and F. Karush, *Biochemistry* 4: 1766 (1965).

54. Y. Masuho and T. Hara, *Gann* 71: 759 (1980).

55. K. L. Carraway and D. E. Koshland, Jr., *Methods Enzymol.* 25: 616 (1972).

56. D. G. Gilliland, R. J. Collier, J. M. Moehring, and T. J. Moehring, *Proc. Natl. Acad. Sci.* 75: 5319 (1978).

57. T. N. Oeltmann and J. T. Forbes, *Arch. Biochem. Biophys.* 209: 362 (1981).

58. F. K. Jansen, H. E. Blythman, D. Carriere, P. Casellas, J. Diaz, P. Gros, J. R. Hennequin, F. Paolucci, B. Pau, P. Poncelet, G. Richer, S. L. Salhi, H. Vidal, and G. A. Voisin, *Immunol. Lett.* 2: 97 (1980).

59. H. E. Blythman, P. Casellas, O. Gros, P. Gros, F. K. Jansen, F. Paolucci, B. Pau, and H. Vidal, *Nature* 290: 145 (1981).

60. G. L. Ellman, *Arch. Biochem. Biophys.* 82: 70 (1959).

61. H. Miyazaki, M. Beppu, T. Terao, and T. Osawa, *Gann* 71: 766 (1980).

62. A. F. S. A. Habeeb, *Methods Enzymol.* 25: 457 (1972).

63. Y. Masuho, T. Hara, and T. Noguchi, *Biochem. Biophys. Res. Commun.* 90: 320 (1979).

64. P. E. Thorpe, W. C. J. Ross, A. N. F. Brown, C. D. Myers, A. J. Cumber, B. M. J. Foxwell, and J. T. Forrester, *Eur. J. Biochem.* 140: 63 (1984).

65. Y. Hamaguchi, S. Yoshitake, E. Ishikawa, Y. Endo, and S. Ohtaki, *J. Biochem.* 85: 1289 (1979).

66. J. W. Goding, *J. Immunol. Methods* 20: 241 (1978).

67. S. Ramakrishnan and L. L. Houston, *Cancer Res.* 44: 201 (1984).

68. D. C. Edwards, W. C. J. Ross, A. J. Cumber, D. McIntosh, A. Smith, P. E. Thorpe, A. Brown, R. H. Williams, and A. J. S. Davies, *Biochin. Biophys. Acta* 717: 272 (1982).

69. Y. Masuho, K. Kishida, M. Saito, N. Umemoto, and T. Hara, *J. Biochem.* 91: 1583 (1982).

70. P. Casellas, J. P. Brown, O. Gros, P. Gros, I. Hellstrom, F. K. Jansen, P. Poncelet, R. Roncucci, H. Vidal, and K. E. Hellstrom, *J. Cancer* 30: 437 (1982).

71. P. Casellas, B. J. P. Bourrie, P. Gros, and F. K. Jansen, *J. Biol. Chem.* 259: 9359 (1984).

72. P. E. Thorpe, A. N. F. Brown, W. C. J. Ross, A. J. Cumber, S. I. Detre, D. C. Edwards, A. J. S. Davies, and F. Stirpe, *Eur. J. Biochem.* 116: 447 (1981).

73. R. J. Youle, G. J. Murray, and D. M. Neville, Jr., *Cell,* 23: 551 (1981).

74. D. Carriere, P. Casellas, G. Richer, P. Gros, and F. K. Jansen, *Exp. Cell Res.* 156: 327 (1985).

75. D. M. Neville, Jr., in *Directed Drug Delivery,* (R. Borchardt, A. Repta, and V. Stella, eds.) Humana Press, New Jersey (1985).

76. D. J. P. FitzGerald, I. S. Trowbridge, I. Pastan, and M. C. Willingham, *Proc. Natl. Acad. Sci.* 80: 4134 (1983).

77. T. N. Oeltman and J. T. Forbes, *Arch. Biochem. Biophys.* 209: 362 (1981).

78. I. S. Trowbridge and D. L. Domingo, *Nature* 294: 171 (1981).

79. M. Seto, N. Umemoto, M. Saito, Y. Masuho, T. Hara, and T. Takahashi, *Cancer Res.* 42: 5209 (1982).

80. T. F. Bumol, Q. C. Wang, R. A. Reisfeld, and N. O. Kaplan, *Proc. Natl. Acad. Sci.* 80: 529 (1983).

81. L. L. Houston and R. C. Nowinski, *Cancer Res.* 41: 3913 (1981).

82. V. Raso and T. Griffin, *J. Immunol.* 125: 2610 (1980).

83. V. Raso, J. Ritz, M. Basala, and S. F. Schlossman, *Cancer Res.* 42: 457 (1982).

12

Tailoring Enzymes for More Effective Use as Therapeutic Agents

Mark J. Poznansky

Department of Physiology
University of Alberta
Edmonton, Alberta, Canada

I. INTRODUCTION

The importance of delivery or carrier systems and specific targeting strategies for drug and/or enzyme therapy stems from the need to alter the pharmacokinetics of these therapeutic agents and to take into consideration their required cellular and subcellular sites of action. For the purposes of this text, we will discuss the concepts of drug and enzyme therapy interchangeably as we consider enzymes as drugs and describe the similarities in the philosophy for designing carrier or delivery systems. The advent of new techniques and understanding in molecular biology, new techniques in protein (including enzyme) purification, advances in drug synthesis and biotechnological approaches to the synthesis of bioactive agents has produced an ever increasing

We are grateful to the Medical Research Council of Canada, to the Alberta Cancer Board and to the Alberta Heritage Foundation for Medical Research for their continued support.

variety of "therapeutic" agents. Many of these, including many new anti-tumor agents, are superior to their predecessors, and this has stimulated the search for more effective drug delivery systems. While this is most evident in the need to direct highly potent anti-cancer agents to tumor cells and away from normal but rapidly dividing cell populations such as those of the digestive tract and bone marrow, it is equally true for other types of drug therapy, including enzyme therapy where the enzyme is only effective if it gains access to a specific site. Nowhere is this more evident than in attempts to use enzyme replacement therapy to treat lysosomal storage diseases, manifested as enzyme deficiencies where substrate accumulation is limited to lysosomal compartments within specific cell and tissue types.

In considering enzyme therapy in general and enzyme replacement therapy in particular, several limitations immediately come to mind:

1. What is the source and availability of the enzyme?
2. Will the enzyme, being a foreign protein, be immunogenic? This is an important question, and in the case of human enzyme deficiency diseases, even a human source of the deficient enzyme may prove immunogenic if the recipient is not genetically coded properly for the enzyme.
3. Will the enzyme be rapidly degraded by endogenous proteases necessitating repeated injections, and
4. What is the specific site of action of the enzyme?

This last point is largely dependent on the availability and permeability of the accumulating substrate. For example, in the case of the inborn error of metabolism phenylketonuria (PKU), while the defective enzyme (phenylalanine hydroxylase) is clearly intracellular, the resultant increase in substrate (phenylalanine) occurs throughout the body, including the plasma and urine. As such, the site of enzyme replacement might not be crucial since lowering substrate in a single compartment, say the plasma, would be expected to create substrate concentration gradients sufficient to draw substrate from other compartments for degradation in the plasma. This would not be the case for many of the enzyme deficiency diseases manifested as "Lipid Storage Diseases" where deficient or defective lysosomal enzymes result in substrate accumulation within lysosomes. In the case of Tay Sachs disease where the enzyme hexosaminidase is deficient the substrate, ganglioside (GM_2) accumulates within lysosomes which with time literally "chokes" the cell, leading to severe cellular damage and eventually organ death. The example of Tay Sachs disease is both important and extreme since the site of action for administered enzyme would not only be lysosomal but in cells of the central nervous system. This then necessitates a crossing of the blood brain barrier prior to the enzyme gaining access to the lysosomal sites of the affected cells.

The range of drug delivery systems involving carriers and targeting agents today seems almost endless, related primarily to the limits of the various investigators' imaginations. The systems devised range from synthetic polymers (poly-L-lysine), to lipoproteins, to intact viable cells (leukocytes) as drug or enzyme carriers (see 1-4 for recent reviews). Another class of compounds has been put forward to act as both carrier and targeting agent or to act as targeting agent in conjunction with some other carrier. Examples of such potential compounds include: antibodies (both mono- and polyclonal), lectins, bioactive ligands with specific cellular receptors such as insulin, epidermal growth factor and the serum lipoproteins, specific sugar residue sequences possessing cell surface receptors, and a range of other ligands including certain plasma proteins. The use of combinations of carrier systems and targeting agents has become popular with the use of antibodies (as the targeting agent) conjugated to liposomes or lipid vesicles (as the drug encapsulating carrier). Much of this material has been reviewed at great length and the author refers the reader to a number of different review articles (1-5). See also example chapters on chimeric toxins and immunotoxins in this text.

It has become obvious over the past few years that no one drug delivery system will prove to be the panacea applicable to a wide range of diseases and problems. More to the point, delivery and carrier systems will have to be devised to suit the circumstances dependent on questions such as accessibility to appropriate sites of action, toxicity of the drug or therapeutic agent to normal tissue, rates of biodegradation, etc. For example, the delivery of toxins such as ricin or diphtheria toxin to tumor cells (6) requires that the toxin be targeted and internalized in order to realize its therapeutic effect, and furthermore the targeting efficiency must be extraordinarily high due to the high toxicity of the toxin to normal tissue. On the other hand, the use of L-asparaginase in a carrier form does not require highly efficient targeting to tumor cells (it has merely to lower substrate asparagine levels in the media bathing the tumor cells), and furthermore the enzyme is not especially toxic to normal tissue (7).

In considering the efficacy of any given drug or enzyme delivery system, we (see Ref. 1) have put forward a number of parameters or questions which might be used to evaluate a given carrier or targeting system.

1. **Selectivity.** To what extent is the delivery system able to discriminate target from non-target tissue?
2. **Load Factor.** What is the capacity of the carrier system and does the ratio of carrier to therapeutic agent make sense?
3. **Immunology.** Does the production of the carrier-drug complex introduce any new antigenic determinants which result in possible immunological complications?
4. **Toxicity.** Is the drug carrier complex non-toxic in the contemplated dosages?
5. **Pharmaceutic Feasibility.** Does the drug delivery system make practical sense in terms of its use in the "field" away from a highly sophisticated experimental clinical setting?

6. **Scope of Diseases.** What diseases might be amenable to treatment with a given drug delivery system?

A major limitation in the applicability of carrier and targeting systems to drug delivery pertains to the routes of administration and the various barriers which have to be crossed in reaching the target site. These have recently been reviewed by Poznansky and Juliano (1). Briefly, if the vascular system is to be used as the initial site of administration, then the drug-carrier system must traverse a particular endothelial barrier to gain access to the underlying "target" tissue. Even after passing this barrier, it still must pass the basement membrane, the plasma membrane and even certain intracellular membranes within the target cell. Clearly encapsulating a drug within some structure or attaching it to a larger carrier system may serve to make these barriers all the more formidable. In addition to these physical barriers, consideration must be given to cells of the reticuloendothelial system (RES) which function to clear the circulation of foreign particulate and macromolecular structures. In devising drug delivery systems, it might therefore be important to consider targeting the drug-carrier complex "away" from the RES as much as targeting it "to" a particular site of action.

In 1974 we became aware of a publication from Daniel Thomas' laboratory in Compiegne, France (8), in which they described the production of "soluble" macromolecular conjugates of the enzymes uricase and L-asparaginase with an excess of albumin. They demonstrated an increased resistance of the conjugate to heat denaturation. We have extended these studies to include eight other enzymes and initiated a number of experiments to determine the *in vivo* or clinical potential of these enzyme-albumin conjugates. Our overall objective has been to make enzymes more amenable to use in medicine as therapeutic agents in light of the limitations described above. The choice of albumin stems from a number of specific advantages that albumin affords as a drug or enzyme carrier system. 1. Albumin is a natural and the most abundant plasma protein. 2. Albumin has a relatively long circulation time with a turnover rate in the order of 30 hours. 3. Albumin probably functions normally as a carrier molecule being responsible for the transport of fatty acids as well as both steroid and polypeptide hormones in the plasma. 4. Albumin is relatively stable, it is inexpensive and readily available and it has a multiple of reactive sites on which to attach therapeutic agents. Our initial interest was not so much in the area of drug therapy, but more in the potential use of enzymes in medicine with particular emphasis on the treatment of a number of "Inborn Errors of Metabolism" by Enzyme Replacement Therapy.

Our early objectives, some of which have been realized, were to produce conjugates of enzyme with a molar excess (in the range of 1:10) of albumin. We hypothesized that this conjugate, following intravenous administration, would a) retain some of the characteristics of native albumin, resulting in a prolonged circulation time, b) mask the antigenic determinants on the enzyme, thus reducing immunologic reactivity and possible hypersensitivity reactions following repeated injection, and c) conjugation of the enzyme with albumin might be expected to increase the resistance of the enzyme to proteolysis and denaturation.

Our early experiments showed indeed that albumin could change a number of *in vitro* and *in vivo* characteristics of the immobilized or conjugated enzyme. These include resistance to proteolytic inactivation and altered circulatory half-lives (2, 9, 10). The most important finding dealt with the immunogenicity of hog liver uricase in rabbits following conjugation with an excess of homologous and heterologous albumins (11). We demonstrated that, at least for uricase, it was possible to either mask the antigenic determinants using albumin or reduce the immunogenicity of the enzyme, which in its free form is a strong antigen. At that time we were careful to point out that the reduced antigenicity following chemical cross-linking with homologous albumin was limited to the enzyme uricase. We have now demonstrated that this reduced immunogenicity can be produced in a similar manner using a number of different enzymes including: alpha-glucosidase from both yeast and human placenta, superoxide dismutase, catalase, L-asparaginase, glucose-6-phosphatase and cholesterol esterase. In the following pages, we will describe some of the more recent experiments using enzyme-albumin conjugates and attempts to target the enzyme to specific tissues, cells and intracellular locales.

The experiments performed in our laboratory using enzyme-albumin conjugates closely parallel a series of experiments by Abuchowski and coworkers (12, 13). In place of albumin, these investigators have used the synthetic polymer polyethylene glycol (PEG) in an effort to promote the potential use of enzymes in medicine while protecting the enzyme from biodegradation and immunologic reactivity. Surprisingly, in spite of the advantages that albumin might offer as a potential drug or enzyme carrier, there are relatively few publications describing albumin as a potential drug or enzyme carrier. The group of Fiume and coworkers (14) has utilized albumin conjugated with fungal toxins in a "lysosomotropic" approach to antiviral chemotherapy, the rationale being that the albumin-toxin conjugates would be rapidly localized in cells with high rates of protein uptake, such as macrophages where DNA viruses may also be concentrated.

A. *PROPERTIES OF ALBUMIN-ENZYME AND ALBUMIN-DRUG CONJUGATES*

In the following pages we will describe some of the physical, biochemical and *in vivo* data for a number of different enzymes and one drug (the iron chelating agent desferrioxamine). While the results of conjugating an excess of albumin to an enzyme produces some common properties, these are often difficult to predict and so each enzyme usually has to be worked up independently in some detail, there often being significant differences in the modes of cross-linking, the degree of protection that has to be afforded to the active site during the cross-linking step, and the properties being sought from the resultant conjugates. The following is a list of enzymes which we have successfully conjugated with albumin retaining better than 50% of the original activity:

Uricase (hog liver)
L-Asparaginase (*Aspergillus niger*)
Alpha-1,4-glucosidase (yeast)
Alpha-1,4-glucosidase (human placenta)
Superoxide dismutase (hog liver)
Catalase (bovine liver)
Glucose-6-phosphatase (rabbit liver)
Cholesterol esterase (porcine pancreas)
Cholesterol esterase (*Pseudomonas fluorescens*)

The following represents a cross-linking recipe for producing a conjugate of alpha-1,4-glucosidase derived from human placenta with an excess of albumin:

2 mg alpha-1,4-glucosidase (14 units)
20 mg human albumin
6 mg maltose
3 ml buffer (PBS at pH 5.5)
50 µl glutaraldehyde (25%)
Stir at 4°C for 3-4 h
Add 300 mg glycine
Dialyze against 1% NaCl and 1% glycine
Pressure ultrafiltration on Amicon XM300, or
Molecular sieve chromatography on Sepharose 4B

The resultant conjugates (Table 1) retain as much as 70% of the native activity towards maltose and as high as 58% of its activity towards the natural substrate. The conjugate (following XM300 pressure ultrafiltration) has an average molecular weight (as determined by molecular sieve chromatography) of $8.5x10^5$ with approximately 80% of the protein eluting with molecular weights between $6x10^5$ and $1x10^6$. Producing the conjugate in the absence of substrate (maltose) in the cross-linking medium results in a conjugate with similar molecular weight characteristics but with little activity (less than 5%).

TABLE 1 *Enzyme recovery rates following cross-linking with albumin*

Enzyme	Cross-linking conditions	Substrate	Recovery %
Uricase (hog liver)	pH 6.5 + uric acid	Uric acid	40
Uricase (hog liver)	pH 9.1 + uric acid	Uric acid	66
Uricase (hog liver)	pH 9.1 + uric acid	Uric acid	4
α-Glucosidase (yeast)	pH 6.8 + PnPG	PnPG[a]	65
α-Glucosidase (yeast)	pH 6.8 + PnPG	Glycogen[b]	2
α-Glucosidase (human)	pH 6.8 + maltose	PnPG[a]	0
α-Glucosidase (human)	pH 6.8 + maltose	Maltose[b]	69
α-Glucosidase (human)	pH 6.8 + maltose	Glycogen[b]	58

The yeast α-glucosidase has a ph optimum of 7.0 and almost no activity at pH 4.5, whereas the alpha-glucosidase from human placenta is a lysosomal enzyme with an optimum pH of 4.8 and virtually no activity at normal pH.

[a] Enzyme assay run at pH 6.8
[b] Enzyme assay run at pH 4.8

The alpha-glucosidase-albumin polymers are produced by straightforward cross-linking of a mixture of enzyme and albumin (in excess) using primary amino groups as the reactive sites and the bifunctional dialdehyde glutaraldehyde to produce a Schiff base linkage. In some cases it is necessary to reduce the Schiff base using sodium cyanoborohydride. In other instances it may be necessary to use other cross-linking strategies (see Poznansky & Juliano (1) for a recent review) such as carbodiimides to cross-link a reactive carboxyl group from one protein to the other. We consider that unless one has a good idea of the chemistry of the two molecules to be conjugated and the nature of the available reactive groups, then the conjugation procedure is a matter of trial and error. It is, however, essential that the cross-linking conditions be as gentle as possible so as to preserve the physiological characteristics of both enzyme and albumin. It was interesting to determine that in the case of uricase-albumin conjugates (9), it was necessary to decrease the turnover rate of the enzyme (by carrying out the reaction in the absence of O_2, a cofactor) in order to achieve good recovery of enzyme activity.

Figure 1 demonstrates the increased resistance to both proteolytic degradation and heat denaturation of the enzyme L-asparaginase conjugated to a ten-fold molar excess of human albumin compared to an equivalent amount of free or non-conjugated enzyme. The degree of protection afforded other enzymes may be even greater as uricase-albumin conjugates have half-lives at 37°C in excess of five days, whereas the native or free enzyme has an activity half-life of only 8 h under the same conditions (9, 10). Similar differences to proteolytic enzymes present in fresh serum can also be determined between unconjugated enzyme and enzyme-albumin complexes.

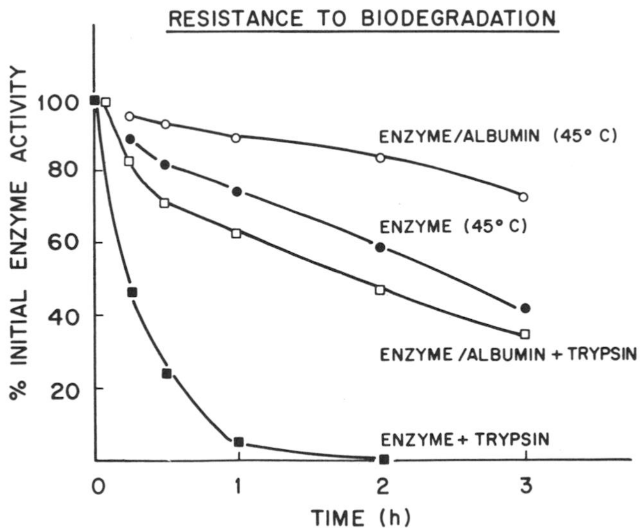

FIG. 1. *L-Asparaginase in its free, unconjugated form or conjugated to a ten-fold molar excess of albumin was incubated at 45°C in buffer (PBS, pH 7.4) or at 37°C in buffer containing the proteolytic enzyme trypsin (10 units/ml). The L-asparaginase activity was then monitored as a function of time. In the same buffer at 37°C without trypsin, the enzyme has a half-life of 12 h which increases to 48 h or more upon conjugation with albumin.*

Figure 2 demonstrates the increased survival in the circulation of the enzyme superoxide dismutase following conjugation with a five-fold or a ten-fold molar excess of albumin. This is the most dramatic alteration in circulation half-times we have seen as the half-life of the free enzyme is less than 15 min, whereas the enzyme-albumin conjugate has a half-life in the order of 15 h. The work on superoxide dismutase (15) represented an effort to determine the possibility of using the enzyme in the treatment of rheumatoid arthritis. In addition to its increased lifetime in the circulation, the conjugate is a much more effective anti-inflammatory agent than equivalent amounts of the enzyme introduced *in vivo* in the uncomplexed form. Other enzymes such as cholesterol esterase and alpha-glucosidase complexed with albumin also show increased circulation, heading for but never reaching that for free albumin. Producing conjugates that are too big (20:1 albumin-enzyme) results in a more rapid clearance of the complex from the circulation.

Although the increased circulation times for the enzymes conjugated with albumin migh be advantageous in certain cases (e.g. the use of L-asparaginase as an anti-tumor agent and the general decrease in L-asparagine levels), the more concerned areas where enzyme therapy might be used for enzyme deficiency diseases involve substrate accumulation in specific tissue and often in intracellular organelles such as lysosomes. We have for the past few years been concerned with the possibility of targeting enzyme-albumin polymers to specific cells to be delivered to lysosomes. Continuing with our work on alpha-glucosidase, we are concerned with the disease, type II glycogenosis or Pompe's disease, a condition whereby glycogen storage results in cellular dysfunction, especially in cells of the liver and cardiac and respiratory muscle (16). This usually results in death within the first year of life. We have utilized two separate targeting agents. Mono- or polyclonal antibodies (12) directed against cell surface antigens are used to home or direct the enzyme-albumin complex to specific cells in anticipation that following

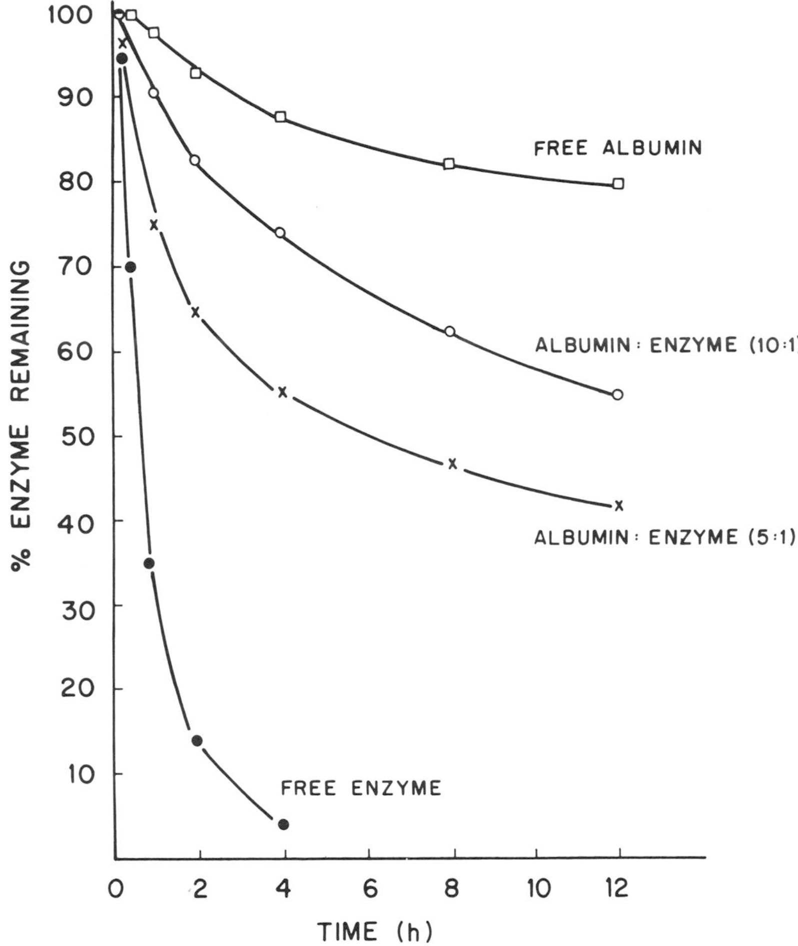

FIG. 2. *Superoxide dismutase (from hog liver), rat albumin and various conjugates of the two were radiolabelled (^{125}I) and blood clearance measurements were made following intravenous administration in rats. While the clearance rates for the albumin and albumin conjugates were fairly reproducible, the clearance rates for free superoxide dismutase varied from half-times between 30 sec and 30 min. The free enzyme is cleared almost exclusively by the spleen and liver.*

binding the entire complex will be internalized and directed towards a lysosome compartment in the manner of receptor mediated endocytosis (17). The second approach we have taken is the use of the small peptide hormone insulin to direct the conjugate to muscle cells rich in insulin receptors (18). Here we were theoretically on safer ground since the internalization of the insulin-insulin receptor complex following binding of the ligand to the receptor is an established observation (19). We consider that it is important to retain the albumin coat on the enzyme (instead of simply using the targeting agent directly attached to the enzyme) in order to avoid the rapid clearance of the conjugate from the circulation via the reticuloendothelial system (RES). In the recent review by Poznansky and Juliano (1), a section is devoted to the question of what barriers a drug-carrier complex must cross in its journey from the site of administration to the site of drug action. We consider the RES to be a sort of negative barrier which must be avoided in order to enhance the possibility of our drug (enzyme)-carrier complex reaching its target. Thus, in considering strategies for drug or enzyme delivery, it might be almost as important to consider how to target the therapeutic agent away from a given area (say the RES) as it is in targeting to a specific site of action. This may not be a simple task since the cells of the RES are expert as recognizing and removing foreign macromolecular structures from the circulation.

Table 2 and Figure 3 demonstrate our attempts to target alpha-glucosidase-albumin conjugates to liver cells using antibodies directed against hepatocytes. Table 2 indicates that while the conjugate in the absence of a specific antibody is directed primarily to Kupffer cells, the presence of a conjugated antibody directed against the hepatocyte population of liver cells results in a redirection of the enzyme to hepatocytes, the cell type which exhibits very severe glycogen storage and dysfunction in Pompe's disease. Figure 3 demonstrates that within

TABLE 2 *Targeting enzyme-albumin polymers to hepatocytes*

^{125}I-labelled protein preparation	^{125}I radioactivty, cpm/mg of cell protein					
	Hepatocytes		Kupffer cells		Ratio	
α-1,4-Glucosidase	108	22	1030	131	0.10	0.01
Albumin	42	20	177	49	0.25	0.05
Enzyme-albumin polymer	95	31	508	133	0.20	0.03
Anti-F IgG	109	15	629	120	0.16	0.03
Anti-H IgG	233	32	280	40	0.85	0.09*
Anti-F IgG polymer	182		1067	189	0.17	0.02
Anti-H IgG polymer	317	80	259	11	1.23	0.15*

Various enzyme or antibody preparations were injected intravenously. When 80-90% of the injected label had cleared from the circulation, the rats were anesthetized and the livers were perfused with collagenase to separate hepatocytes from Kupffer cells. The ratio represents the counts/mg cell protein in the hepatocytes over that in the Kupffer cells. The anti-H IgG preparation was derived from the serum of a rabbit that had received repeated injections of isolated rat hepatocytes. The antiserum was absorbed with rat spleen cells prior to conjugation.

F = fibroblasts, H = hepatocytes

* Significantly different from all other preparations at $P < 0.001$.

the liver cells the enzyme is targeted to a lysosomal fraction. The liver cells were homogenized following a 20 min or an 80 min incubation of the antibody-albumin-enzyme complex (radiolabelled in the enzyme) and subjected to subcellular fractionation on Percoll (20). The upper panel describes the fractionation of marker enzymes: 5'-nucleotidase for a membrane fraction and acid phosphatase for the lysosomal fraction. The lower panel shows that at 20 min most of the label is still associated with a membrane fraction, whereas at 80 min a much more significant fraction of the label has reached the lysosomal fraction. In separate experiments (21), we have verified that not only does the radiolabel appear in the lysosomal fraction with time, but in fact an increase in lysosomal enzyme activity (in this case alpha-glucosidase from yeast) could also be identified with a lysosomal fraction as a function of time.

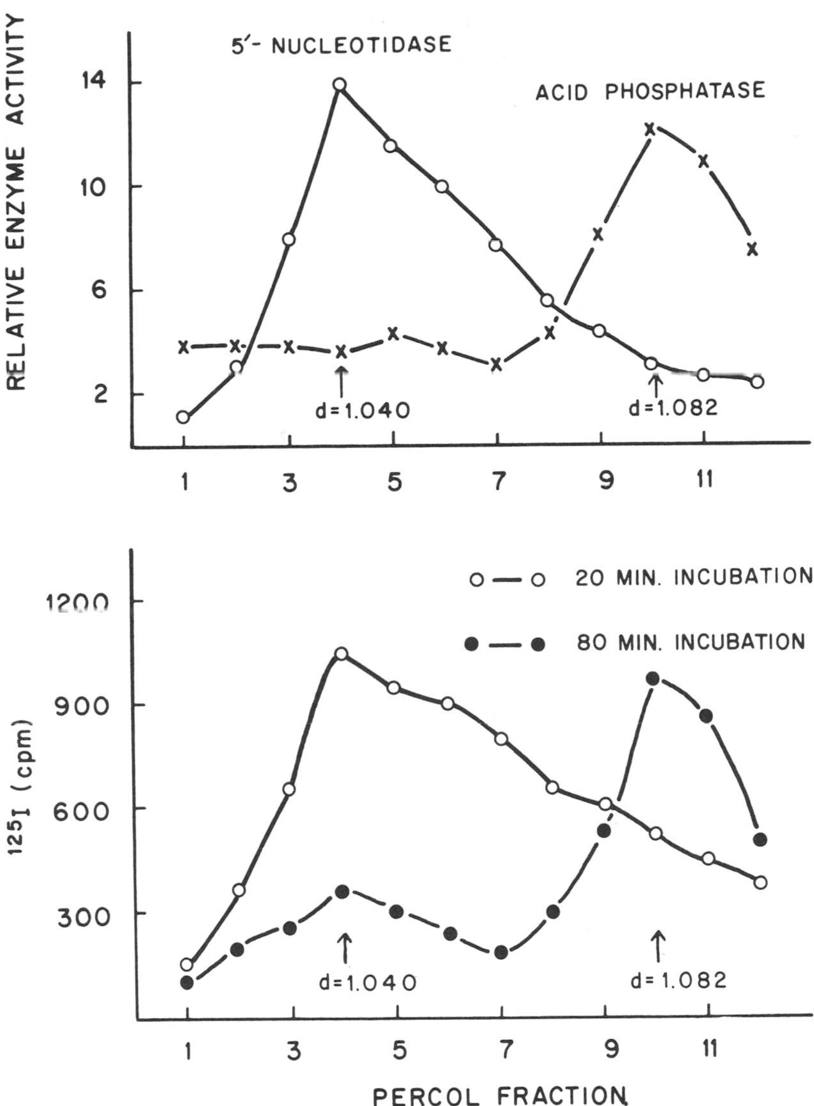

FIG. 3. *Rat liver homogenates (following removal of nuclear material and cell debris by a single low speed spin; 5 min at 500 x g) were subjected to density gradient centrifugation (see 18 and 20) using 35% Percoll in .25 M sucrose. The top panel identifies the plasma membrane (5'-nucleotidase as marker with d=1.04) and the lysosomal fractions (acid phosphatase as a marker with d=1.082) for the rat liver preparation under these conditions. The bottom panel identifies the fate of* ^{125}*I-labelled alpha-glucosidase-albumin-antihepatocyte antibody conjugated following a 20 min or 80 min incubation with the liver cells.*

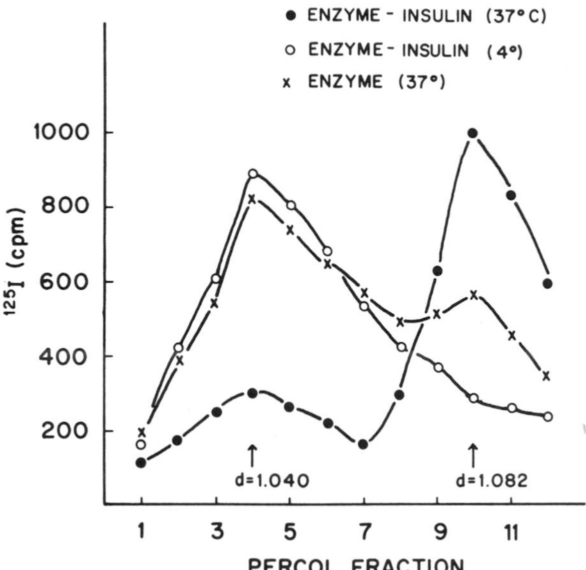

FIG. 4. *This figure demonstrates the subcellular fractionation of chick myoblasts grown in tissue culture for several days. The plasma membrane fraction (d = 1.04) and lysosomal fraction (d=1.082) are identified as in Figure 3. The cells were incubated in the presence of enzyme or enzyme-insulin conjugates for 2 h prior to homogenization and subcellular fractionation. The enzyme in this case is alpha-glucosidase from yeast, which in of itself binds well to cells in culture but which is not internalized (x--x). Enzyme-insulin conjugates are found associated with a lysosomal fraction (d=1.082) only above 20°C, the temperature below which the process of receptor mediated endocytosis works very poorly, if at all. Similar results are seen when enzyme-albumin conjugates are used with insulin except very little binding of the enzyme-albumin alone to the cells is observed.*

In the case of Pompe's disease, the immediate cause of death is failure of cardiac and respiratory muscle as a result of glycogen accumulation, and if enzyme therapy is to be effective, then the enzyme must be delivered to muscle tissue. We have attempted to utilize the insulin receptor as a target since muscle cells are especially rich in these receptors and subsequent internalization of the ligand (insulin) has been demonstrated (18). We have recently demonstrated the potential of insulin as a targeting agent to direct alpha-glucosidase-albumin polymers to muscle cells in tissue culture and to muscle tissue *in vivo*. Figure 4 examines targeting of enzyme and enzyme-insulin conjugates to chick myoblasts in tissue culture and the subsequent subcellular fractionation. Enzyme is shown to reach a lysosomal fraction only when insulin is conjugated to the enzyme and incubated with the cells at 37°C. At 4°C, although binding occurs, no internalization is observed as the enzyme remains associated with a membrane fraction. This figure describes the use of alpha-glucosidase, which itself binds fairly well to the muscle cells (but is not internalized). We have carried out similar experiments using L-asparaginase where both the binding of the enzyme to muscle and its subsequent internalization (at 37°C) is very strongly dependent on the attachment of insulin.

B. *IMMUNOLOGY*

Our initial objective in conjugating homologous albumin to enzymes was to reduce the immunogenicity of foreign enzymes. In 1978 we (11) demonstrated that hog liver uricase could be rendered non-immunogenic (unable to elicit an antibody response) and non-antigenic (unable to react with preformed anti-uricase antibodies). We have now demonstrated that this can be considered a possibility for a range of different enzymes (see Table 3). Yagura and colleagues (24) have also shown that L-asparaginase can be rendered non-immunogenic by conjugating it to an excess of homologous albumin using our cross-linking procedures.

In a separate set of experiments initiated in an attempt to produce a more effective iron chelating agent, we have cross-linked the iron chelating agent desferrioxamine (DF) to albumin. We have shown that the conjugated DF (Bhardwaj *et al.*, 1986, submitted for publication) is a more effective iron chelator when compared to equivalent amounts of free DF. We believe that the albumin allows the DF a greater circulation time and hence chance to chelate iron compared to free DF, which is rapidly cleared from the circulation via the kidneys. Figure 5 demonstrates

TABLE 3 *Immunological properties of enzyme-albumin polymers*

Enzyme	Immune response
Uricase (hog liver)	+++
Uricase-albumin (1:10)	-
α-Glucosidase (yeast)	+++
α-Glucosidase-albumin (1:10)	-
α-Glucosidase (human placental)	+++
α-Glucosidase-albumin	-
Superoxide dismutase (bovine)	+++
Superoxide dismutase-albumin (1:5)	+
Superoxide dismutase —albumin (1:10)	-
L-Asparaginase (*E coli*)	+++
L-Asparaginase-albumin (1:5)	+
L-Asparaginase-albumin (1:10)	-

The immunogenicity of these enzymes and enzyme-albumin conjugates was examined in mice and antibody response was determined by RIA.

	ANTIGEN INJECTED	ANTIBODIES PRODUCED TO: DF	MSA	BSA	MSA-DF	BSA-DF
1	DF	—	—	—	—	—
2	Ficoll · DF	X	—	—	X	X
3	BSA · DF	X	—	X	X	X
4	MSA · DF	—	—	—	—	—
5	MSA · DF + DF	—	—	—	—	—
6	MSA · DF + BSA · DF	—	—	X	—	X
7	BSA · DF + MSA · DF	X	—	X	X	X
8	Ficoll · DF + MSA · DF	X	—	—	X	X

FIG. 5. *Groups of mice (five/group) were injected on three successive days followed by seven days rest. This regimen was followed for three separate sets of injections. The mice were then bled and the antisera tested by radioimmunoassay against DF (desferrioxamine), MSA (mouse serum albumin), BSA (bovine serum albumin), MSA-DF (conjugates of MSA and an excess of DF) and BSA-DF (conjugates of BSA and an excess of DF). --- represents no response whereas x represents an immune response. Groups 5 through 8 received two sets of three series of injections each, the first representing the "toleragen" and the second representing the challenging antigen.*

some of the immunological properties of DF attached to albumin or Ficoll. The concern is that while DF itself as a small molecule is non-immunogenic (mouse 1), following attachment to a larger carrier molecule it may work as an effective hapten. This is clearly so, as is seen in mice 2 and 3, where attaching DF to either Ficoll or BSA (bovine serum albumin) causes antibodies to be formed to both the carrier and drug hapten. This is not so for mouse serum albumin (MSA), which does not cause the attached DF to be immunogenic (mouse 4). We assume initially that since the MSA as a "self protein" is not recognized as foreign, the mouse immune system decides that the attached hapten (DF) is self as well. Subsequent injection of mice receiving MSA-DF with straight DF (5) produces no response, as expected, but what is surprising (mouse 6) is that mice injected with BSA-DF following a series of injection of MSA-DF do not produce antibodies against the DF (although they do against the BSA since it is still a foreign protein). We explain this by suggesting that the self albumin is working as a toleragen to render the subsequent injections of DF non-immunogenic as the mouse is now programmed to recognize the DF as self. The concept of tolerance induction using small haptens attached to carrier molecules has been previously described as the self immunoglobulin, IgE, can behave as a toleragen to attached hapten (22) and carboxymethyl cellulose (CMC) has been shown also to be an effective toleragen (23). Mouse groups 7 and 8 show that in order to induce tolerance to the attached DF drug molecule, the toleragen (MSA-DF) must be given first. When the mice are first injected with either BSA-DF or with Ficoll-DF followed with a series of injections of MSA-DF, then the ability to induce tolerance to the DF molecule is destroyed. The assumption here would be that the mouse having received the drug molecule in an immunogenic form has "learnt" to recognize DF as foreign and will continue to do so. We do not as yet have any strong evidence to suggest that in the case of the enzyme-albumin conjugates that the albumin molecule is not simply acting to mask the antigenic sites on the enzyme (see 2 and 11), but also acting as an effective toleragen for the attached enzyme (foreign protein).

II. CONCLUSION

The primary objective of this work is the production of enzymes which are more presentable to clinical medicine. The ability to reduce or eliminate the immunogenicity of foreign proteins has been our most important finding. We feel that albumin has some important advantages over other carrier systems because even in its conjugated form with enzyme and other albumin molecules the organism, and especially the reticuloendothelial system, appear to be less able to recognize the complex as foreign. This may also have important advantages in allowing the complex sufficient time in the circulation to at least start to seek out its target. The fact that a ligand such as insulin can be used to target the enzyme-albumin conjugate to specific cells and intracellular organelles is a hopeful sign for the potential treatment of a wide range of enzyme deficiency diseases manifested as lysosomal storage diseases.

REFERENCES

1. M.J. Poznansky and R.L. Juliano, *Pharm. Rev.*, in press (1985).
2. M.J. Poznansky, *Pharmac. Ther. 21*: 53-76 (1983).
3. G. Gregoriadis, *The Lancet 2*: 241-246 (1981).
4. G. Poste and R. Kirsh, *Biotechnology 1*: 869-878 (1983).
5. F.J. Martin and D. Papahadjopoulos, *J. Biol. Chem. 257*: 286-288 (1983).
6. E.S. Vitetta, K.A. Krolock, M. Miyama-Lanab, W. Cushley, W. and J.W. Uhr, *Science 219*: 644-650 (1983).
7. M.J. Poznansky, M. Shandling, M.S. Salkie, J. Elliott and E. Lau, *Cancer Res. 42*: 1020-1025 (1982).
8. B. Paillot, M-H. Remy, D. Thomas and G. Broune, *Pathol. Biol. 22*: 491-495 (1974).
9. M.J. Poznansky, in *Biological Applications of Immobilized Enzymes and Proteins* (ed. T.M.S. Chang) Vol. 2, pp. 341-354, Plenum Press, New York (1977).
10. M.J. Poznansky and D. Bhardwaj, *Can. J. Physiol. Pharm. 58*: 322-325 (1980).

11. M-H. Remy and M.J. Poznansky, *The Lancet ii*: 68-70 (1978).

12. A. Abuchowski, T. van Es, N.C. Palczuk and F.F. Davis, *J. Biol. Chem. 252*: 3578-3581 (1977).

13. A. Abuchowski and F.F. Davis, in *Enzymes as Drugs* (eds. J.S. Holcenberg and J. Roberts) pp. 367-384, Wiley, New York (1981).

14. L. Fiume, A. Mattioli, P.G. Balboni and G. Barbanti-Brodano, in *Drug Carriers in Biology and Medicine* (ed. G. Gregoriadis) pp. 3-22, Academic Press, New York (1979).

15. K. Wong, L.G. Cleland and M.J. Poznansky, *Agents & Actions 10*: 231-244 (1980).

16. R.R. Howell and J.C. Williams, in *The Metabolic Basis of Inherited Disease* (eds. J.B. Stanbury, J.B. Wyngaarden, D.S. Fredrickson, J.L. Goldstein and M.S. Brown) 5th Ed., pp. 141-166, McGraw-Hill, New York (1983).

17. J.L. Goldstein, R.G.W. Anderson and M.S. Brown, *Nature 279*: 679-685 (1979).

18. M.J. Poznansky, R. Singh, B. Singh and G. Fantus, *Science 223*: 1304-1406 (1984).

19. J. Schlessinger, Y. Schechter, M.C. Willingham and I. Pastan, *Proc. Natl. Acad. Sci. U.S.A. 75*: 2659-2664 (1978).

20. H. Pertoft, B. Warmegard and M. Hook, *Biochem. J. 174*: 309-317 (1978).

21. M.J. Poznansky and R. Singh, in *Advances in the Treatment of Inborn Errors of Metabolism* (eds. M.d'A. Crawfurd, D.A. Gibbs and R.W.E. Watts) pp. 161-174, John Wiley, New York (1982).

22. Y. Borel, *Immunological Rev. 50*: 71-104 (1980).

23. U.E. Diner, D. Kunimoto and E. Diener, *J. Immunol. 122*: 1886-1891 (1979).

24. T. Yagura, Y. Kamisaki, H. Wada and Y. Yamamura, *Int. Arch. Allergy App. Immunol. 64*: 11-18 (1981).

13

Attaching Metal Ions to Antibodies

Claude F. Meares

Department of Chemistry
University of California
Davis, California

I. INTRODUCTION

Attachment of metal ions to antibodies by means of bifunctional chelating agents can add the nuclear, physical, and chemical properties of the metallic elements to target-selective proteins. The products may be used for immunoassays *in vitro* employing either radioactive or luminescent metals (1), or for *in vivo* imaging of human patients with radioactive or paramagnetic metals (2), or for therapy with radioactive metals (3). In the future, many other applications may be envisioned; in particular, the photochemical (4) and catalytic (5) properties of chelated metals are likely to be important. For several years we have been involved in the development of new bifunctional chelating agents which may be applied to this general purpose, though we have turned to monoclonal antibodies relatively recently. Here I will briefly

The work described here was supported principally by Research Grant CA 16861 from the National Cancer Institute.

review the important principles and practical aspects of bifunctional chelate chemistry, and describe some recent developments.

II. INORGANIC CHEMICAL CONSIDERATIONS

A remarkably large number of metallic ions form "stable" chelate complexes. For example, diligent - but probably incomplete - searching of the literature reveals references to the preparation and/or thermodynamic stability constants of the chelates of about *70* different elements with the chelator EDTA! (6,7) Those 70 metallic elements have a great range of properties, and many would not be stable *in vivo* as EDTA chelates, so there are good reasons to investigate new chelators which will permit particular metals to be used as probes or cell-killing agents in living systems.

The structures of metal chelates depend on both the metal and the chelator. For example, (Figure 1) the structure of Fe(III)-EDTA (8) is similar to those of indium(III)EDTA, cadmium(II)-EDTA, cobalt(II)EDTA, and manganese(II)EDTA (9,10,11,12). The presence of metal-bound water molecules, which are often easily displaceable by competing ligands in solution, is not an uncommon feature of metal chelates. The *rates* at which metal-bound groups (ligands) exchange with other ligands are exquisitely dependent on the metal involved, but they also are strongly influenced by the structure of the chelator (13).

It is easy to become confused over the relative significance of the equilibrium, versus the kinetic properties of metal chelates, when the goal is application to a living system. The thermodynamic equilibrium (stability) constant of a metal chelate is often readily available (6,7), and it gives a gross indication of the suitability of the chelate for further investigation. Chelators that contain highly basic groups (such as EDTA, DTPA, and related polyaminocarboxylate compounds) will bind H^+ avidly, and this can reduce the effective value of the stability constant by a large factor (14).

All reactions proceed toward equilibrium, so it helps to know where the equilibrium point is located; however, *rates* are often

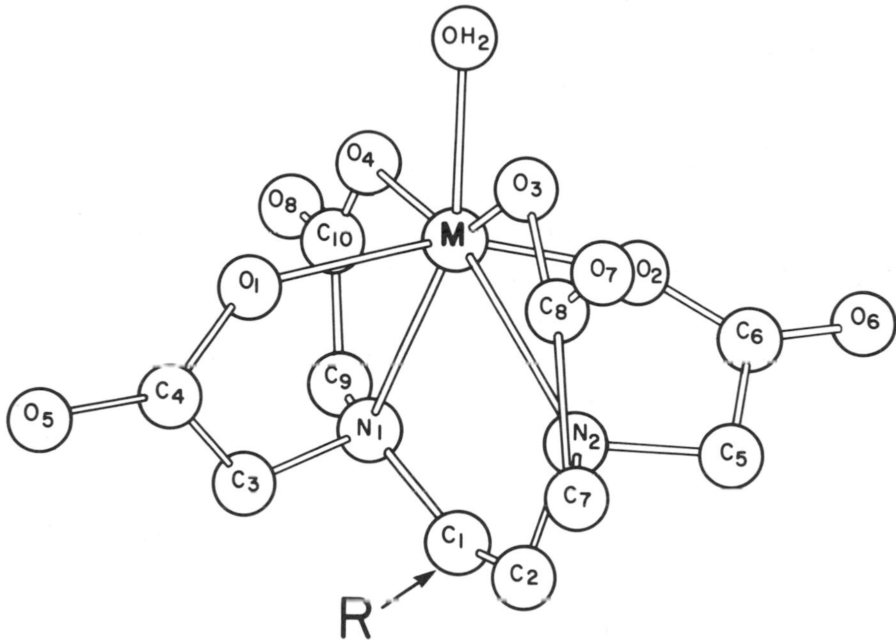

FIG. 1 *The structure of Fe(III)EDTA, which is similar to that of several other EDTA chelates. The phenyl- and benzyl-EDTAs discussed in the text have a substituent at carbon-1, indicated here by R. Taken from ref. 33.*

the key feature in practice. Another troublesome matter is the presence of contaminating metal ions; this makes it practically impossible to prepare a metal chelate by mixing 10^{-9}M metal ion with 10^{-9}M chelator, no matter how large the stability constant! Container surfaces, "pure" water, buffer salts, protein samples, and carrier-free radionuclides are all potential sources of trace metal impurities. Keeping the concentration of chelator as high as possible (e.g. > 10 μM) is one way to minimize problems due to slow rates or to contaminating metal ions.

A. *THE PRESENCE OF PROTEINS*

A conjugate formed by a chelating agent and an antibody contains a very large number of potential metal-binding sites; besides the chelator site, there may be dozens of weaker metal-binding sites on the native protein. It is often the case that metal ions bound to those weaker sites dissociate rather slowly. This can cause practical problems when the radiometal-conjugate is placed in a living system; metal ions bound to weak sites are soon transferred to other proteins and so provide nonuniform "background" interference. Only in rare cases are metal ions bound irreversibly to proteins after simple mixing. One way to prepare a radiometal-tagged antibody with the metal bound only to the attached chelating groups is to use a buffer with weak metal-binding properties (e.g. citrate) in an appropriate concentration so that the desired metal chelate is more stable than the metal citrate, which is in turn more stable than the complexes formed between the metal and the native protein.

For applications which do not require the use of short-lived radionuclides, the metal chelate can be prepared *before* conjugation to the protein; this greatly simplifies matters!

In cases where limited time or limited chelator concentration prevent quantitative binding of the radiometal to the desired chelator, unbound metal ions can be scavenged by adding excess EDTA (or DTPA, etc.). The radiolabeled protein conjugate can then be quickly isolated by centrifugation through a small gel-filtration column (15).

B. *CONDITIONS IN VIVO*

After radioactive metal chelates are injected into living systems, they are extremely dilute. Blood contains substantial concentrations of molecules that bind metals very strongly (e.g. albumin [$\approx$ 1 mM] and transferrin [$\approx$ 0.01 mM]), and it also contains chelatable metals (Ca^{++}, Mg^{++}) that may compete for available chelating groups. Thus if a radioactive metal ion dissociates from its chelating group, it will almost certainly not return. The *rate of loss* of the metal from the chelate *in vivo* is generally the critical feature. Redox reactions involving metals are also possible, since blood contains dissolved ascorbate and O_2 (in addition to that bound to hemoglobin).

In order for a metal chelate to be stable *in vivo*, a number of possible routes to decomposition must be avoided. In our experience, the best way to accomplish this is by carefully analyzing each step of the process of preparing a new conjugate, from organic synthesis all the way to tests in living systems.

III. SYNTHESIS OF BIFUNCTIONAL CHELATING AGENTS

The first step in preparing a stable metal-protein conjugate is preparation of a bifunctional chelator. The usual tools of organic synthesis apply to chelator preparation, including structural characterization by spectroscopic methods (16). The most useful molecules of this type so far have been polyamino-carboxylates such as EDTA and DTPA. These molecules tend to form metal chelates rapidly under very mild conditions (room temperature, pH > 5) so that short-lived radiometals may be used conveniently. Also there are well-studied cases in which the (inevitable) loss of metal from the chelator occurs extremely slowly (< 1% per day) under physiological conditions.

As more results become available in this area, it appears that no single chelator will stably bind all interesting metal ions. For example, the cyclic anhydride of DTPA leads to a product which binds indium well enough for many *in vivo* applications (17), but does not hold cobalt or copper well (18).

In designing bifunctional chelating agents, we find it useful to put the metal-binding and protein-modifying functions on the opposite sides of a benzene ring. The metal-binding group may be EDTA, DTPA, a macrocycle like TETA, or some other polyatomic group which is expected to bind the desired metal. The protein-modifying group may be an isothiocyanate, bromoacetamide, diazonium, or other reactive group chosen to attach itself to the protein without causing loss of biological activity. We have recently begun to experiment with an additional feature; a linker arm between these two groups which may be susceptible to cleavage under certain conditions. This would permit removal of the metal chelate from the protein under certain circumstances; as

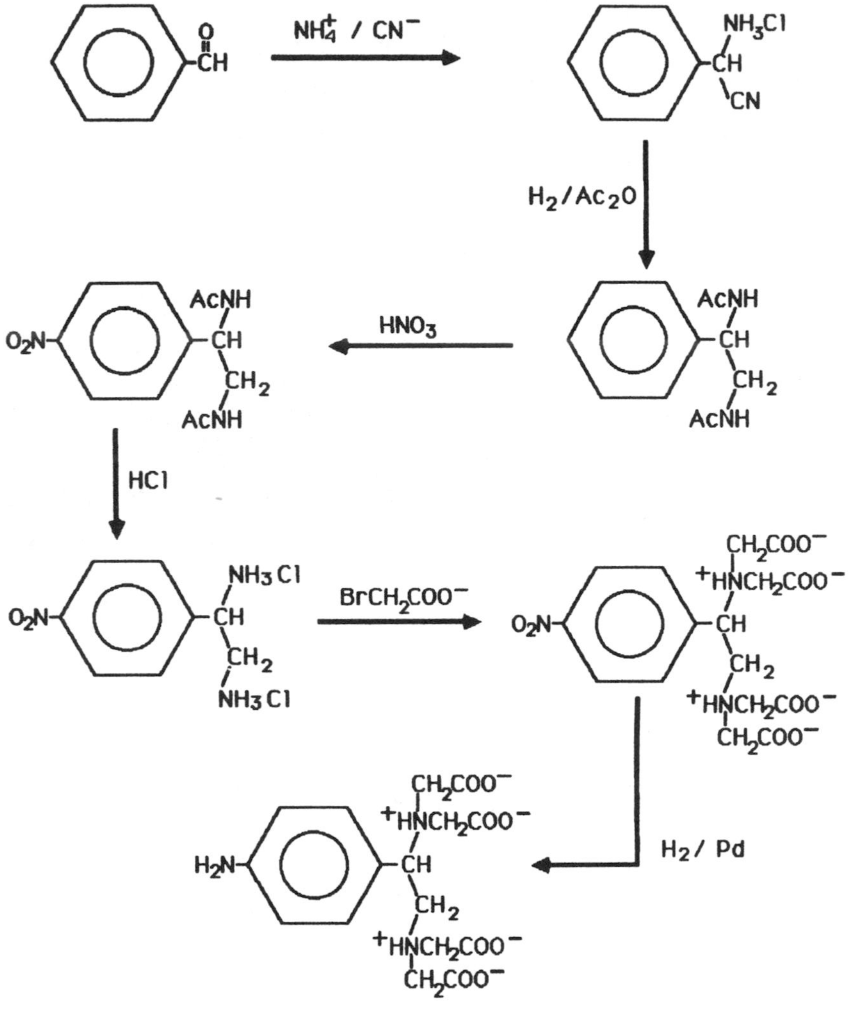

FIG. 2 *The synthesis of aminophenyl-EDTA (19).*

discussed below, a *reversible* radiolabel can exhibit improved biological distributions.

Our first bifunctional chelator (19) (Figure 2) was prepared from benzaldehyde by a seven-step procedure. The steps involved a Strecker amine synthesis (20), followed by reduction of the 1-phenylglycinonitrile product to a protected 1-phenyl-ethylenediamine, nitration of the phenyl ring, deprotection, and carboxymethylation to form 1-[p-nitrophenyl]-EDTA in 30% yield. (Altman *et al.* (21) have investigated the latter reaction and suggested an explanation for the low yield.) At this point, the aromatic nitro group could be reduced to an amine and diazotized or otherwise activated for attachment to a protein.

The next generation of reagents arrived following the insight that naturally occurring amino acids could be converted to bifunctional chelating agents (16). In this case the analogy with the first generation is best shown by starting with

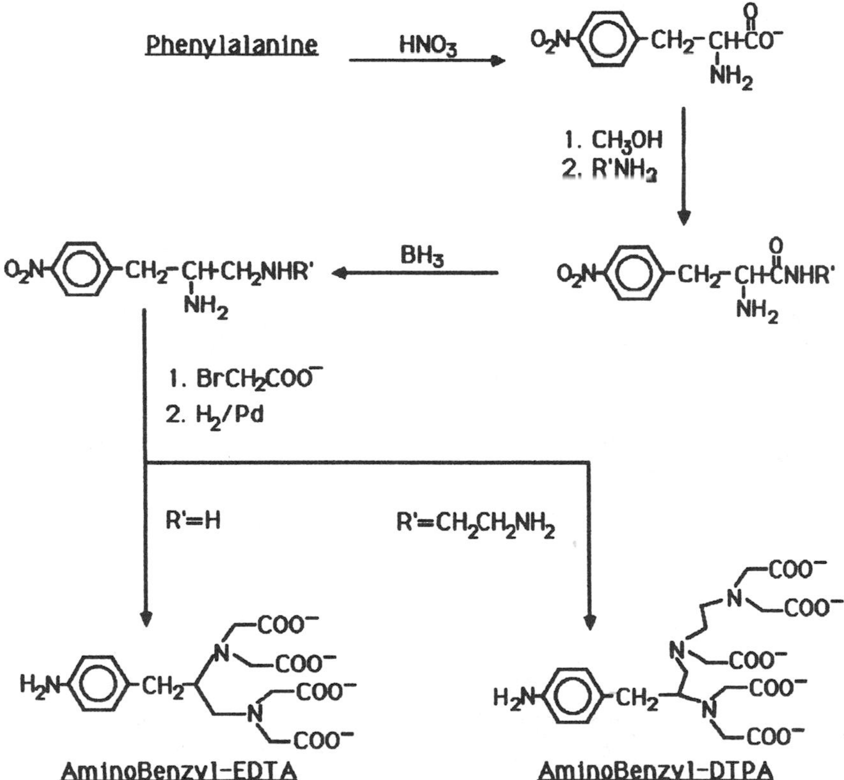

FIG. 3 *The conversion of phenylalanine to aminobenzyl-EDTA (16). Many other α-amino acids may be carried through this procedure, with minor modifications, and many different amines $R'NH_2$ may be used to form the amide.*

L-phenylalanine (22) (Figure 3). First a para nitro group is introduced into the phenyl ring, then the amino acid is converted to an ester, then an amide. The amide is reduced to an amine with borane, then carboxymethylated to give 1-[p-nitro*benzyl*]-EDTA. The aromatic nitro group is then reduced to an amine, which may be further derivatized as shown in Figure 4. Amino*benzyl*-EDTA has an obvious similarity to amino*phenyl*-EDTA, but it has the advantages of easier chemistry, generally high yields (> 70%), and the provision of a single stereoisomer rather than a racemate. The metal-binding properties of phenyl- and benzyl-EDTA seem to be quite similar (23). Furthermore, while in our experience the original synthesis was not readily amenable to small changes in reactants (e.g. nitrobenzaldehyde vs benzaldehyde), the new

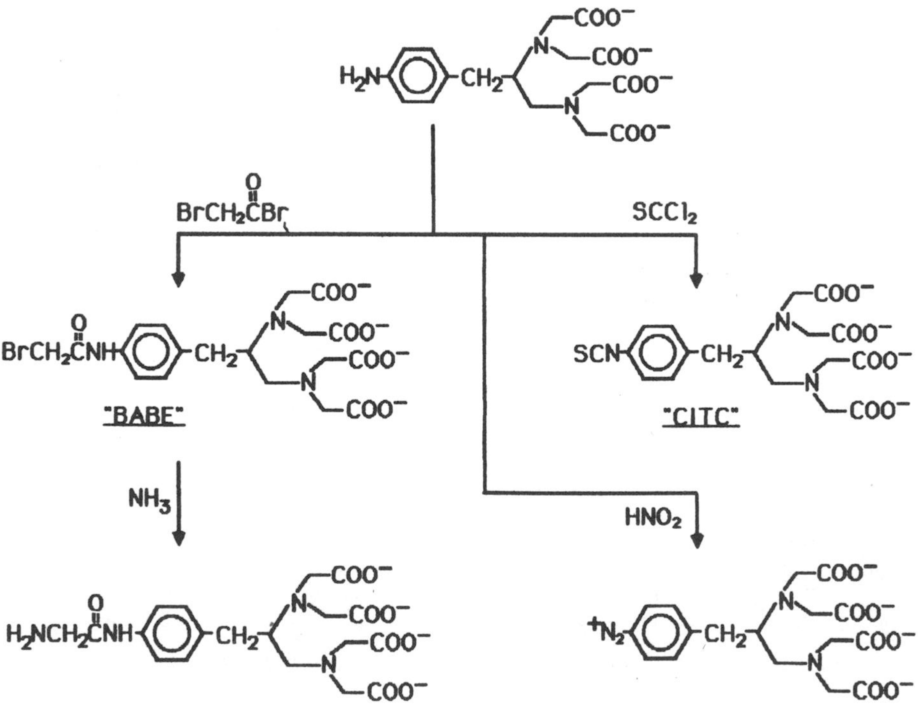

FIG. 4 *Examples of the conversion of the aromatic amino group to protein-labeling reagents. The glycinamide product in the lower left is useful for attachment to a peptide chain.*

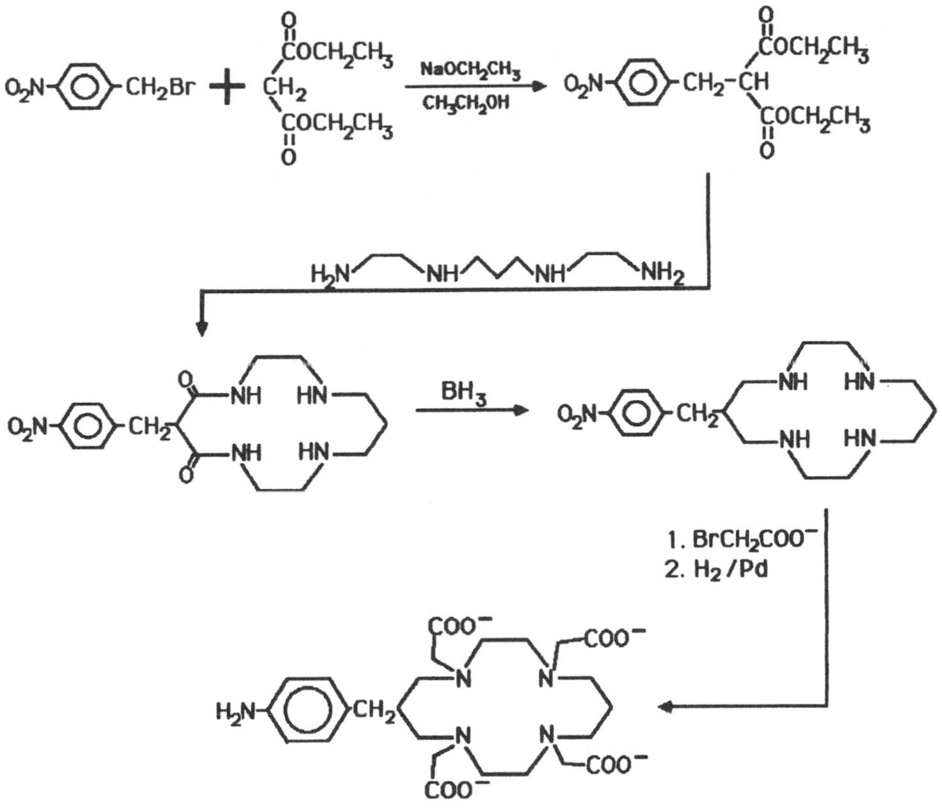

FIG. 5 *Synthesis of aminobenzyl-TETA, a macrocyclic bifunctional chelating agent which binds copper stably in vivo (26).*

synthesis is remarkably flexible. The use of different amino acids as starting materials gives quite a variety of sidechains in the product (24). Also, the use of different amines to form the amide gives products with many useful chelating groups such as EDTA, DTPA (Figure 3), HED3A, etc. (25). This flexibility is very important in practice.

The first macrocyclic bifunctional chelator (Figure 5) was prepared in response to the need for a reagent which would stably bind copper to monoclonal antibodies for *in vivo* use (26). Here, malonic ester was alkylated by nitrobenzyl bromide, the product was condensed with a tetraamine, and familiar reactions were then performed to give the desired bifunctional chelator. This reagent fulfilled the requirement of binding copper extremely well. The macrocyclic TETA group avidly binds Cu^{++} and Co^{++}, but gives multiple products with In^{+++}; apparently this is due to the limited size of the macrocyclic cavity and the large number of amine and carboxylate groups present in the TETA structure.

IV. CONJUGATION TO PROTEINS

Each of the bifunctional chelating agents described above contains an aromatic amino group; in each case, this is activated by conversion to a protein-tagging reagent. Many choices are available for activation; some of these include diazotization, formation of an isothiocyanate, formation of a bromoacetamide, reaction with cyanuric chloride to form an alkylating agent, construction of peptide chains connected to the amino group, use of bifunctional cross-linking reagents such as ethylene glycol bis-[succinimidyl succinate], and *many* others.

The usual strategy with monoclonal antibodies is to use a reagent that will not block the antibody combining site. Because an IgG antibody molecule may well contain >50 lysine residues (of which only a small number are likely to be near the combining site), lysine ε-amino groups are usually good targets (27). Since it is necessary to modify only a few groups (< 5) per protein molecule, it is usually possible to produce a conjugate with high biological activity (> 80% by *in vitro* assay). Many of the activation procedures mentioned above yield products which attach well to protein amino groups. Comparisons of the *in vivo* metabolic properties of the various products are at an early stage, but differences are clearly evident (D. A. Goodwin *et al.*, unpublished).

Typical conjugation reaction conditions now in use in our laboratory are incubation of isothiocyanate (≈ 0.5 mM) with protein (≈ 20 mg/ml if possible) in sodium phosphate (0.15 M, $9.0 < pH < 9.5$) at 37°C for 2 hours (15). A centrifuged gel-filtration column (in a 1 ml tuberculin syringe) is used for rapid separation of protein from small molecules. Few multivalent metal ions are soluble in the phosphate buffer, helping to minimize contamination of the chelator. A quick change to citrate buffer provides an environment suitable for labeling the conjugate with indium^{3+}.

Conjugation of the chelator to the protein is followed by titration to quantitate protein-bound chelating groups and by analysis of biological activity (15). Conditions are developed for which specific addition of short-lived radiometals to protein-bound chelating groups can be performed quickly. Citrate (0.1 M, 5 < pH < 7) is not a universal buffer for this procedure; it works well with indium, copper, and cobalt, but less well with gallium and yttrium. It is important to use the highest convenient *concentration* of protein-chelator conjugate, since the rate of binding of a carrier-free radiometal will be directly proportional to chelator concentration. Also it is important to remember that contaminating metal ions are always present in everything; our "metal-free" solutions typically behave as though they contain 1 μM to 5 μM chelatable metal ions. Upon storage in liquid form for more than a few hours, the metal-binding capacity of a protein-chelator conjugate may drop measurably!

Appropriate tests and control experiments in living systems are used to explore the properties of the metal-tagged product *in vivo*, and in particular to compare it with the separate components (unchelated metal, or metal chelate, or metal plus *unmodified* protein) which might contribute to unwanted interference. In developing a new system, several iterations may be needed before the desired result is obtained; this is illustrated in another article which tells something of the history of this topic (28).

V. REVERSIBLE RADIOLABELING

Simply injecting radiolabeled antibodies intravenously does not often lead to their accumulation in a target site at the desired rate and with the desired selectivity. The use of antibody *fragments* which still contain the antigen-binding site has been found to give improved biological distributions (29,30,31,32) of radiolabel, but at the cost of less uptake in the target. We are now exploring alternative chemical approaches to improve the target/background ratio of labeled antibodies *in vivo*. One approach involves the use of a metabolizable link between antibody and chelate, such as a disulfide or an oligopeptide. A primary goal of this strategy is to reduce the radiation dose to the liver during therapy, by providing for rapid metabolic release of the metal chelates from antibodies taken up by the liver.

A novel method for removing circulating antibody-bound chelates from the circulation is to bind the chelate reversibly to an *anti-chelate* antibody (33). The antibody carries the chelate with it until a competing, nonradioactive chelate is injected; at that point, practically all of the bound chelate remaining in circulation is displaced and rapidly excreted (34). Since under normal circumstances substantial quantities of radiolabeled antibodies remain in the circulation for many days, this procedure promises to be useful in cases where circulating "background" cannot be tolerated.

For the future, it appears likely that a combination of chemical and biological innovations will be necessary in order to realize the potential of radiolabeled antibodies as general tools for *in vivo* diagnosis and therapy.

ACKNOWLEDGMENT

The contributions of many collaborators, including David Goodwin, Michael Sundberg, Charles Leung, Simon Yeh, David Sherman, Leslie Anderson, Gary David, Molly Stone, Michael McCall, Sally DeNardo, Min Moi, and others listed in the references, is gratefully acknowledged.

REFERENCES

1. E. Soini and H. Kojola, *Clinical Chemistry 29:* 65 (1983).
2. J. A. Koutcher, C. T. Burt, R. B. Lauffer, and T. J. Brady, *J. Nucl. Med. 25:* 506 (1984).

3. S. E. Order, personal communication (1985).

4. C. H. Chang and C. F. Meares, *Biochemistry 23:* 2268 (1984).

5. G. B. Dreyer and P. B. Dervan, *Proc. Natl. Acad. Sci. USA 82:* 968 (1985).

6. L. G. Sillen and A. E. Martell, eds. Stability Constants of Metal Ion Complexes (Chemical Society, London, 1964).

7. A. E. Martell and R. M. Smith, eds. Critical Stability Constants (Plenum, New York, 1974).

8. J. L. Hoard, M. Lind, and J. V. Silverton, *J. Am. Chem. Soc. 83*: 2770 (1961).

9. Ya. M. Nesterova and M. A. Porai-Koshits, *Koord. Khim. 8:* 994 (1982).

10. S. Richards, B. Pederson, J. V. Silverton, and J. L. Hoard, *Inorg. Chem. 3:* 27 (1964).

11. X. Solans, S. Gali, M. Font-Alba, J. Oliva, and J. Herrera, *Acta Crystallogr. C39:* 438 (1983).

12. V. M. Agre, N. P. Kozlova, V. K. Trunov, and S. D. Ershova, *Zh. Strik. Khim. 22:* 138 (1981).

13. D. W. Margerum, G. R. Cayley, D. C. Weatherburn, and G. K. Pagenkopf In *Coordination Chemistry, vol II,* A. E. Martell, ed. (American Chemical Society Monograph 174, Washington, D.C., 1978), Chapter 1.

14. A. Ringbom, *Complexation in Analytical Chemistry* (Interscience, New York, 1963).

15. C. F. Meares, M. J. McCall, D. T. Reardan, D. A. Goodwin, C. I. Diamanti, and M. McTigue, *Anal. Biochem. 142:* 68 (1984).

16. S. M. Yeh, D. G. Sherman, and C. F. Meares, *Anal. Biochem. 100:* 152 (1979).

17. D. J. Hnatowich, W. W. Layne, R. L. Childs, D. Lanteigne, M. A. Davis, T. W. Griffin, and P. W. Doherty, *Science, 220:* 613 (1983).

18. W. C. Cole, S. J. DeNardo, M. K. Moi, and C. F. Meares, submitted for publication.

19. M. W. Sundberg, C. F. Meares, D. A. Goodwin, and C. I. Diamanti, *Nature 250:* 587 (1974); *J. Med. Chem. 17:* 1304 (1974).

20. R. D. Steiger, "Organic Syntheses," collective volume III, Wiley, New York, 1955, p. 84.

21. J. Altman, N. Shoef, M. Wilchek, and A. Warshawshky, *J. Chem. Soc., Perkin Trans. 1:* 365 (1983).

22. L. H. DeRiemer, C. F. Meares, D. A. Goodwin, and C. I. Diamanti, *J. Labelled Compounds and Radiopharmaceuticals 18:* 1517 (1981).

23. S. M. Yeh, C. F. Meares, and D. A. Goodwin, *J. Radioanal. Chem. 53:* 327 (1979).

24. A. Abusaleh and C. F. Meares, *Photochem. Photobiol. 39:* 763 (1984).

25. C. F. Meares and T. G. Wensel, *Accts. Chem. Res. 17:* 202 (1984).

26. M. K. Moi, C. F. Meares, M. J. McCall, W. C. Cole, and S. J. DeNardo, *Analyt. Biochem. 148:* 249 (1985).

27. C.-H. Chang, C. F. Meares, and D. A. Goodwin, In *Applications of Nuclear and Radiochemistry*, R. M. Lambrecht and C. H. Morcos, eds. (Pergamon, New York, 1982) p. 103.

28. Meares, C. F., *Intl. J. Nucl. Med. Biol.*, in press (1986).

29. J. P. Mach, J. F. Chatal and J. D. Lumbroso, et al., *Cancer Res. 43:* 5593 (1983).

30. F. Buchegger, C. M. Haskell, and M. Schreyer, et al., *J. Exp. Med. 158:* 413 (1983).

31. A. B. Wilbanks, J. A. Peterson, and S. Miller, *Cancer 48:* 1768 (1981).

32. R. L. Wahl, C. W. Parker, and G. W. Philpott, *J. Nucl. Med. 24:* 316 (1983).

33. D. T. Reardan, C. F. Meares, D. A. Goodwin, M. McTigue, G. S. David, M. R. Stong, J. P. Leung, R. M. Bartholomew, and J. M. Frincke, *Nature 316:* 265 (1985).

34. D. A. Goodwin, C. F. Meares, G. S. David, M. McTigue, M. J. McCall, J. M. Frincke, M. R. Stone, R. M. Bartholomew, and J. P. Leung, *Intl. J. Nucl. Med. Biol.*, in press (1986).

14

Cancer Selective Macromolecular Therapeusis: Tailoring of an Antitumor Protein Drug

Hiroshi Maeda, Yasuhiro Matsumura,
Tatsuya Oda, and Kazumi Sasamoto

Department of Microbiology
Kumamoto University Medical School
Kumamoto, Japan

I. INTRODUCTION

Cancer selective drug targeting is an important tool in cancer chemotherapy for the treatment of solid tumors. If this objective is achieved, one can foresee substantial reduction in toxicity, more of a systemic one such as bone marrow suppression of nausea, anorexia, loss of hair, bleeding, cardiac, liver, and renal toxicity, whereas therapeutic efficacy will increase.

Investigations using monoclonal antibodies have been increasing in the past decade toward the above objective, and a few chapters in this book will deal with that approach. In that method we anticipate several drawbacks to be overcome: 1)

Supported in part by Grants-in-Aid for Cancer Research from the Japanese Ministry of Education, Science and Culture, by the Princess Takamatsu Award for Cancer Research, and by a Research Award from the Sapporo Bioscience Foundation to H.M., for 1985.

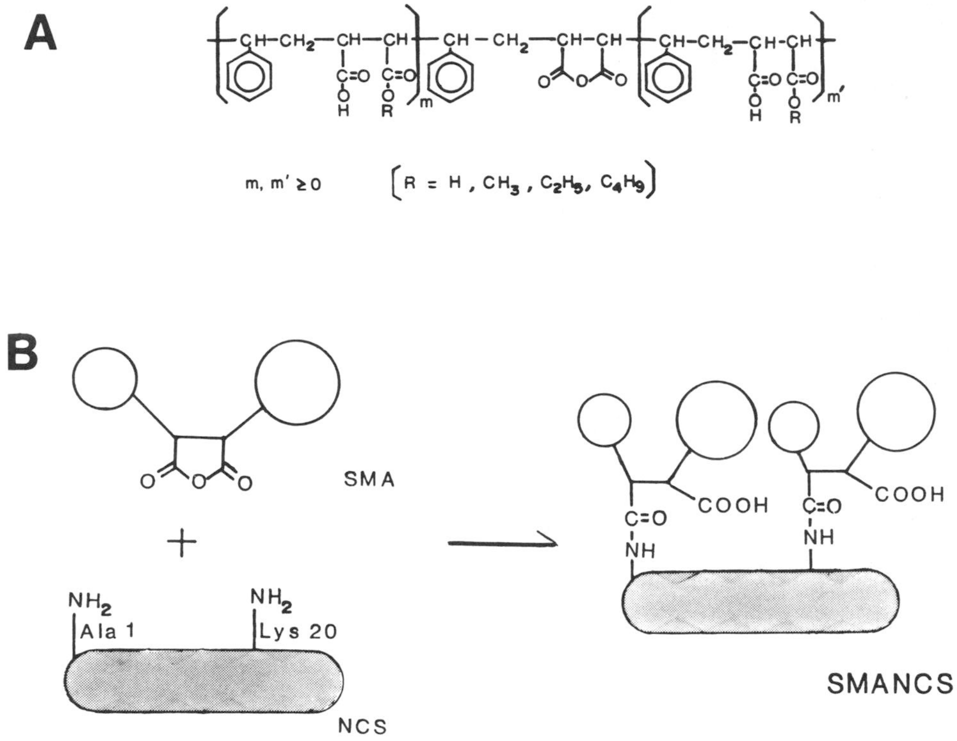

FIG. 1 *(A) Chemical structure of polystyrene-co-maleic acid (anhydride) SMA. (B) Diagrammatic representation of conjugation of SMA with antitumor protein (neocarzinostatin; NCS). The reaction proceeds stoichiometrically (see Ref. 6). The conjugate was designated as smancs.*

problems of using non-human IgG; 2) stability of the chemical bond between the drug and IgG, such as disulfide or ester bonds which are labile *in vivo*; 3) unwanted interaction with the humoral antigen or unpredictable tissues which might have a common antigenicity to the tumor antigen; 4) overwhelming divergence of mosaic tumor antigens, which may not be covered by a single clonal antibody; 5) inefficient accessibility or penetrability of the drug tagged-IgG to the extracapillary tumor tissues; 6) difficulties in the internalization into the cytosol from the tumor cell surface and subsequent liberation of the active component of the drug from IgG in the endosomes.

The use of protein drugs is usually accompanied by several inherited drawbacks such as the production of antibodies, rapid clearance from the circulation, proteolytic breakdown, accumulation in certain normal organs or target organs if possible, and handling instability *in vitro* as well as *in vivo*. In the present article we have described remarkable improvements in these points with the prototype drug, *smancs*.

We approached the above objective by using a new methodology: tailoring a macromolecular drug to utilize the architectural uniqueness of the blood vessels of tumors at the tissue level. Consequently one can accomplish tumor selective targeting. For this purpose we synthesized a conjugate of an antitumor protein neocarzinostatin (NCS) with a synthetic copolymer of styrene maleic acid (SMA)(see Fig. 1). NCS is an antitumor agent (M_r ~11,700) produced by Streptmyces carzinostaticus var F-41 into the culture medium. It is a single chain polypeptide (1-4). Details of the synthesis and purification of SMA, its conjugation with NCS and purification of the conjugate smancs have been published (5,6,7).

Using smancs as a prototype macromolecular drug, we carried out a number of basic and clinical investigations which revealed that a great degree of difference in effects exists between tumor and normal tissues at the physiological, anatomical and pathological levels. This difference becomes much clearer when one examines the behavior of large molecular weight substances and lipids. In this article we elucidate a unique behavior of native and tailored proteins *in vivo*, and discuss applicability of tumor targeting for cancer treatment and diagnosis. The prototype model protein discussed is mainly smancs.

II. ARCHITECTURAL DIFFERENCE OF TUMOR AND NORMAL TISSUES: MACROMOLECULES AND LIPIDS BEHAVE DIFFERENTLY FROM SMALL MOLECULES

A. *BASIC CONCEPT OF TUMOR VASCULATURE AND PERMEABILITY*

Figure 2 shows a conceptual presentation of tumor and normal tissue emphasizing blood and lymphatic capillaries, and the

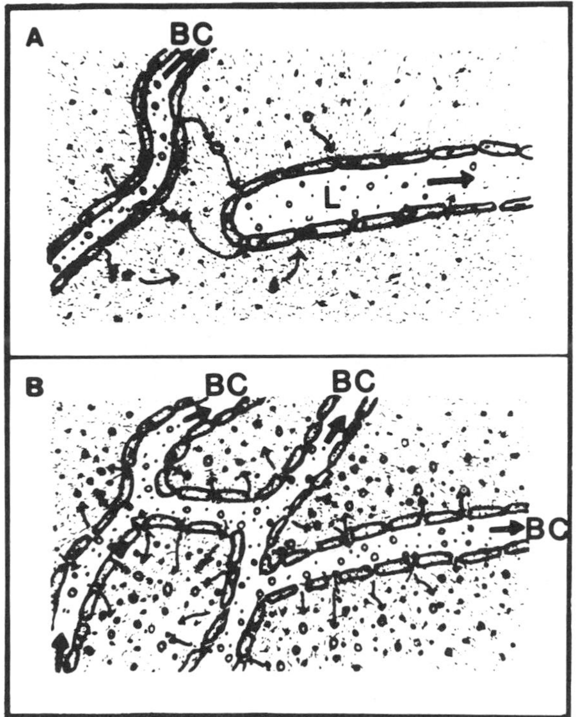

FIG. 2 *Architectural difference between normal and tumor tissues. (A) Diagrammatic presentation of normal tissue which has sparse blood capillary (BC) and lymphatic capillary (L). (B) Tumor tissue. Very few lymphatic vessels but highly developed new blood vessels are seen. In the tumor tissue (B), large molecules (proteins) and lipids are highly permeable and leak out into the extravascular space, but that is not so in the normal tissue (A). Large dots represent proteins or lipid particles.*

transfer of small, large or lipid molecules. It is now well known that tumor cells secrete tumor angiogenesis factor (angiogenin) when the foci of the tumor becomes more than 2-3 mm in diameter, and hence the extensive tumor neovasculature (hypervasculature) develops in the tumor (8,9). The massively developed tumor vasculature can be seen by angiography (10). When the tumor mass becomes more than a certain limit, 10-20 mm in diameter, the center of the tumor tissue may become hypovascular, while the periphery remains more hypervascular. The hypervasculature has another unique character; vascular permeability is highly enhanced. This vascular permeability is also mediated by a factor that is secreted from tumor cells (11,12). This enhanced permeability or leakings of the blood vessels in the tumor becomes more prominent for macromolecules and lipids than for low molecular weight compounds. Furthermore, there is almost no recovery system (the lymphatic system) for the lipids or macromolecules in the tumor, as will be shown later (Fig. 2). Normally macromolecules would not leak from blood vessels so easily but would remain in the circulating blood. Small molecules, on the other hand, diffuse and traverse freely into or from the blood vessels in the normal or the tumor tissues. In

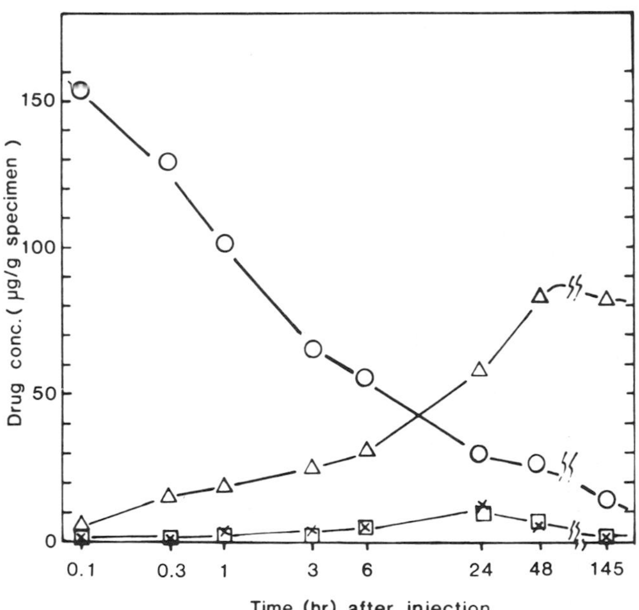

FIG. 3 *In vivo fate of Evans blue-albumin complex. A prototype model of macromolecular drug (70K) after intravenous injection (10 mg/kg, mouse). o, plasma concentration; Δ, tumor tissue; □, normal muscle; and X, normal skin. Note that the intratumor concentration of a model drug (albumin) is more than ten times that of normal tissue or 6-7 times more than that of the plasma level in 24 hrs or later. No free dye exists in this condition.*

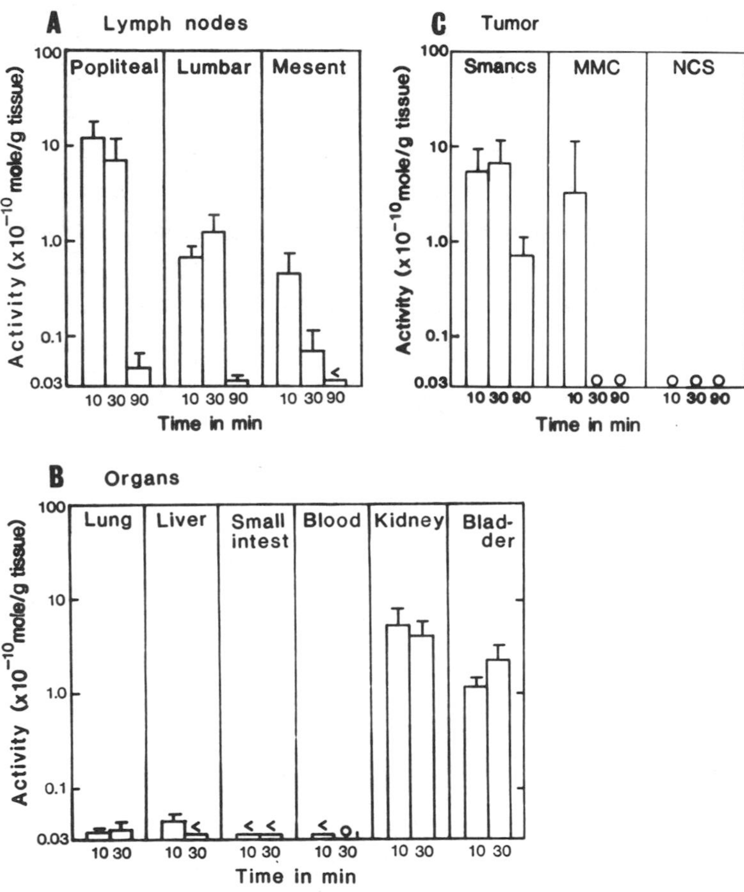

FIG. 4 *Tissue distribution and intratumor concentration of smancs, mitomycin C and neocarzinostatin. Accumulation of smancs in various lymph nodes (A), organs (B) and tumor tissue (C). Drugs were injected intravenously at a dose of 10 mg/ml in rats. In data (C), mitomycin C (MMC) was injected at a dose of 29 μmol/kg, while that of smancs and neocarzinostatin (NCS) were 0.42 μmol/kg and 0.85 μmol/kg, respectively (7).*

contrast to the normal tissues, large molecules like plasma proteins or lipid particles can leak out of the capillaries in the inflammatory tissues due to the action of a permeability factor such as bradykinin. Furthermore, they can not get back to the circulating blood directly, but are recovered into the lymphatic systems, which contrasts with the tumor tissue (13). An experiment using Evans blue-bovine albumin complex in tumor-bearing mice demonstrates this (Fig. 3).

Several workers have shown previously that radiolabeled plasma proteins accumulate more in tumor tissues than in normal tissues (14-16)*. We have demonstrated that the polymer conjugated antitumor protein, smancs, does indeed show remarkable tumoritropic accumulation when given intravenously (7) (Fig. 4C). This is explained by the above mechanism.

Figure 4A shows high concentrations in the various lymph nodes. This indicates that small proteins like smancs as well as NCS can leak from normal blood vessels and be recovered more via the lymphatics, as shown in Fig. 2A, thus resulting in higher concentrations in the lymph nodes (19). This was demonstrated when NCS was injected subcutaneously; it was recovered in high concentration in the primary lymph node (20).

It has been known for a long time that several radioactive metal ions such as Ga^{2+} and Ru^{2+}, free of carrier, accumulated in the tumor and also in the inflammatory tissues. This phenomenon has been utilized for the diagnosis of solid tumor as radio scintigraphic technique. The mechanism of tumortropic accumulation of these metal ions is reported lately that the citrate salt of metals injected intravenously formed a complex with transferrin in the blood plasma (21-23). Thus, the radioemitting complex accumulates in the tumor.

**Since the quantitation in vivo was done using radioiodine labeled proteins, the results now seem only qualitative (see Ref. 17 and 18).*

Therefore, the different behavior of the macromolecules in the tumor and normal tissue can be attributed to the difference in the respective anatomical architecture, which includes hypervasculature with enhanced permeability and a less developed lymphatic recovery system in the tumor tissue, as described below.

B. *TUMOR TROPIC ACCUMULATION OF A LIPID AND MACROMOLECULAR ANTICANCER AGENT*

Lipid behaves similarly to macromolecules at the tissue level. We have previously reported that when the oily lymphographic agent Lipiodol (or Ethiodol)* is injected intraarterially in the tumor feeding artery, it leaked out much more predominantly in the tumor area and remained there for a longer period of time (24) than the normal counterpart. Since the Lipiodol is a lymphographic agent, it is known that this agent is taken specifically into the lymphatic vessels. Thus, the lack of its recovery from the tumor tissue can be interpreted as a lack or underdevelopedness of the lymphatic systems in the tumor tissue (7,24). Further experiments using the Evans blue-albumin complex clearly demonstrated the accumulation and retention of these macromolecules much more in the tumor tissue than in the normal tissue, as shown in Fig. 3 and Fig. 5. Similar results were observed for other lipids (25). This will be discussed further.

To summarize, four unique characteristics in the tumor tissue that can be utilized for preferential accumulation of macromolecular drugs for tumor targeting are: 1) hypervasculature, 2) enhanced leakage of macromolecules, 3) lack of reabsorption into the tumor blood vessels, and 4) lack of an efficient recovery system for macromolecules or lipids from the extracapillary space, namely, the lymphatic systems.

**Lipiodol[R] is an iodinated fatty acid ester of poppy-seed oil (a product of Gelbe, Paris, France). It has an iodine content of 38% (w/w) and a density of 1.38 (g/ml).*

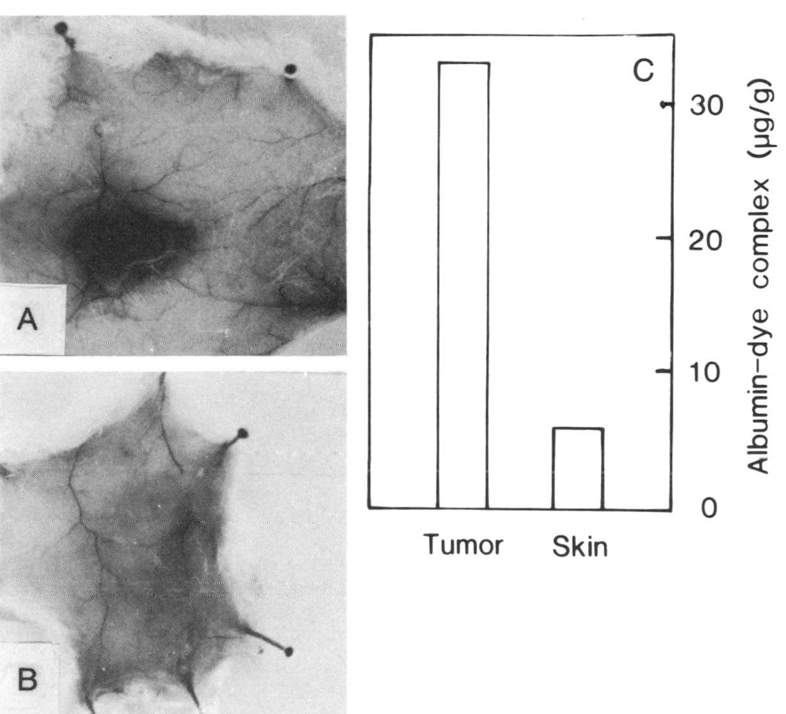

FIG. 5 *Clearance of bovine serum albumin dye complex from the tumor (A) or from the normal skin tissue (B). The albumin-dye complex is a prototype drug of 70K dalton which was injected into the tumor (A) or into the non-tumorous skin (B). The picture was taken 48 hr after the injection. In the normal tissue the macromolecules were cleared via the lymphatic (reticulo-endotherial) system and the dye became undetectable, whereas it was retained for more than 100 hr in the tumor. This indicates no, or very little, presence of the lymphatic recovery system in the tumor tissue. In (C) the dye (Evans blue) was extracted by formamide for 48 hr and quantified spectrophotometrically. Experiments were done in mice with S-180 tumor.*

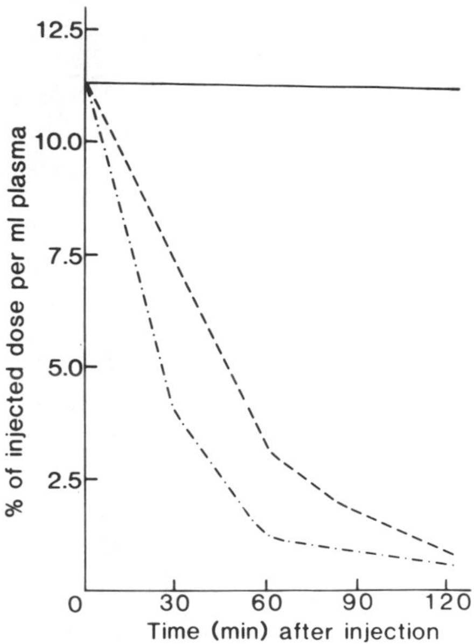

FIG. 6 *Clearance of native and modified albumin from the plasma in mice. The solid line shows native albumin and the broken and dashed lines indicate formaldehyde treated and nitroguanidiated albumin, respectively. Extensively chemically modified albumin is cleared very rapidly from circulation while native albumin, which has a much higher biocompatibility, is retained for a much longer period of time in the circulation. [Data was modified and redrawn based on the work of C. H. C. M. Buys et al. (32)].*

III. BIOCOMPATABILITY OF MODIFIED OR FOREIGN PROTEINS

A. GENERAL MECHANISM OF PROTEIN CLEARANCE FROM CIRCULATION

When a foreign protein or protein conjugate is injected into the blood circulation, it is cleared very rapidly. There are at least four different clearance mechanisms. The first one involves sugar residues of glycoproteins which are mediated by an asialoglycoprotein receptor of the liver cells (parenchymal hepatocyte) and the other organs. An example of this is desialidated serum glycoproteins (26-28) which are rapidly cleared from the circulation by the receptor. The second mechanism, which seems more relevant in the present line of research, is the so-called scavenger-receptor mediated clearance mainly in macrophage, Kupper cells in the liver, or hepatocytes (29-31). An example is shown in Fig. 6 in which formaldehyde-treated or nitroguanidinated bovine serum albumin is cleared much more rapidly than the native albumin. This and other data indicated that more modification of amino and other functional groups results in more efficient uptake; if more than 20% of an amino group is modified, clearance is extensive through the scavenger receptor of macrophages and endothelial cells (31,33,34). Similarly, this trend has been reported previously with the other modified protein (acetyl low density lipoprotein) (33-35). Macrophage and other cells are also known to have an α_2-macroglobulin receptor. Gonias and Pizzo (35) have shown that when α_2-macroglobulin was complexed with plasmin it was cleared from the plasma much more rapidly than when it was not complexed. Furthermore, separated half molecules of α_2-macroglobulin are cleared more rapidly than the intact α_2-macroglobulin (35,36). These results indicate that a subtle difference in the conformation of proteins is recognized by these cell surface receptors, and this may play an important role in the clearance of endogeneously modified proteins or heterologous proteins (Table 1).

A third mechanism of clearance was studied in the earlier days to investigate systemically the protein half-life *in vivo* with regard to proteolytic degradation. Those studies showed that the half-life is a function of amino acid composition. Acidic residues or polarity of proteins render them more vulnerable to proteolytic degradation or even deamidation of asparagine or glutamine to aspartic acid or glutamic acid, respectively* (45-47). Neocarzinostatin also seems to go through this process (48). A fourth line of clearance is via renal or urinary clearance. This appears to be more important for the small proteins like neocarzinostatin and ribonuclease (see Table 1) (49).

**A hypothesis that claims amide residues, thus DNA, determine the protein half-life in vivo (45).*

TABLE 1 *Plasma clearance time of various modified and native proteins in vivo*

Protein	Nature/origin test animal	pI	Modification/Probe	M.W. x 10^3	$t_{1/2}$[a]	$t_{1/10}$	Reference
Neocarzinostatin	*Streptomyces/ mouse hydrophilic*	3.4	*DTPA/NH_2/^{51}Cr*[b]	12	1.8 min	15 min	This work
Ribonuclease	*bovine/mouse*	9.5	*None/enzyme*	13.7	5 min	~30 min	38
Ribonuclease dimer	*bovine/mouse*		*Cross-linked/enzyme*	27	18 min	5 hrs	38
Smancs	*acidic polymer hydrophobic/mouse*	~3.0	*DDC/Lys*[c] / *HN_2/DTPA/^{51}Cr*	17	19 min	5 hrs	This work
Ovomucoid	*chick, rich in sialic acid/mouse*	2.8	*DTPA/NH_2/^{51}Cr*	29	5 min	34 min	This work
Serum albumin	*mouse/mouse*	4.8	*None*	68	3-4 days[d]	--	37
	"	--	*Evans blue/dye-binding*	--	2 hr	30 hr	This work
	"	--	*DTPA/NH_2/^{51}Cr*	--	6 hr	30 hr	"
	bovine/mouse	--	*DTPA/NH_2/^{51}Cr*	--	1 hr	24 hr	"
	bovine/rat	--	*Iodination/^{125}I*	--	4.5 hr	65 hr	39
Formaldehyde conjugated HSA	*human/rat*	--	*Formaldehyde/^{125}I*	--	25 min	4 hr	29
Transferrin	*human/rat*	5.2/5.5	*Iodination/^{125}I*	87	8 days[e]	--	40

L-asparaginase	*E. coli/rat*	*4.5-5.5*	*None/enzyme*	*65* x *(2-8)*	*1.5-3.4 hr*	--	*41*
	"	--	*DL-alanylation*	--	*13 hr*	--	*42*
	"	--	*PEG_2-linked*[f]	--	*56 hr*	*11 days*	*41*
IgG	*mouse/mouse*	*6.8*	*DTPA*	*150*	*60 hr*		*43*
IgG	"		*Iodination/^{131}I*	*150*	*45.6 hr*		*44*
α_2 *Macroglobulin*	*human/mouse*	--	*Iodination/^{125}I*	*180* x *4*	*140 hr*	*~22 days*	*35*
α_2 *Macroglobulin (Half molecule)*	"	--	"	*180* x *2*	*36 hr*	--	*35*
α_2 *Macroglobulin/ plasmin*	"	--	"	*180* x *2*	*2.5 min*	*~20 min*	*35*
α_2 *Macroglobulin/ plasmin*	"	--	"	*180* x *4*	*5.0 min*		*35*

[a] $t_{1/2}$ indicates here an initial decline, α phase in pharmacokinetics.

[b]DTPA: diethylene triamine pentaacetic acid; (pK_{as} of ^{51}Cr=24).

[c]DCC: dicyclohexylcarbodiimide.

[d]Human albumin in man, 19 days (37).

[e]Endogeneous (37).

[f]PEG: polyethylene glycol, biantenary.

B. *RADIOLABELING OF PROTEINS FOR THE STUDY OF CLEARANCE OF VARIOUS PROTEINS IN VIVO AND ACCUMULATION IN THE TUMOR*

Studies on the fate of protein *in vivo* have been carried out more frequently using radioactive iodine. The iodinated proteins have, however, serious drawbacks for the study *in vivo* due to the labile character of labeled iodine on the protein (17). Therefore, we prepared a number of protein derivatives tagged by the chelating agent DTPA (diethylenetriaminepentaacetic acid) (18,50), with radioactive ^{51}Cr from different animals and polymer-conjugated proteins. By using these radiolabeled protein derivatives, we have analyzed the accumulation of these proteins in the tumor and other organs, clearance from the plasma, and the biocompatability based on plasma half-time. DTPA derivatives are stable in both metal binding and the formation of a chemical bond which withstands hydrolytic breakdown during circulation *in vivo*. As described by Hnatowich et al. using Hepes buffer at pH 7.0 for 5 min (50), the anhydride of DTPA reacts with the primary amino group on proteins very readily, forming carbamide bonds through primary amino groups of lysine residues or amino-terminus.

In the case of smancs, in which no free amino groups were available (Fig. 1) (5-7), we have introduced lysine to which radioactive metal was chelated through DTPA. The first step involves attachment of lysine using carbodiimide (51), and the second step utilizes the amino group of lysine residues to which DTPA is attached (see Fig. 7). In the present derivative of smancs, about 3.7 moles of lysine and about 0.6-1.0 mole of DTPA per mole of protein were incorporated after the reaction. In the case of albumin, which has about 60 free amino groups per mole of protein, no significant decrease in amino group was noted after coupling with DTPA although detectable by radioactivity. Table 1 shows the plasma clearance time of various proteins in which the molecular weights range from 12×10^3 to more than 360×10^3.

The clearance of small foreign proteins appears rapid (eg. neocarzinostatin and RNase). The data for albumin in Table 1 show

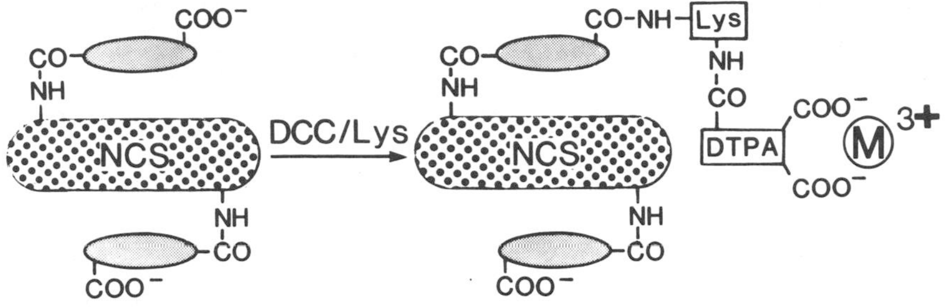

: styrene maleic acid copolymer
DCC : dicyclohexylcarbodiimide
Lys : lysine
DTPA : diethylenetriaminepentaacetic acid
M^{3+} : radioactive ^{51}Cr

FIG. 7 *Scheme of attachment of DTPA to smancs for radio-labeling. There is no amino group in smancs, thus L-lysine was coupled with DCC at first, then DTPA (anhydride) was conjugated as described by Hnatowich et al. (50). It chelated with radio-active ^{51}Cr and became a stable probe. The half-life of ^{51}Cr was 27.8 days and it was obtained as $CrCl_3$.*

that the same modification of mouse and bovine serum albumin with DTPA revealed a much longer half-life in the homologous system (mouse albumin/mouse), than in the heterologous (bovine albumin/mouse) one. As shown in Table 1, mouse albumin with DTPA exhibited a $t_{1/2}$ of 6 hr, while bovine albumin showed onle one hr. When chicken ovomucoid was injected, it was cleared much more rapidly than the bovine albumin (Table 1). These data indicate that the biocompatibility of the injected molecules to the injected host is an important matter. Molecules homologous to the host animal seem to exhibit longer $t_{1/2}$ naturally. The difference appears to be due to species.

As was described, human α_2-macroglobulin had a $t_{1/2}$ of 140 min in mice, while its intact form or dissected half molecule (which was prepared by cleavage of disulfide bonds) became only 5.0 or 2.5 min, respectively, after complexing with plasmin

(35,36) (Table 1). This is a clear example of how protein modification or conformation plays an important role in determining the half-life *in vivo*.

C. *INCREASE AND DECREASE OF PLASMA HALF-LIFE IN VIVO AFTER POLYMER CONJUGATION*

The plasma concentration of a drug has a great pharmacological significance, particularly when one is targeting a drug to the tumor tissue based on the vascular and lymphatic difference as described, or depleting L-asparagine in the plasma by L-asparaginase. Usually, a longer plasma life can ensure more prolonged pharmacological effect and less frequent administration of the drug. In the case of large proteins, if they are biocompatible and maintained high in plasma level, then a high intratumor concentration is attained after a long period of administration (see Fig. 3 and Table 2).

In the case of neocarzinostatin and smancs (5-7,52), ribonuclease and its dimer (38), and L-asparaginase and its polyethylene glycol conjugates (41), the plasma half lives of their conjugates and their $t_{1/10}$ become many times longer than those of the nonconjugates (Table 1).

A similar result was obtained earlier on a peptide hormone LHRF (luteinizing hormone releasing factor), whose plasma half-life was prolonged at least 3-fold by the attachment of polyglutamate (MW 100,000) (53). More recently, a conjugate of dextran and the antitumor agent mitomycin C, or pyran conjugates with other low molecular weight anticancer agents, polyamino acid, albumin with adriamycin and others, were also found to exhibit great pharmacological improvements resulting in clinical benefits (see reviews in 54-56). L-asparaginase, whose clinical use greatly decreased due to a number of drawbacks, now appears to be revived. Among them, dextran-mitomycin C conjugates (55) and albumin-daunorubicin conjugates (57) seem to have some clinical successes. In general all conjugates seem to have superior antitumor activity with less toxicity.

TABLE 2 *Intratumor accumulation of ^{51}Cr labeled proteins in mice*

Proteins[a]	Approx. M.W. x 10^{-3}	Time to reach tumor/blood ratio of	
		1	10
Neocarzinostatin	*12*	*5 hr*	*not attained*
Smancs	*16*	*5 hr*	*2 days*
Ovomucoid	*29*	*1 hr*	*2 days*
Albumin (bovine)	*68*	*17 hr*	*4 days*
Albumin (mouse)	*68*	*20 hr*	*5 days*

a ^{51}Cr is chelated by DTPA.

Accordingly, the tumor/blood ratios of smancs and other macromolecules became much better than neocarzinostatin itself (Table 2, Fig. 3-5). As described, other large molecules, bovine or mouse albmin, took about 4 to 5 days to reach a tumor/blood ratio of 10, whereas ovomucoid (29K) and smancs took about 3 days. Gallium scintigraphy, which depends on transferrin accumulation in the tumor, is usually carried out 3-5 days after intravenous injection of gallium citrate (8,23,58). A profile of the tumor/blood ratio at different times is shown in Table 2. Smancs and ovomucoid exhibited a more progressively efficient accumulation in the tumor, whereas ovomucoid was cleared much more rapidly from the blood. Because of the low biocompatibility of ovomucoid, its plasma level cleared very rapidly; thus, the absolute concentration in the tumor was low. On the other hand, albumin attained a higher tumor/blood ratio, but after a longer period. However, many of these cases with a long half-life suffer inactivations of their biological activity. Most of the present data reflect the radioactivity of the proteins used. Many of these became delineated from their biological activity because their biological half-life was much shorter than their immunological activity or the structural integrity of the modified protein.

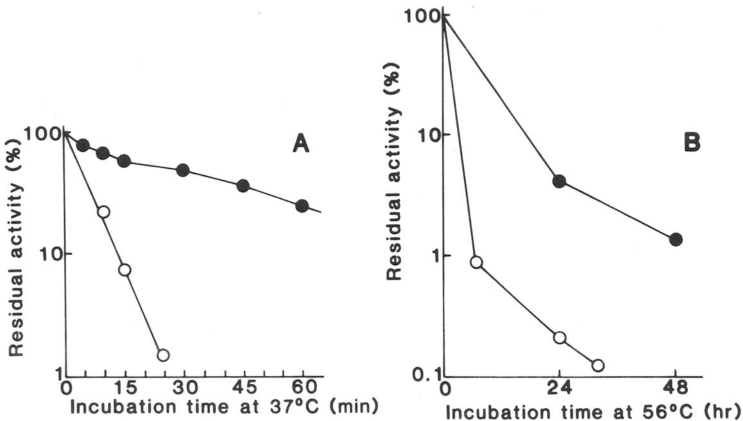

FIG. 8 *Increased stability of smancs in the blood at 37°C (A); or in physiological saline, at 56°C, pH 6.8 (B).* ●, *smancs;* o, *neocarzinostatin. Both final concentrations were 100 μg/ml.*

We have studied the stability of smancs *in vitro*, particularly in blood serum and to various physical exposures (pH, temperature, uv) and found that substantial improvements were obtained (Fig. 8A,B) (7,52). This trend was found with succinylated neocarzinostatin against proteolytic degradation, in which both of the amino groups were modified (48). However, other proteins are known to exhibit more susceptibility to proteases after modification (59,60).

D. *IMMUNOGENIC ALTERATION*

When one utilizes a protein as a drug, it is always possible to elicit antibody and subsequent immunological reaction in the subject host at cellular and humoral levels, including induction of anaphylaxis, or a delayed type reactivity. If this happens these side effects nullify clinical benefits of the protein drug. This was the case with L-asparaginase, once used as an antileukemic agent.

Antigenic determinants of proteins have interested protein chemists and immunologists for a long time (60-65). It is usually important for protein drugs to have no immunogenity but to retain

TABLE 3 *Antigenicity of smancs and neocarzinostatin (NCS): Precipitin reaction in vitro.*

Antigen used	Concentration of antigen (mg/ml) and immune reaction[a]								
	2	1	0.5	0.25	0.125	0.062	0.031	0.015	0.008
Smancs	+	+	+	+	-	-	-	-	-
NCS	+	+	+	+	+	+	+	+	-

[a]Antiserum was raised in rabbit using NCS with complete Freund adjuvant. Plus and minus signs indicate formation of precipitin line in 1% agar gel (Noble Agar, Difco) at pH 7.4. Incubation: 24 hr at 4°C (from Ref. 7 with permission).

a pharmacological effect. It is, however, the opposite for vaccines. A number of instances were reported in the early days in which cellular or humoral immunogenicity of protein antigens were changed by different manipulations. In the classic experiments by Habeeb et al., succinylation of serum albumin, lactoglobulin, and γ-globulin resulted in a considerable alteration in conformation and thus, antigenicity (60). The protein tailoring with poly D-Glu-Ala-Lys or a copolymer such as poly (Glu-Lys) with non-immunogenic polymers of the D-isomer resulted in loss of immunogenicity (63). More recently, a general rule to predict the antigenic determinants of a given protein was proposed (64). In view of the present context, smancs (Table 3) and polyethylene glycol conjugated L-asparaginase (7,41), uricase (66), and other plasma proteins (60) exhibited loss of antigenicity. Thus, their clinical applicabilities, when compared to the parental proteins, have increased considerably (7,67,68). These examples show that immunogenicity can be diminished in protein drugs by tailoring with appropriate polymers.

IV. ALTERATION TO LIPID SOLUBLE PROTEIN

As exemplified by smancs, and the conjugate with polyethylene glycol, lipid solubility can be increased considerably.

TABLE 4 *Accumulation of [^{14}C]-iodinated fatty acid*[a]

Organs/Tissues	Radioactivity DPM/g ($x10^3$) 15 min	3 days
Tumor (in liver)	*1252.58*	*130.94*
Liver (adjacent)[b]	*566.25*	*17.02*
Liver (remote)	*28.95*	*6.89*
Small intestine	*1.06*	*4.44*
Lung	*2.66*	*2.02*
Kidney	*1.61*	*2.57*
Stomach	*10.97*	--
Heart	*2.65*	*1.72*
Large intestine	*0.35*	*1.06*
Spleen	*2.39*	*3.28*
Bladder	*0.28*	*1.31*
Brain	*<0.1*	*0.38*
Muscle	*<0.1*	*0.46*
Skin	*<0.1*	*1.42*
Mes. lymph node	*0.15*	*2.21*
Cer. lymph nodes	*0.22*	*1.61*
Thymus	*0.22*	*0.93*
Serum	*0.58*	*1.03*
Plasma cells	*0.86*	*1.57*
Bone marrow	*<0.1*	*2.97*
Urine (exc)	--	*1.14*
Urine (vesical)	*<0.1*	*1.06*
Bile	*70.91*	*1.78*

[a]Intrahepatic arterial dose, 0.3 ml (from Ref. 24 with permission). Tumor VX2 was transplanted in the liver of a rabbit.

[b]Specimen location to tumor.

This made smancs a more beneficial drug for an injection together with a lipid contrast medium for arterial injection and hence becoming an extremely cancer selective drug delivered to the target tumor (25,68,69). Table 4 shows the results of drug (lipid) targeting to the tumor by this method in rabbits with VX-2 tumor that was inoculated at subcapsular parenchyma space of the liver. These results made this drug useful therapeutically and also diagnostically for small tumors less than 4 mm (67,68). This benefit will be described in section VI.

In other examples, conjugates of lipoprotein lipase and horseradish peroxidase both became soluble in benzene and water. Free ester synthesis was demonstrated for lipase in benzene. The peroxidase also retained the enzyme activity. Further explorations to other proteins and industrial applications appear interesting (70-72).

V. ENHANCEMENT OF CELL BINDING AND INTRACELLULAR INCORPORATION BY TAILORING OF PROTEIN

The binding of neocarzinostatin and smancs to cultured cells was studied, and the latter was found to exhibit a remarkably increased association constant to cells. From time course analyses, it was found that the binding of the fluorescein-labeled protein drugs to HeLa cells was time-dependent and proceeded rapidly at 37°C in the initial 30 min, reaching equilibrium after 2 hr. The result is shown in Fig. 9.

In the presence of 100-fold molar excess of unlabeled NCS, the binding of smancs was inhibited, as was that of NCS. These results suggested that the binding of NCS or smancs to HeLa cells was mediated by a specific receptor or binding site on the cell surface, and most of the smancs also bound to the same receptor. Furthermore, the amount of cell-bound smancs was increased 20-fold as compared to NCS. The increase of binding activity of smancs to cells was examined by the Scatchard plot analyses for the number of binding sites and apparent association constant (Ka) to HeLa

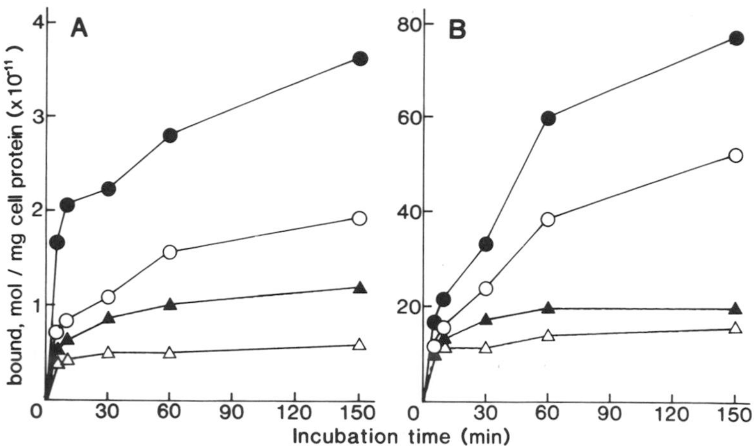

FIG. 9 *Binding of neocarzinostatin (A) and smancs (B) to HeLa cells. Note the 20-fold difference in the ordinate scale between A and B. Neocarzinostatin or smancs were labeled with fluorescein isothiocyanate (approximately one mol per mol of protein). Smancs was derivatized with lysine to introduce amino groups (see Fig. 7). ● and ▲ indicate experiments at 37°C, in the absence or presence of 100-fold excess non-labeled drug, respectively. o and Δ show the similar experiments at 4°C, in the absence or presence of 100-fold excess of non-labeled drug, respectively. The data apparently indicate that almost 20 times more smancs seems to bind to the cell than does neocarzinostatin (See also Table 5).*

cells. The results are summarized in Table 5. No significant differences between NCS and smancs in the number of binding sites could be demonstrated. However, Ka-values of smancs were 13.4 (x 10^4 M^{-1}), while that of neocarzinostatin was 0.5. These facts indicate that the binding affinity and the rate of internalization of smancs into HeLa cells are extensively increased by conjugating an appropriate polymer. This result was also substantiated by the biological activity (inhibition of colony formation) for the conjugate which required a much shorter exposure time (5 min) to the cells than the parental drug (NCS) (more than 80 min) to kill 80% of the tumor cells in culture (73,74). Details will be described elsewhere. Thus, tailoring of proteins has another prominent aspect at the subcellular level.

TABLE 5 *Association constant (Ka) and number of binding sites of neocarzinostatin (NCS) and smancs evaluated from Scatchard plot analysis*

Cells	Drugs	Temp. (^oC)	Ka ($x10^{-4}M^{-1}$)	No. of binding sites per cell ($x10^{-7}$)
HeLa	*NCS*	*0*	*1.97*	*0.68*
		37	*0.53*	*3.20*
	Smancs	*0*	*20.38*	*1.68*
		37	*13.39*	*4.48*

VI. CLINICAL OUTLOOK OF THE TAILOR-MADE DRUG SMANCS

Our preliminary results on the liver and the lung cancer, with mostly primary and some metastatic origins, have shown remarkably good therapeutic effects. Most of them are stage III or IV and/or inoperable, or resistant to other treatments. Our primary treatment utilized lipid contrast medium (Lipiodol; see previous footnote) as a carrier solvent of the drug, and the drug was injected arterially: about 3-4 mg/3-4 ml for hepatoma or 0.5-2.0 mg/0.5-2.0 ml for lung cancer via hepatic or bronchial artery, respectively.

Response rates in the patients were unprecedented. Namely, about 92% of the patients showed a decrease in α-fetoprotein, a marker protein for hepatoma, after one or two times of injection. About 90% of the patients also showed a decrease in tumor size. Similar results were observed in the patients with lung cancer (68,75).

The toxicity in the above procedure is essentially a mild fever (37-39^oC) for a few days in 48% of the subjects. There was no hematosuppression, loss of hair or other adverse effects to cause termination of the therapeutic procedure. These remarkable

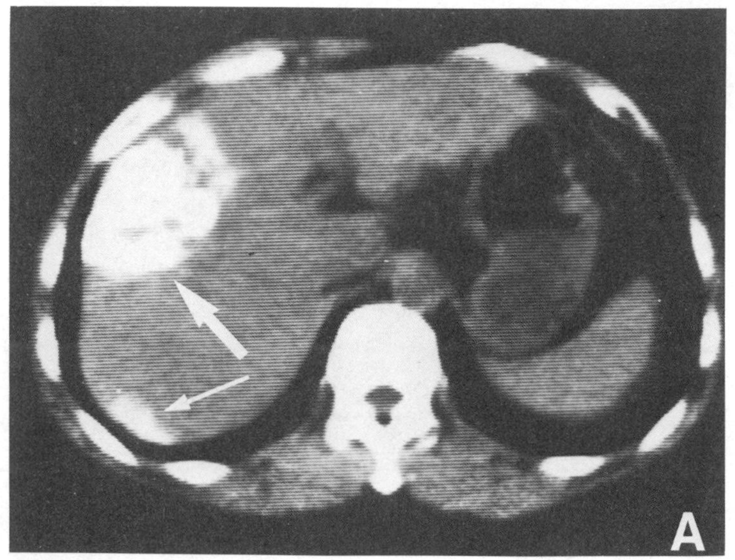

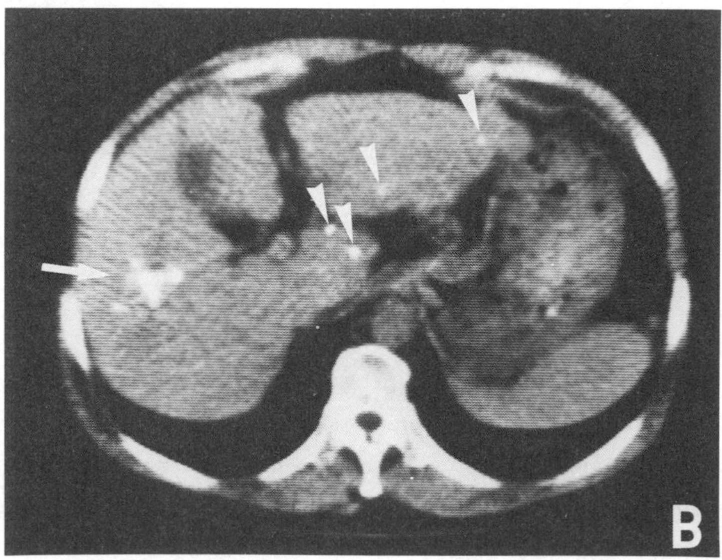

FIG. 10 *CT scan image of patients with liver cancer. The tumorous areas of the massive solitary type (A) and multiple nodular type (B) became high in electron density due to the iodine of Lipiodol (bright area shown by arrows) which were originally low in density (dark). The bright area indicates the prescence and location of tumor in which anticancer agent is solubilized in Lipiodol which is embedded in the tissue.*

TABLE 6 *Comparison of primary tumor lesion and tumor nodules by various methods in the same patients*

	Number of tumor lesions detected		
	Hepatoma, multiple type[a]		
Method	Major lesion	Daughter nodules	Hepatoma, massive type[b]
Angiography	*17*	*5*	*>9*
CT, ordinary	*15*	*0*	*14*
Scintigraphy	*14*	*0*	*0*
Ultrasonic	*18*	*0*	*3*
CT, Smancs/Lipiodol	*23*	*31*	*35*

[a]Number of cases entered = 9.

[b]Number of cases entered = 15.

results can be attributed to the tumor-selective drug targeting and slow and consistent drug release from Lipiodol (7,67,69,76).

Other advantages are in diagnosis and in determination of the successive dose regimen (69). Tumors which took up Lipiodol with the drug became high in electron density, and thus highly detectable by X-ray computed tomography or plain X-ray (Fig. 10, Table 6). Daughter nodules of the tumor as small as 3-4 mm in diameter were found to take up the drug almost selectively and become able to be visualized; they are undetectable otherwise. The normal area remained unchanged. Most of the hepatocellular carcinoma retain smancs/Lipiodol for a much longer period (more than one to six months). Thus, the Lipiodol stain which also indicates the presence of the drug can be a very useful measure for follow-up in the patient prognosis (67-69, 76).

A very good response in animal tumors was found with this arterial injection of aqueous forms of smancs. More importantly, although not published yet, human tumors in the lung, brain, esophagus and stomach seem to respond to smancs very well by this administration (H. Maeda et al., unpublished). Critical evaluation has to await several more years of observations.

VII. CONCLUSIONS

By conjugating synthetic polymers one can alter the pharmacological properties of the drug to a great extent. *In vivo* half-life can be expanded from ten- to hundreds-fold; tumor targeting is well over ten-fold; immunogenicity can be nullified; and overall therapeutic efficacy can be dramatically improved. Another aspect is the increase in lipophilicity. Namely, the protein can be made very amphipathic, soluble both in water and in lipid. This may permit an utterly different drug formulation. In the case of smancs, diagnosis with Lipiodol can be simultaneously therapeutic.

Although not described in this article, we now have ample evidence that smancs has acquired another property, that of biological response modifier, which is beneficial from a therapeutic point of view. Namely, this property was undetectable in the parental neocarzinostatin, but after attaching to the copolymer it became capable of activating tumor combating macrophages (77), and induction of interferon γ, and tumor necrosis factor, or enhancing cellular immunity *in vivo* (78,79). Therefore, protein tailoring can not only improve the drawbacks in proteins but it can also add novel or secondary functions.

REFERENCES

1. N. Ishida, K. Miyazaki, K. Kumagai and M. Rikimaru, *J. Antibiot.* (Tokyo) Ser A. *18*: 68 (1965).
2. H. Maeda, K. Kumagai and N. Ishida, *J. Antibiot.* (Tokyo) Ser. A. *19*: 253 (1966).
3. H. Maeda, C. B. Glaser, K. Kuromizu and J. Meienhofer, *Arch. Biochem. Biophys.* *164*: 369 (1974).
4. B. W. Gibson, W. C. Herlihy, T. S. A. Samy, K.-S. Hahm, H. Maeda, J. Meienhofer and K. Bieman, *J. Biol. Chem.* *259*: 10801 (1984).

5. H. Maeda, J. Takeshita and R. Kanamaru, *Int. J. Peptide Protein Res.* *14*: 81 (1979).

6. H. Maeda, M. Ueda, T. Morinaga and T. Matsumoto, *J. Med. Chem.* *28*: 455 (1985).

7. H. Maeda, T. Matsumoto, T. Konno, T. Morinaga and K. Iwai, *J. Protein Chem.* *3*: 181 (1984).

8. J. Folkman, *Adv. Cancer Res.* *19*: 331 (1974).

9. J. Folkman and H. P. Greenspan, *Biochim. Biophys. Acta* *417*: 211 (1975).

10. R. L. Wright, *Angiology* *18*: 69 (1967).

11. D. R. Senger, S. J. Galli, A. M. Dvorak, C. A. Perrnezzi, V. S. Harrey and H. F. Dvorak, *Science* *219*: 983 (1983).

12. H. F. Dvorak, D. R. Senger, A. M. Dvorak,, V. S. Harvey and J. McDonagh, *Science* *227*: 1059 (1985).

13. F. C. Courtice in *Lymph and the Lymphatic System* (H. S. Mayersen, Chairman), C. C. Thomas, Springfield, IL., p. 89 (1963).

14. H. I. Peterson, L. Appelgren, G. Lundborg and B. Rosengren, *Bibliotheca Anatomica* No. 12, 511 (1973).

15. H. I. Peterson and K. L. Applelgren, *Eur. J. Cancer* *9*: 543 (1973).

16. K. Shibata, H. Okubo, H. Ishibashi, K. Tsuda-Kawamura and T. Yanase, *Br. J. Pathol.* *59*: 601 (1978).

17. S. Halpern, P. Stern, P. Hagan, A. Chen, J. Frincke, R. Bartholomew, G. David and T. Adams, in *Radioimmunoimaging and Radioimmunotherapy*, Elsevier, Amsterdam, p. 197 (1983).

18. C. F. Meares and T. G. Wensel, *Acc. Chem. Res.* *17*: 202 (1984).

19. J. Takeshita, H. Maeda and R. Kanamaru, *Gann.* *73*: 278 (1982).

20. H. Maeda, J. Takeshita and A. Yamashita, *Eur J. Cancer* *16*: 723 (1980).

21. C. L. Edwards and R. L. Hayes, *J. Nucl. Med.* *11*: 103 (1969).

22. P. Som, Z. H. Oster, K. Matsui, G. Guglielmi, B. R. R. Person, M. L. Pellettieri, C. S. Scrivastava, P. Richards, H. L. Atkins and A. B. Brill, *Eur. J. Nucl. Med.* *8*: 491 (1983).

23. S. M. Larson, *Seminar Nucl. Med.* *8*: 193 (1978).

24. K. Iwai, H. Maeda and T. Konno, *Cancer Res.* *44*: 2115 (1984).

25. K. Iwai, H. Maeda, T. Konno and N. Ohtsuka, *J. Jpn. Soc. Cancer Therapy* *19*: 335, Abstr. No. 373 (1983).

26. A. G. Morrel, G. Gregoriadis, I. H. Scheinberg, J. Hickeman and G. Aswell, *J. Biol. Chem.* *246*: 1461 (1971).

27. K. Bridges, J. Harford, G. Ashwell and R. D. Klausner, *Proc. Natl. Acad. Sci. USA 79*: 350 (1982).

28. H. J. Gueze, J. W. Slot, G. J. A. M. Strous, H. F. Lodish and A. L. Schwartz, *Cell 32*: 277 (1983).

29. J. L. Mego and J. D. McQueen, *Biochim. Biophys. Acta 100*: 136 (1965).

30. S. Horiuchi, K. Takata and Y. Morino, *J. Biol. Chem. 260*: 482 (1985).

31. J. L. Goldstein, Y. K. Ho, S. K. Basu and M. S. Brown, *Proc. Natl. Acad. Sci. USA 76*: 333 (1979).

32. C. H. C. M. Buys, A. S. H. DeJong, J. M. W. Bouma and M. Gruber, *Biochim. Biophys. Acta 392*: 95 (1975).

33. M. S. Brown and J. L. Goldstein, *Ann. Rev. Biochem. 52*: 223 (1983).

34. S. Horiuchi, K. Takata, H. Maeda and Y. Morino, *J. Biol. Chem. 260*: 53 (1985).

35. S. L. Gonias and S. V. Pizzo, *Biochemistry 22*: 4933 (1983).

36. K. A. Ney, S. Gidwitz and S. V. Pizzo, *Biochemistry 24*: 4586 (1985).

37. H. E. Schultze and J. F. Heremans, in *Molecular Biology of Human Proteins*, Vol. 1, Elsevier, Amsterdam, p. 321 (1966).

38. J. Barthleyns and S. Moore. *Science 186*: 444 (1974).

39. J. W. Baynes, J. Van Zile, L. A. Henderson and S. R. Thorpe, *Birth Defect. Original Artic. Ser. 16*: 103 (1980).

40. E. H. Morgan and T. Peter, Jr., *J. Biol. Chem. 246*: 3508 (1981).

41. Y. Kamisaki, H. Wada, T. Yagura, A. Matsushima and Y. Inada, *J. Pharmacol. Exp. Therapeut. 216*: 410 (1981).

42. J. R. Uren and R. C. Ragin, *Cancer Res. 39*: 1927 (1979).

43. D. A. Scheinberg, M. Strand and O. A. Gansow, *Science 215*: 1511 (1982).

44. F. J. Dixon, D. W. Talmage, P. H. Maurer and M. Deichmiller, *J. Exp. Med. 96*: 313 (1952).

45. A. B. Robinson and C. J. Rudd, *Current Topics Cell. Regul. 8*: 247 (1974).

46. F. A. Momany, J. J. Aguanno and R. R. Larrabee, *Proc. Natl. Acad. Sci. USA 73*: 3093 (1976).

47. J. F. Dice, E. J. Hess and A. L. Goldberg, *Biochem. J. 178*: 305 (1979).

48. H. Maeda and J. Takeshita, *J. Antibio. 29*: 111 (1976).

49. H. Maeda, S. Sakamoto and J. Ogata, *Antimicrob. Agent. Chemoth. 11*: 941 (1977).

50. D. J. Hnatowich, W. W. Layne and R. L. Childs, *J. Appl. Radiat. Isot. 33*: 327 (1982).

51. D. G. Hoare and D. E. Koshland, Jr., *J. Biol. Chem. 242*: 2447 (1967).

52. S. Hirayama, T. Oda, F. Sato and H. Maeda, *Jpn. J. Antibiotics, 39,* March (1986).

53. M. S. Amoss, Jr., M. W. Monahan and M. S. Verlander, *J. Clin. Endocrinol. Metab. 39*: 187 (1974).

54. C. E. Carraher, Jr. and C. G. Gebelein (eds.), *Biological Activity of Polymers*, Adv. Chem. Ser., Vol. 186, Am. Chem. Soc., Washington, D. C. (1982).

55. H. Sezaki and M. Hashida, *CRC Crit. Rev. Drug Carrier Systems 1*(1): 1 (1984).

56. M. J. Poznansky and R. L. Juliano, *Pharmacol. Rev. 36*: 277 (1984).

57. A. Trauet, M. Masqielier, R. Baurain and D. DeCampeneere, *Proc. Natl. Acad. Sci. USA 79*: 626 (1982).

58. L. J. Anghileri, M. Ottaviani and C. Raynaud, *J. Nucl. Med. Allied Sci. 27*: 17 (1983).

59. N. Yamasaki, K. Hayashi and M. Funatsu, *Agr. Biol. Chem. 32*: 660 (1968).

60. A. F. S. A. Habeeb, H. G. Cassidy and S. J. Singer, *Biochim. Biophys. Acta 29*: 587 (1958).

61. M. Sela, *Science 166*: 1365 (1969).

62. R. J. Brawn and W. B. Dandliker, in *Immunochemistry of Proteins* (M. Z. Atassi, ed.) Vol. 2, Plenum Press, N.Y., p. 45 (1977).

63. V. E. Jones and S. Leskowitz, *Nature 207*: 596 (1965).

64. T. P. Hopp and K. R. Woods, *Proc. Natl. Acad. Sci. USA 78*: 3824 (1981).

65. A. F. S. A. Habeeb, in *Immunochemistry of Proteins* (M. Z. Atassi, ed.) Vol. 1, Plenum Press, N.Y. (1977).

66. H. Nishimura, Y. Ashihara, A. Matsushima and Y. Inada, *Enzyme 24*: 261 (1971).

67. T. Konno, H. Maeda, K. Iwai, S. Tashiro, S. Maki, T. Morinaga, M. Mochinaga, T. Hiraoka and I. Yokoyama, *Eur. J. Cancer Clin. Oncol. 19*: 1053 (1983).

68. T. Konno, H. Maeda, K. Iwai, S. Maki, S. Tashiro, M. Uchida and Y. Miyauchi, *Cancer 54*: 2367 (1984).

69. S. Maki, T. Konno and H. Maeda, *Cancer 56*: 751 (1985).

70. K. Takahashi, H. Nishimura, T. Yoshimoto, Y. Saito and Y. Inada, *Biochem. Biophys. Res. Comm. 122*: 845 (1984).

71. Y. Inada, H. Nishimura, K. Takahashi, T. Yoshimoto, A. R. Saha and Y. Saito, *Biochem. Biophys. Res. Comm. 122*: 845 (1984).

72. A. Ajima, T. Yoshimoto, K. Takahashi, T. Tamaura, Y. Saito and Y. Inada, *Biotech. Lett. 183*: 170 (1985).

73. T. Oda and H. Maeda, *Proc. Jpn. Cancer Assoc.* Abstr. No. 1051 (Fukuoka) (1984).

74. T. Oda and H. Maeda, *Proc. Jpn. Cancer Assoc.,* Abstr. No. 960 (Tokyo) (1985).

75. H. Maeda and T. Konno, *Cancer Chemother.* (Gan to Kagakuryoho) 12: 773 (1985) (in Japanese).

76. Y. Yumoto, K. Jinno, K. Tokuyama, Y. Araki, T. Ishimitsu, H. Maeda, T. Konno, S. Iwamoto, K. Ohnishi and K. Okuda, *Radiol. 154*: 19 (1985).

77. T. Oda, T. Morinaga and H. Maeda, *Proc. Soc. Exp. Biol. Med. 181*: in press (1986).

78. S. Uchimura, F. Suzuki and H. Maeda, *Proc. Jpn. Cancer Assoc.,* Abstr. No. 1220 (Tokyo) (1985).

79. H. Maeda and F. Suzuki, *Proc. Jpn. Cancer Assoc.,* Abstr. No. 1221 (Tokyo) (1985).

Index

D